T0338183

Processing and Properties of Advanced Ceramics and Composites V

Ceramic Transactions, Volume 240

Edited by

Narottam P. Bansal
J. P. Singh
Song Won Ko
Ricardo H. R. Castro
Gary Pickrell
Navin Jose Manjooran
K. M. Nair
Gurpreet Singh

Published by John Wiley & Sons, Inc., Hoboken, New Jersey.
Published simultaneously in Canada.

For general information on our other products and services or for technical support, please contact our
Customer Care Department within the United States at (800) 762-2974, outside the United States at
(317) 572-3993 or fax (317) 572-4002.

Wiley also publishes its books in a variety of electronic formats. Some content that appears in print may
not be available in electronic formats. For more information about Wiley products, visit our web site at
www.wiley.com.

Library of Congress Cataloging-in-Publication Data is available.

ISBN: 978-1-118-74409-3
ISSN: 1042-1122

Printed in the United States of America.

10 9 8 7 6 5 4 3 2 1

Contents

INNOVATIVE PROCESSING

NANOTECHNOLOGY

ELECTRONIC AND FUNCTIONAL CERAMICS

Preface

This volume contains papers presented at seven international symposia held during the Materials Science & Technology 2012 Conference (MS&T'12), Pittsburgh, PA, October 7–12, 2012. The symposia include: Innovative Processing and Synthesis of Ceramics, Glasses and Composites; Advances in Ceramic Matrix Composites; Solution-Based Processing for Ceramic Materials; Novel Sintering Processes; Nanotechnology for Energy, Healthcare and Industry; Dielectric Ceramic Materials and Electronic Devices; and Controlled Synthesis, Processing, and Applications of Structural and Functional Nanomaterials.

These conference symposia provided a forum for scientists, engineers, and technologists to discuss and exchange state-of-the-art ideas, information, and technology on advanced methods and approaches for processing, synthesis, characterization, and applications of ceramics, glasses, and composites.

Thirty six papers that were discussed at these symposia are included in this proceeding volume. Each manuscript was peer-reviewed using The American Ceramic Society's review process.

The editors wish to extend their gratitude and appreciation to all the authors for their submissions and revisions of manuscripts, to all the participants and session chairs for their time and effort, and to all the reviewers for their valuable comments and suggestions.

We hope that this volume will serve as a useful reference for the professionals working in the field of synthesis and processing of ceramics and composites as well as their properties.

NAROTTAM P. BANSAL
J. P. SINGH
SONG WON KO
RICARDO H. R. CASTRO
GARY PICKRELL
NAVIN JOSE MANJOORAN
K. M. NAIR
GURPREET SINGH

Ceramic Matrix Composites

DEVELOPMENT OF CONTINUOUS SiC FIBER REINFORCED HfB$_2$-SiC COMPOSITES FOR AEROSPACE APPLICATIONS

Clifford J. Leslie[1,2], Emmanuel E. Boakye[1,3], Kristin A. Keller[1,3], Michael K. Cinibulk[1]

[1] Air Force Research Laboratory, Wright-Patterson Air Force Base, Ohio 45433-7817, USA
[2] NRC Research Associateship Program, Washington, DC 20001, USA
[3] UES Inc., Dayton, Ohio 45432, USA

ABSTRACT

The fabrication of SiC/HfB$_2$-SiC continuous fiber reinforced composites by polymer impregnation and slurry infiltration of partially densified SiC-SiC laminates was investigated. A polymer-based HfB$_2$ precursor was synthesized and used to impregnate the preform panels via vacuum infiltration. Subsequent heat treatment at 1600°C in argon carbothermally reduced the polymer to HfB$_2$ and residual carbon. Slurry infiltration was performed via vacuum infiltration, pressure infiltration, vibration infiltration, vibration-assisted vacuum infiltration and vibration-assisted pressure infiltration using 5 vol % HfB$_2$ slurries in ethanol. An additional pressure infiltration experiment was performed using a 15 vol % slurry. Subsequent heat treatment at 1600°C was sufficient to initiate sintering of HfB$_2$ particles, but not densification. Polymer impregnation resulted in poor infiltration due to the low ceramic yield of the polymer and poor adhesion to the preform. SEM analysis, Archimedes density measurements and computational image analysis showed some success for the slurry infiltration techniques, with the best results observed for pressure infiltration using the 5 vol % slurry.

I. INTRODUCTION

Ultra-high temperature ceramics (UHTCs) are a small group of transition metal borides, carbide and nitrides possessing superior environmental resistance, excellent strength retention at high temperatures, and melting points above 3000°C. Interest in these materials has greatly increased recently due to developments in hypersonic flight and high speed propulsion. Among the materials capable of performing in the extreme environments required for such applications are HfB$_2$, ZrB$_2$, HfC, TaC, HfN, TaN and ZrN. [1] Much of the most recent work has focused on processing and characterizing dense, high strength monoliths and coatings of HfB$_2$ or ZrB$_2$ alloyed with SiC. [2,3,4,5,6,7,8] Although these diboride monoliths show good oxidation resistance and high-temperature strength retention, especially with additions of 20-30 vol % SiC, they are severely limited by low fracture toughness.

A common approach to increasing the fracture toughness of brittle ceramics is to reinforce the matrix with fibers to form a composite (CMCs). Thus, in response to the need for UHTCs with increased toughness, there have been numerous efforts in recent years to develop fiber reinforced UHTC matrix composites. Wang et al. developed a carbon fiber reinforced ZrC (C$_f$-ZrC) composite through slurry infiltration and in-situ formation of ZrC in a three-dimensional needle-punched carbon fiber preform with a PyC/SiC interface coating. Although these composites, tested in flexure, were found to have low strength relative to conventional C$_f$/SiC composites, they showed the desired non-brittle fracture behavior.[9] Li et al., took the work of Wang a step further by incorporating SiC into the ZrC matrix. In this work, three-dimensional carbon fiber preforms were infiltrated with a mixture of polycarbosilane and a ZrC precursor and processed via polymer infiltration and pyrolysis (PIP). Three-point flexure tests indicated that the C$_f$/ZrC-SiC composites had higher strength than C$_f$/ZrC composites and also displayed non-brittle fracture behavior. These composites were thermally tested at 1800°C by oxy-propane torch flame test and retained 52 to 91% of their strength after exposure.[10] Hu et al., developed a carbon fiber reinforced SiC laminate with 10 to 24.6 vol % ZrB$_2$ matrix additive (C$_f$/SiC-ZrB$_2$) using a combination of polymer impregnation of polycarbosilane and ZrB$_2$ slurry infiltration. The reinforcement improved the fracture toughness by >50% over typical ZrB$_2$ and ZrB$_2$-SiC monoliths.[11] Sciti et al., developed a chopped SiC fiber reinforced ZrB$_2$ composite (SiC/ZrB$_2$) by hot pressing a mixture of up to 30 vol % fiber and ZrB$_2$ powder with various sintering additives. Incorporation of the chopped fibers improved the fracture toughness of ZrB$_2$ by 43%, with the greatest increase being observed for composites with 20 vol% chopped fibers. These composites also showed nearly 100% strength retention, and increased flexure strength compared to the unreinforced material, after exposure to 1200°C in air.[12] Finally, Nicholas et al., has developed a continuous SiC fiber reinforced ZrB$_2$-SiC matrix CMC fabricated by impregnation of a polycarbosilane-based polymer (Starfire Systems SMP-10) loaded with ZrB$_2$ particles using a low-temperature PIP technique with a final high-temperature pyrolysis at 1600°C. Three point flexure tests showed that, although their strength was lower than monolithic ZrB$_2$, the composites did not fail catastrophically.[13]

Based on the literature, much of the work on fiber reinforced UHTCs has been focused on Zr-based matrices, with little consideration of Hf-based matrix materials. This article discusses efforts to develop a fabrication process for continuous SiC fiber reinforced HfB$_2$-SiC (SiC$_f$/HfB$_2$-SiC) composite laminates. A polymer impregnation technique using a HfB$_2$ polymer-based precursor as well as numerous slurry infiltration techniques were used to infiltrate partially densified SiC$_f$/SiC laminates with HfB$_2$. The impregnation/infiltration step was followed by heat treatment in an inert environment. In the case of polymer impregnation, the heat treatment promoted the in-situ carbothermal formation of HfB$_2$, while in the case of slurry infiltration, it acted to partially sinter the HfB$_2$ matrix. The densities and microstructures of the resulting composites were evaluated and used to determine the most effective processing technique.

II.   EXPERIMENTAL

Partially densified SiC/SiC panels, provided by Goodrich (now UTC Aerospace Systems), were used as preforms for the composites developed under this effort. These preforms consisted of porous two and four-ply laminates of 5-harness satin weave Hi-Nicalon S fabric with a boron nitride fiber coating, partially infiltrated with SiC via chemical vapor infiltration. The preform fiber volume percentage was reported as 37%, while the BN and SiC coatings added an additional 5 and 21 vol %, respectively. A total porosity of 37 vol% was also reported. Commercially available HfB$_2$ powder (-325 mesh, Materion Advanced Chemicals, Milwaukee, WI) was used in the as-received condition for slurry infiltration, while HfCl$_4$ (Sigma-Aldrich, St. Louis, MO), boric acid (Matheson Coleman & Bell, Cincinnati, OH) and Cellobond J2027L liquid phenolic resin (Momentive Specialty Chemicals, Columbus, OH) were used as the precursors for the polymer-based HfB$_2$ synthesis. A representative batch of the HfB$_2$ powder, measured using a laser diffraction particle size analyzer (LS230, Beckman Coulter, Brea, CA), was found to have an average particle size of 2.56 μm.

II.1   Slurry Infiltration

Slurries of 5 and 15 vol% HfB$_2$ in ethanol were formulated for preform infiltration. 2 wt% polyethyleneimine (PEI) was added as both a dispersant and binder. The slurries were stirred for a minimum of two hours at room temperature and subsequently used to infiltrate two-ply preform panels. Prior to infiltration the preforms were cut into approximately one inch squares and weighed. The infiltration techniques evaluated in this work were: a) vacuum infiltration, b) pressure infiltration, c) vibration infiltration, d) vibration-assisted vacuum infiltration and e) vibration-assisted pressure infiltration. After infiltration, the panels were heat treated for five hours in 99.9999% pure argon at 1600°C to partially consolidate the HfB$_2$ matrix. The infiltration techniques considered are described below.

In vacuum infiltration (VI), the preform was submerged in a volume of the slurry which was subsequently degassed in a Chamber evacuated using house vacuum. The degassing step was used to evacuate the voids between the tows of the preform weave, allowing the slurry to infiltrate the panel. The submerged panel was held under vacuum for two hours, removed and dried in a drying oven.

For pressure infiltration (PI), the preform panels were inserted into a small pressurizable filtration system (Millipore YT30 142 HW, EMD Millipore) and a vacuum was pulled on the panels. The slurry was poured into the chamber and the chamber was subsequently sealed and pressurized to 80-100 psi using air. The combination of vacuum and pressure forced the slurry into the voids in the preform, infiltrating the panels with the HfB$_2$ particles. A 1.2 μm filter membrane was positioned beneath the preforms in order to allow the extraction of ethanol with minimal HfB$_2$ loss while retaining the particles in the preforms. The extracted ethanol was captured by a liquid nitrogen cooled cold trap installed in the vacuum line.

The vibration infiltration technique was based on the report by Yang et al., for oxide composite densification.[14] In this case, the panel was submerged in a small volume. Here, only the 5 vol% slurry was used. The submerged panel was placed on a vibrating table for approximately 2 hours. After infiltration, the panel was removed from the slurry and dried.

Finally, the vibration-assisted vacuum infiltration (VVI) and vibration-assisted pressure infiltration (VPI) techniques were combinations of the techniques described above. In both cases, only the 5 vol% slurry was used.

II.2   Polymer Impregnation

For the polymer impregnation technique, a phenolic-based polymer, developed under a separate research effort and consisting of a hafnium precursor (HfCl₄), a boron precursor (H₃BO₃, boric acid) and phenolic resin (carbon source), all dissolved in ethanol, was utilized.[15] Based on the known carbon yield of phenolic resin, sufficient quantities of each precursor needed to achieve a range of desired B:Hf and C:Hf molar ratios were added to ethanol and heated separately. These solutions were then combined and heated at 120°C until the liquid was clear and amber in color, yielding a polymer-based HfB₂-C precursor. This precursor was used to impregnate preform panels via a process akin to polymer infiltration and pyrolysis (PIP). The impregnated panels were subjected to a 1600°C heat treatment for five hours during which the precursor was carbothermally converted to HfB₂ with varying amounts of residual carbon. In the case of the slurry infiltrated panels, this heat treatment also served to further the matrix densification.

II.3   Characterization

After heat treatment, the mass gains of the composite panels were measured and their densities were calculated using the Archimedes technique. Sample panels representing each infiltration technique were then cut, polished and evaluated using scanning electron microscopy (SEM, Quanta, FEI, Hillsboro, OR) to assess the extent of infiltration and matrix densification. The phase volume fractions were subsequently estimated from the micrographs using the EM/MPM (BlueQuartz Software, Springboro, OH) image segmentation software and a MATLAB volume fraction calculation routine developed internally.

III.   RESULTS AND DISCUSSION

SEM backscatter images of composite panels processed by vacuum, vibration and pressure infiltration using a 5 vol % HfB₂ slurry, and by pressure infiltration using a 15 vol % slurry are shown in Fig. 1. In all cases, there is a significant amount of porosity present within the fiber tows. This porosity is intrinsic in that it is present in the preforms prior to infiltration as a result of the initial CVI processing and is closed and thus inaccessible to the slurry. The vacuum and vibration infiltration panels, Fig. 1a and b, show only partial infiltration, with large portions of the regions between fiber tows being unfilled or partially filled with HfB₂. In these specimens, infiltration through the entire thickness of the panels was not observed in any region of the specimen. The panel pressure infiltrated with 5 vol % slurry (Fig. 1c), however, showed excellent infiltration through the entire thickness, with porosity present predominately within tows. This improvement was the result of active slurry infiltration through the preform driven by the pressure. Increasing the HfB₂ slurry concentration to 15 vol % resulted in a poorer pressure infiltration. As shown in Fig. 1d, the broad regions between fiber tows are filled; however, the narrower regions between fiber tows are not filled. This observation suggests that the narrow regions between fiber tows were inaccessible to the slurry, possibly due to increased viscosity resulting from the higher solids loading and increased binder volume fraction.

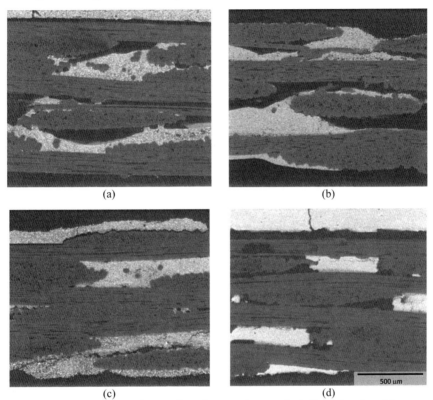

     (a)   (b)

     (c)   (d)

Fig. 1. SEM backscatter micrographs of composite panels infiltrated using (a) vacuum infiltration, (b) vibration infiltration and (c) pressure infiltration with 5 vol % HfB$_2$ slurries and (d) pressure infiltration with a 15 vol % slurry. The bright phase represents HfB$_2$ and the dark phase is SiC.

       In an attempt to improve the density of the panels processed by vacuum and pressure infiltration techniques, vibration-assisted vacuum infiltration (VVI) and vibration-assisted pressure infiltration (VPI) techniques were developed. These techniques were a combination of methods in which vacuum and pressure infiltration were performed on a vibrating table. Micrographs of the resulting panels are shown in Fig. 2. Based on Fig. 2a, the VVI process offered a considerable improvement in infiltration over either vacuum or vibration infiltration alone. Although the edges of the panel retained significant amounts of porosity, nearly complete infiltration through the entire thickness of the panel was observed. In the case of VPI processing (Fig. 2b), the extent of infiltration was similar to using just the pressure infiltration technique (Fig. 1c).

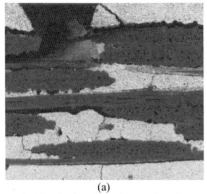 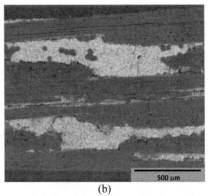

(a)   (b)

Fig. 2. SEM backscatter micrographs of composite panels infiltrated using (a) vibration-assisted vacuum infiltration (VVI) and (b) vibration-assisted pressure infiltration (VPI) with 5 vol % HfB$_2$ slurries.

The sintered matrix microstructure of the pressure infiltrated specimens is shown in Fig. 3b. Although heat treatment was performed at only approximately 0.5T$_m$, this temperature was sufficient to initiate sintering, with extensive necking between particles evident after heating. Subsequent heat treatments were found to slightly increase the matrix density, although matrix cracking was induced and extensive porosity was retained.

(a)   (b)

Fig. 3. SEM secondary electron images of the HfB$_2$ powder (a) before sintering and (b) after infiltration and sintering for five hours at 1600°C. Extensive neck formation between particles is observed after sintering.

In addition to the slurry infiltration techniques discussed above, a polymer-based precursor impregnation process was used to infiltrate SiC/SiC panels with HfB$_2$. The panels were impregnated with the precursor using the vacuum infiltration process described above. Subsequent heat treatment converted the polymer to HfB$_2$. SEM analysis of the panels showed that the polymer impregnation technique formed localized sparse coatings of poorly adhered

HfB$_2$ particles along the internal surfaces of the preform with no true infiltration. Much of this coating detached during subsequent handling and SEM preparation. The lack of HfB$_2$ infiltration with this technique is attributed to a low ceramic yield of the polymer, measured to be 1 vol %. The sparse coverage and poor adhesion is thought to be associated with poor wetting of the SiC preform surface by the polymer and the low processing temperature relative to the melting point of HfB$_2$.

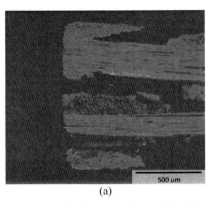 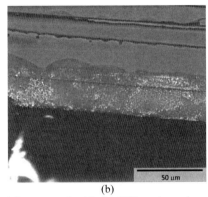

(a)   (b)

Fig. 4. SEM backscatter micrographs of a panel impregnated with the HfB$_2$ polymer-based precursor and subsequently heat treated at 1600°C.

After infiltration and heat treatment, the densities of the panels were measured using the Archimedes method. The calculated densities and open pore volumes (percentage) were averaged for each infiltration technique and are reported in Table I. Based on these calculations, it is evident that the pressure infiltration and VPI techniques yielded the best results, in agreement with the SEM analysis discussed above. Surprisingly, the density of the vibration infiltration panel was higher than expected relative to the pressure infiltrated panel densities, with both techniques resulting in panel densities near 3.3 g/cm^3. Fig. 1 suggests that the pressure infiltrated panels are considerably more dense then the vibration infiltration specimen. However, vibration during the infiltration process is expected to induce higher particle packing efficiency in the green state, ultimately resulting in increased density in the matrix phase. This phenomenon also explains the increased density of the VPI panels.

Table I. Average Mass Gain, Density and Open Porosity

	% Mass Gain	Density, g/cm^3	% Open Porosity
Preform	--	1.90	37.1 (Total Porosity)
Polymer Impregnation	--	2.14	27.3
Vacuum Infiltration	35.3	2.67	26.2
Pressure Infiltration	80.5	3.36	22.8
Vibration Infiltration	64.7	3.29	21.3
VVI	--	--	--
VPI	71.2	3.47	20.9

Finally, in order to quantify the extent of HfB$_2$ infiltration for each slurry infiltration technique, representative backscatter micrographs of panels fabricated using each technique were evaluated using the EM/MPM image segmentation program. This program allows the user to select multiple regions within the image and assign them to different classes or phases based on their range of grayscale values. Once each phase is assigned, the image is processed and a single grey scale value is assigned to each phase. The resulting image consists of clearly identified phases with distinct grayscale values. This image is then input into an internally developed MATLAB routine that calculates the areal fraction of each phase in the image. The result is an estimated volume fraction of each phase present in that particular image. This analysis is somewhat limited in that it takes into account only the larger intertow porosity and a portion of the finer closed porosity within the tows. In this evaluation, multiple images for each panel were analyzed. The resulting volume fraction values were averaged and are reported in Table II.

Table II. Estimated Average Phase Volume Fraction, Porosity and Matrix-to-Porosity Ratios

	HfB$_2$ (Matrix)	SiC (Fibers)	Porosity, %	Matrix:Porosity
Preform	--	--	--	--
Polymer Impregnation	--	--	--	--
Vacuum Infiltration	0.16	0.72	12	1.27
Pressure Infiltration	0.29	0.68	2	13.28
Vibration Infiltration	0.15	0.68	17	0.87
VVI	0.27	0.59	14	1.96
VPI	0.21	0.76	4	5.28
VPI (Abridged)	0.27	0.70	4	7.12

From Table II, it is evident that pressure infiltration technique gave the best result with the highest degree of infiltration. The HfB$_2$ matrix phase comprised nearly 30% of the total volume while porosity was 2% of the volume. As discussed earlier, the matrix is only lightly sintered so the porosity is expected to be greater than 2%. However, so far as comparing the relative porosities for the various methods, the porosity of panels processed by the pressure infiltration technique and VPI method are relatively less, with values between 2 and 4% compared to the other methods, with porosities greater than 12% (Table II). Therefore, in terms of the effectiveness of the infiltration techniques, it is inferred that the pressure infiltration process gave the greatest HfB$_2$ infiltration into the preform. Based on Table II, the vibration infiltration technique was the least effective slurry infiltration method. This observation is reasonable, since vibration infiltration was the only technique in which there was no applied presure to promote slurry infiltration. However, the use of vibration in combination with vacuum infiltration resulted in improved infiltration compared to either vacuum or vibration infiltration alone.

An important observation from the phase volume fraction calculations that cannot go unmentioned is the high SiC volume fraction. As stated earlier, the preform supplier reported that the SiC coating accounted for 21% of the total preform volume. According to the values in Table II, the SiC fibers and CVI preform matrix comprise upward of 59% of the total composite volume on average, consistent with the supplier data. In the case the VPI panel, this value exceeds 75%. Such high SiC volume fractions severely limit the volume of HfB$_2$ matrix that can be added to the panels. The 80 vol % HfB$_2$ matrix required for optimal thermal and oxidation resistance cannot be achieved with the current preform design, thus a viable initial processing technique must be developed. With an appropriate preform design, pressure infiltration is expected to be adequate as a primary infiltration technique for composite fabrication.

IV.   CONCLUSION

SiC/HfB$_2$-SiC composite panels were fabricated via polymer impregnation and slurry infiltration of partially densified SiC fabric laminates. In the case of polymer impregnation, the preform was vacuum infiltrated with a polymer-based HfB$_2$ precursor and subsequently heat treated at 1600°C to carbothermally convert the polymer. For the slurry infiltration, a) vacuum infiltration, b) pressure infiltration, c) vibration infiltration, d) vibration-assisted vacuum infiltration (VVI) and e) vibration-assisted pressure infiltration (VPI) techniques were developed. The slurry infiltrated specimens were heat treated at 1600°C to partially consolidate the matrix. These heat treatments were sufficient to initiate HfB$_2$ matrix sintering, but only negligible densification. Polymer impregnation was not successful due to low ceramic yield of the polymer and poor adhesion to the preform. Of the slurry infiltration techniques, pressure infiltration and VPI using 5 vol % slurries achieved the highest panel densities and nearly complete infiltration of the open porosity. Vacuum infiltration, vibration infiltration and VVI, as well as pressure infiltration using a 15 vol % slurry, resulted in incomplete infiltration of the inter-tow regions of the preforms. Based on computational analyses of the phase volume fractions, pressure infiltration was found to be the optimum technique. However, the rather high SiC volume fraction resulting from the initial preform fabrication illuminates a need to develop an improved preform processing technique.

ACKNOWLEDGEMENTS

The authors wish to acknowledge M. Aherns (UES, Inc.) for preform sample preparation and metallography assistance, C. Przybyla (AFRL/RXCC) for assistance with computational analysis, and R. Corns (UES, Inc.) for technical support.

[1] E. Wuchina, E. Opila, M. Opeka, W. Fahrenholtz and I. Talmy, UHTCs: Ultra-High Temperature Ceramic Materials for Extreme Environment Applications, *Electrochem. Soc. Interface*, Winter, 30-36 (2007).

[2] F. Monteverde, The thermal stability in air of hot-pressed diboride matrix composites for use at ultra-high temperatures, *Corros. Sci.*, **47**, 2020-33 (2005).

[3] W. G. Fahrenholtz, G. E. Hilmas, I. G. Talmy and J. A. Zaykoski, Refractory Diborides of Zirconium and Hafnium, *J. Am. Ceram. Soc.*, **90**, 1347-64 (2007).

[4] Y. D. Blum, J. Marschell, D. Hui and S. Young, Thick Protective UHTC Coatings for SiC-Based Structures: Process Establishment, *J. Am. Ceram. Soc.*, **91**, 1453-60 (2008).

[5] M. M. Opeka, I. G. Talmy and J. A. Zaykoski, Oxidation-based materials selection for 2000°C+ hypersonic aerosurfaces: Theoretical considerations and historical experience, *J. Mater. Sci.*, 39 5887-5904 (2004).

[6] C. M. Carney, Oxidation resistance of hafnium diboride-silicon carbide from 1400 to 2000°C, *J. Mater. Sci.*, **44**, 5673-81 (2009).

[7] E. Eakins, D. D. Jayaseelan and W. E. Lee, Toward Oxidation-Resistant ZrB$_2$-SiC Ultra High Temperature Ceramics, *Metall. Mater. Trans. A*, **42A**, 878-87 (2011).

[8] C. M. Carney, T. A. Parthasarathy and M. K. Cinibulk, Oxidation Resistance of Hafnium Diboride Ceramics with Additions of Silicon Carbide and Tungsten Boride or Tungsten Carbide, *J. Am. Ceram. Soc.*, **94**, 2600-7 (2011).

[9] Z. Wang, S. Dong, L. Gao, X. Zhang, Y. Ding and P. He, Fabrication of Carbon Fiber Reinforced Ultrahigh Temperature Ceramics (UHTCs) Matrix Composite, *Processing and Properties of Advanced Ceramics and Composites II*, 69-75 (2010).

[10] Q. Li, S. Dong, Z. Wang, P. He, H. Zhou, J. Yang, B. Wu and J Hu, Fabrication and Properties of 3-D C$_f$/SiC-ZrC Composites, Using ZrC Precursor and Polycarbosilane, *J. Am . Ceram. Soc.*, **95**, 1216-9 (2012).

[11] H. Hu, Q. Wang, Z. Chen, C. Zhang, Y. Zhang and J. Wang, Preparation and characterization of C/SiC-ZrB$_2$ composites by precursor infiltration and pyrolysis process, *Ceram. Int.*, **36**, 1011-16 (2010).

[12] D. Sciti and L. Silvestroni, Processing, sintering and oxidation behavior of SiC fibers reinforced ZrB$_2$ composites, *J. Euro. Ceram. Soc.*, **32**, 1933-40 (2012).

[13] J. Nicholas, V. G. K. Menta, K. Chandrashekhara, J. Watts, B. Lai, G. Hilmas and W. Fahrenholtz, Processing of continuous fiber reinforced ceramic composites for ultra high temperature applications using polymer precursors, *International SAMPE Technical Conference 2012, SAMPE 2012 Conference and Exhibition*, Baltimore, Maryland, May 21-24, 2012.

[14] J. Y. Yang, J. H. Weaver, F. W. Zok and J. J. Mack, Processing of Oxide Composites with Three-Dimensional Fiber Architectures, *J. Am. Ceram. Soc.*, **92**, 1087-92 (2009).

[15] S. Venugopal, E. E. Boakye, A. Paul, K. Keller, P. Mogilevsky, B. Vaidhyanathan, J. G. P. Binner, A. Katz and P. M. Brown, Low Temperature Synthesis and Formation Mechanism of HfB$_2$ Powder, unpublished work.

EFFECT OF PRIMARY GRAIN SIZE OF SrZrO$_3$/ZrO$_2$ NANO-DISPERSED COMPOSITE ABRASIVE ON GLASS POLISHING PROPERTIES

Takayuki HONMA[1], Koichi KAWAHARA[1], Seiichi SUDA[1] and Masasuke TAKATA[1,2]

[1]Japan Fine Ceramics Center,   [2]Nagaoka University of Technology

[1]Nagoya, Aichi, Japan,   [2]Nagaoka, Niigata, Japan

ABSTRACT

We studied glass polishing properties with SrZrO$_3$/ZrO$_2$ nano-dispersed composite abrasive with different primary grain size. The nano-composite particle was synthesized by spray pyrolysis technique and calcined at temperatures of 800-1200°C in order to control primary grain size. The high chemical mechanical polishing (CMP) performance generated using the abrasive with primary grain size of less than 80 nm. On the other hand, the CMP effect decreased in the case of the grain size of more than 100 nm. The distance between operating points of chemical and mechanical interaction was of consequence for CMP performance in the nano-composite abrasive.

INTRODUCTION

Precisely polished glasses are used as substrates for hard disk drives and flat panel displays. The polishing with a polishing pad and an abrasive requires higher removal rate as well as smooth surface. Therefore it is important to develop the abrasive with high polishing performance. Cerium oxide based abrasives have the excellent polishing properties generated by chemical mechanical polishing (CMP)[1-4] which is attributed to a high chemical reactivity with glass surface and a suitable hardness. Though the mechanism of the chemical reactivity is still vague, both adequate chemical reactivity with glass and suitable mechanical strength would be indispensable for glass abrasives.

Some novel glass abrasives of the core-shell composite were developed[5-7]. However, polishing properties of the core-shell abrasive were dominated by the properties of the shell material which contacted with glass surface. The core material contributed a little to glass polishing properties. Recently, we focused on the nano-dispersed abrasives composed of two kinds of materials; one has a high chemical reactivity with glass and the other has a high mechanical strength. We have demonstrated that an excellent CMP effect could be originated using the SrZrO$_3$/ZrO$_2$ nano-dispersed composite abrasives[8]. The high removal rate and the smooth surface were obtained with the SrZrO$_3$/ZrO$_2$ nano-composite abrasives. From the polishing property investigation with different compositions of SrZrO$_3$/ZrO$_2$ nano-composites, the optimum composition of the abrasive for glass polishing was determined as SrZrO$_3$:ZrO$_2$=7:3 in molar fraction. The distance between operating points of chemical and mechanical interaction was suggested to be one of the important factors for the CMP properties with the nano-dispersed composite abrasives. Thus, the primary grain size would have a significant effect on the CMP properties because the distance between operating points of chemical and mechanical interaction is determined by the primary grain size. In this study, we investigated effects of the primary grain size of $SrZrO_3/ZrO_2$ nano-dispersed composites on glass polishing properties.

EXPERIMENTAL PROCEDURE

Preparation of $SrZrO_3/ZrO_2$ nano-composite particles

The $SrZrO_3/ZrO_2$ nano-dispersed composite particles were synthesized by spray pyrolysis method. Figure 1 shows a schematic illustration of the spray pyrolysis apparatus. Aqueous solutions dissolving strontium nitrates ($Sr(NO_3)_2$, 2N5) and zirconyl nitrates ($ZrO(NO_3)_2 \cdot 2H_2O$, 2N5) were used as a starting solution. The composition of the solution was set to $SrZrO_3:ZrO_2=7:3$ by molar ratio. The concentration of nitrates in the solutions was 0.4 mol L^{-1} in resulting oxides. Solution droplets were generated by an ultrasonic atomizer and were introduced by carrier gas (air) into a tube furnace which had four heating zones controlled at temperatures of 200, 400, 800 and 1000°C. The particles were captured by a membrane filter and then calcined at 800-1200°C for 4 h in air.

Characterization of $SrZrO_3/ZrO_2$ nano-composite particles

The morphology of synthesized particles was observed by scanning electron microscopy (SEM; S-4500, HITACHI). The average sizes of the particles and the primary grains were measured from the SEM images containing more than 200 particles and grains, respectively. The crystalline phases in particles were identified by the X-ray diffractometer (XRD; RINT-2000, RIGAKU).

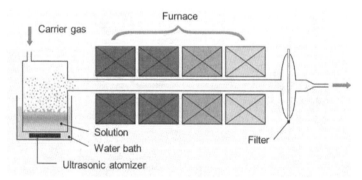

Fig. 1. Schematic illustration of spray pyrolysis apparatus.

Glass polishing properties

The synthesized abrasives were dispersed in ion-exchanged water. The total amount of slurry was 500 g containing 5 mass% abrasive. The slurry was cycled by a tubing pomp at a slurry flow rate of 0.1 L min^{-1}. A workpiece and a polishing pad were an aluminoborosilicate glass (AN100, AGC) of 37.5 mm x 30.0 mm and a formed polyurethane (MH-C15A, Nitta Haas), respectively. The three workpieces were simultaneously polished using a polishing machine (Tegra system, Struers) with the prepared slurry for 30 min under a pressure of 10 kPa. Relative velocity between the workpiece and the polishing pad was 1.2 m s^{-1}. The glass removal rate was calculated from the weight loss of glass during polishing.

Surface morphology of the polished glasses was evaluated by atomic force microscopy (AFM; SPM-9600, SHIMADZU). Average surface roughness Ra and maximum valley depth Rv were evaluated with the AFM profile. The scanning area was 30 μm x 30 μm, and the average surface roughness was calculated by averaging the surface roughness from five different scanning areas.

RESULTS AND DISCUSSION

Characterization of SrZrO₃/ZrO₂ nano-composite particles

Figure 2 shows SEM images of synthesized particles by spray pyrolysis method. The particles were spherical shapes which reflected the solution droplet generated using an ultrasonic atomizer. Figure 3 shows the calcination temperature dependence of particle sizes. The particle sizes were 0.9-1 μm in diameter, irrespective of the calcination temperature. The particle surface was rather rough and primary grains were clearly recognized from the enlarged SEM images (Inset of Fig. 2). Figure 4 shows the relation between the primary grain size evaluated from SEM images and the calcination temperature. The primary grain size increased with increasing the calcination temperature. Figures 2-4 shows that the morphology of the particles except for the primary grain size (the shape and the size of particles) was almost unchanged by calcination. Therefore we can evaluate the relation between the primary grain size and the polishing properties using the particles synthesized by spray pyrolysis method.

Figure 5 shows XRD patterns of the synthesized particles after calcination at different temperatures. The particles calcined at 800-1000°C were composed of perovskite SrZrO₃ (ICDD #74-2231) and tetragonal ZrO₂ (ICDD #80-0784). On the other hand, monoclinic ZrO₂ (ICDD #37-1484) was clearly identified in the particles calcined above 1100°C. No impurity phase was observed in the XRD patterns of all particles. The broad peak in the particles calcined at 800-900°C would be attributed to the small size of the primary grains. With increasing the calcination temperature, the peak intensity from tetragonal ZrO₂ decreased while that from monoclinic ZrO₂ increased. The formation of tetragonal ZrO₂ instead of monoclinic ZrO₂ would be attributed to strontium dissolution into ZrO₂ [9] and/or the small size of primary

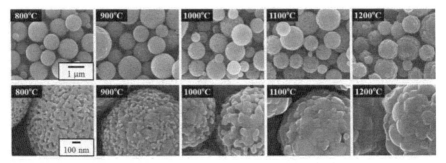

Fig. 2.   SEM images of nano-composite particles calcined at various temperatures.

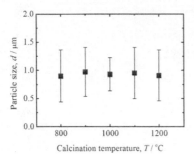

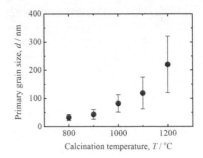

Fig. 3.   Relationship between particle size and calcination temperature.

Fig. 4.   Size dependence of primary grain on calcination temperature.

grains. The dissolution of Sr into ZrO_2 would stabilize tetragonal ZrO_2, however, the Sr dissolution would occur more easily at higher temperature than lower temperature. Therefore, the experimental results that the volume fraction of monoclinic ZrO_2 increased with increasing the calcination temperature would be explained by the effect of the primary grain size rather than the effect of Sr dissolution. In order to clarify the effect of the primary grain size on the crystal structure, the pure ZrO_2 particles were synthesized by spray pyrolysis method and were calcined at various temperatures. Figure 6 shows XRD patterns of the pure ZrO_2 particles before and after calcination at 600-900°C. The primary grain size of pure ZrO_2 calcined at various temperatures varied from about 20 nm to about 70 nm depending on the calcination temperature. The XRD pattern of the as-synthesized particles showed that the dominant peaks were from the tetragonal phase. The XRD pattern of the particles calcined at 700°C was consisted of both tetragonal and monoclinic phases. The peaks from tetragonal phase scarcely observed in the XRD patterns of the particles calcined above 800°C. These results indicated that the crystal

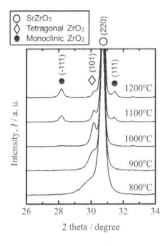

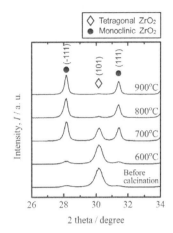

Fig. 5. XRD patterns of nano-composite particles calcined at various temperatures.

Fig. 6. XRD patterns of ZrO₂ particles calcined at various temperatures.

structure of ZrO_2 particle was obviously transformed from tetragonal into monoclinic by increasing the calcination temperature. XRD analyses revealed that the crystal structure of pure ZrO_2 changed depending on the primary grain size. Therefore, the formation of tetragonal ZrO_2 phase in nano-composite particles (Fig. 5) was concluded to be attributed to the small primary grain size.

Glass polishing properties

Glass polishing tests were carried out with the abrasives prepared by spray pyrolysis method. Figure 7 shows relationship between glass removal rate and the calcination temperature of abrasives. The removal rate of nano-composite abrasive calcined at 800°C was 230 nm min⁻¹. The removal rate increased with increasing the calcination temperature, and reached the maximum removal rate of 450 nm min⁻¹ at 1000°C. With further increase in the calcination temperature, the removal rate decreased with increasing temperature. The maximum removal rate with nano-composite abrasive was approximately 80% of the removal rate with commercial ceria-based abrasive. The removal rates evaluated using mixed abrasive of $SrZrO_3$ and ZrO_2 particles prepared by spray pyrolysis method were also shown in Fig. 7. The removal rate with mixed abrasives gradually increased with increasing the calcination temperature in the range of 800-1200°C. In the case of the calcination temperature of 1000°C, the removal rate with nano-composite abrasive was twice as high as that with the mixed abrasive. According to our previous study, the mixed abrasive did not show noticeable CMP effect[8].

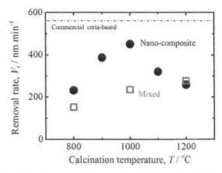

Fig. 7. Relationship between glass removal rate and calcination temperature for nano-composite and mixed abrasives.

Therefore, the high removal rate with the nano-composite would be brought by the CMP effect. The CMP effect was significant using the nano-composite particles calcined at 1000°C.

Figure 8 shows AFM images of the glass surface after polishing with nano-composite abrasives calcined at various temperatures. The surfaces were relatively smooth and had few scratches polished with nano-composite abrasives calcined at 800-1000°C. On the other hand, deep scratches and pits were distinct on the glass surfaces polished with the abrasives calcined at 1100-1200°C. Figure 9 shows the average surface roughness Ra and the maximum valley depth Rv evaluated on the glass surface after polishing as a function of the calcination temperature. Polishing with nano-composite abrasive calcined at 800°C resulted in the surface roughness of 0.7 nm Ra. The roughness slightly decreased with increasing the calcination temperature and reached the minimum Ra of 0.5 nm at 1000°C. The polished glass surface with nano-composite abrasive calcined at 1000°C was smoother than that with commercial ceria-based abrasive (Ra with the nano-composite abrasive was approximately 30% lower than that with the commercial ceria-based abrasive). With further increase in the calcination temperature, the surface roughness increased with increasing the calcination temperature. The valley depth Rv had a similar tendency of the surface roughness. The glass surfaces had less than 8 nm Rv after polishing with nano-composite abrasive calcined at temperature below 1000°C, however, the valley depth increased with increasing the calcination temperature on the glass surface polished with nano-composite calcined at 1000-1200°C. These results indicated that nano-composite composite abrasives calcined below 1000°C had an advantage in the surface smoothness and that the advantage decreased with increasing the calcination temperature above 1000°C.

The removal rate and the surface smoothness varied depending on the calcination temperature. The calcination treatment would affect three factors. The first was increase of particle strength by sintering. Focusing on the dependence of removal rate on the calcination

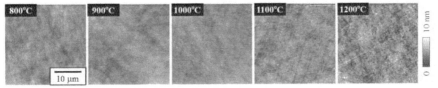

Fig. 8.   AFM images of polished glass surface with abrasives calcined at various temperatures.

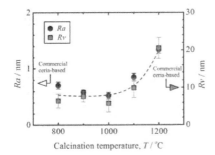

Fig. 9.   Average surface roughness, Ra, and maximum valley depth, Rv, as a function of calcination temperature.

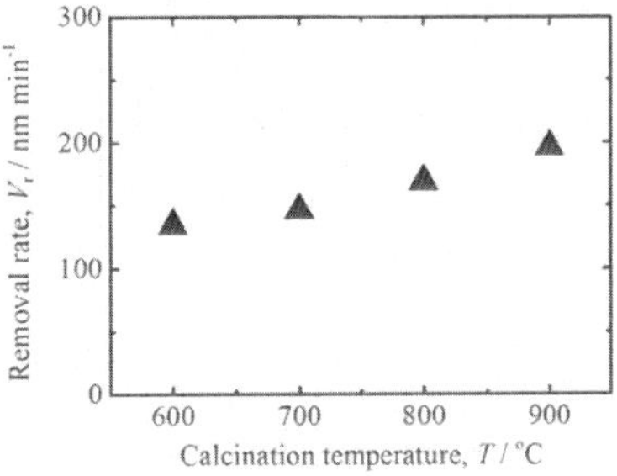

Fig. 10.   Relationship between glass removal rate and calcination temperature for ZrO$_2$ abrasive.

temperature with the mixed abrasive, the tendency of the removal rate would reflect the increase in the particle strength with increasing the calcination temperature.   Therefore, the increase of the removal rate with the nano-composite abrasive in the calcination temperature range of 800-1000°C would be due to increase in particle strength.   The second factor was the distance between SrZrO$_3$ and ZrO$_2$ grains.   The mechanism of CMP was considered that Si-O bond in the glass was weakened by a chemical reaction between the glass and the abrasive, and then glass surface was mechanically removed.   Therefore, it would be necessary that ZrO$_2$ has to exist at the softened glass area where SrZrO$_3$ chemically reacted with the glass.   The experimental results indicated that the distance between SrZrO$_3$ and ZrO$_2$ of 80 nm would be appropriate while the distance above 200 nm would be too large to originate CMP effect under the present polishing conditions.   The third factor was crystal structure change of ZrO$_2$.   In order to clarify the effect of crystal structure change of ZrO$_2$ on the removal rate, we conducted the polishing tests using pure ZrO$_2$ abrasives calcined with different temperatures. Figure 10 shows relationship between the calcination temperature and the removal rate using pure ZrO$_2$ abrasives.   Although phase transformation occurred around 700°C (Fig. 6), the removal rate did not show discontinuous change around 700°C.   Therefore, there would be little influence of crystal structure change of ZrO_2 on the removal rate.

The distance between $SrZrO_3$ and ZrO_2 grains below 80 nm was also appropriate for achieving smooth surface. In the case of primary grain size above 100 nm, deep scratches and pits (10-20 nm depth) generated on the glass surface would be attributed to large grain size. These results indicated that the microstructure control of particles was of great importance for achieving high CMP performance using the $SrZrO_3/ZrO_2$ nano-composite abrasives.

CONCLUSION

We studied the glass polishing properties using $SrZrO_3/ZrO_2$ nano-dispersed composite abrasives with different primary grain size to clarify the effects of the primary grain size on glass polishing properties. The nano-composite abrasives were synthesized by spray pyrolysis method and calcined at temperatures of 800-1200°C.

The optimum removal rate and smooth surface were obtained by calcination at 1000°C. The polishing properties with the abrasives calcined at a temperature of below 1000°C were explained by increase in particle strength and appropriate distance between operating points of chemical and mechanical interaction. In the case of calcination temperature of above 1000°C, the polishing properties were explained by increase in the distance between operating points due to coarsening of primary grain.

ACKNOWLEDGEMENT

This work was supported by New Energy and Industrial Technology Development Organization (NEDO), Japan as part of the Rare Metal Substitute Materials Development Project.

REFERENCES
[1]L. M. Cook, *J. Non-Cryst. Solids*, **120**, 152-171 (1990).
[2]R. Sabia and H. J. Stevens, *Mach. Sci. Tech.*, **4**, 235-251 (2000).
[3]T. Hoshino, Y. Kurata, Y. Terasaki, K. Susa, *J. Non-Cryst. Solids*, **283**, 129-136 (2001).
[4]P. Suphantharida and K. Osseo-Asare, *J. Electrochem. Soc.*, 151, G658-G662 (2004).
[5]H. Lei and P. Zhang, *Applied Surf. Sci.*, **253**, 8754–8761 (2007)
[6]S. Armini, C. M. Whelan, K. Maex, J. L. Hernandez and M. Moinpour, *J. Electrochem. Soc.*, **154**, H667-H671 (2007).
[7]Q. Yan, H. Lei, Y. Chen and Y. Zhu, "Advanced Tribology", Ed. by J. Luo, Y. Meng, T. Shao and Q. Zhao, Springer, Berlin (2010) pp. 625-629.
[8]T. Honma, K. Kawahara, S. Suda and K. Kinoshita, *J. Ceram. Soc. Japan*, **120**, 295-299 (2012).
[9]T. Noguchi, T. Okubo and O. Yonemochi, *J. Am. Ceram. Soc.*, **52**, 178-181 (1969).

THERMAL EFFECT STUDIES ON FLEXURAL STRENGTH OF SiCf/C/SiC COMPOSITES FOR TYPICAL AERO ENGINE APPLICATION

Vijay Petley[1*], Shweta Verma[1], Shankar[1], S.N. Ashritha [2], S. N. Narendra Babu[1], S. Ramachandra[1]

1. Gas Turbine Research Establishment, DRDO, Bangalore, Karnataka, India
2. Student, BMS College of Engineering, VTU-Bangalore, Karnataka, India

ABSTRACT

The high temperature thermal stability of the CVI generated SiCf/C/SiC composites under cyclic heat exposure with different quenching rates has been studied. One set of samples was heat exposed at 1073, 1273 K followed by air cooling and water quenching. Second set of samples was exposed to thermal cycling simulating typical aeroengine exhaust conditions in the burner-rig. The degradation of the composites was evaluated by a three point flexure test at room temperature. Severe degradation in the flexural strength was observed for the air cooled as well as water quenched samples exposed at 1273 K and minimum degradation for burner rig tested samples. Matrix micro cracks are observed predominantly in these samples and the fiber matrix interface has been oxidized. The quenching rate was found to have less significance on the flexural strength retention as compared to the total time of exposure at high temperature. The failure patterns of these samples are correlated with their load-displacement response and the flexural strength levels.

1. INTRODUCTION

The requirement of high temperature materials for the aeroengine application has seen the rise in the development of materials from wrought to cast with engineered grain orientations, high temperature coatings and high temperature composites. The high temperature properties of these materials are characterized at various levels including engine level testing. Of these materials, the continuous fiber reinforced ceramic matrix composite materials such as SiCf/SiC offers high temperature capabilities with high specific strength, high specific modulus, damage tolerance and are under active consideration for usage in the exhaust cone, convergent-divergent (CD)-nozzle flaps of the aero engine.[1-4] The characterization of these composites for thermal stability under the oxidative environment is important for practical application. The effect of long term cyclic heat exposure on degradation behavior and the residual flexural strength of the SiCf/SiC composites with different interfaces have been studied.[5] The oxidation mechanism of these composites at high temperature have been reported.[6-8] A number of work has been published on the nature of the load-deflection curve, oxidation mechanism, glass formation and its effect for enhancement of the properties, correlation of the properties with fracture surfaces, embrittlement of the fibers as a result of thermal exposure, etc.[9-16] The developmental efforts for realizing the low oxygen content and a highly stoichiometric SiC fiber with functionally graded interface coating playing a major role for imparting toughness are being studied, but the cost for these are going to be on a higher side. The inverted T-section of the typical component like flaps has two anchors equidistant from the major axis with a span of 30 mm and the flap is supposed to experience bending force during operation across the major axis. A study was undertaken to characterize the flexural behavior of the Isothermal Chemical Vapor Infiltration (ICVI) processed SiCf/C/SiC composites when subjected to thermal soak and thermal fatigue cycle for short duration.

2. EXPERIMENTAL PROCEDURE:

2.1 Specimen preparation:

The ICVI generated 2D-SiC$_f$/C/SiC laminates (0/90° fiber orientation) with carbon (C) coating was applied for the fiber bundle. These composite laminates were received from M/s ILN Tech. Inc. USA and the complete details of the processing parameters are not available. However, the density and porosity measured as per standard BS EN 1389:2003 was found to be 2.4 gm/cc and 8% for as-fabricated specimen. The laminates dimensions were like length of 300 mm, width of 150 mm and thickness of ~3.7 mm. Flexural test specimens with dimensions of length 75 mm, width 6.7 mm and thickness of ~3.7 mm were prepared from the laminates through abrasive water jet cutting and the same were used for three-point flexure tests. The file sand of 80 mesh size was used as the abrasive material and the pressure of abrasive water jet was maintained at 3600 bar. The flexural test specimens were not SiC seal coated as an additional protection of the composites. For all the specimens the major axis of the specimen was maintained along the longitudinal direction of the laminate. The macrostructure of the SiC$_f$/C/SiC composite, parallel to the plane of plies and perpendicular to the plane of plies (side lateral plane) are shown in Fig. 1.

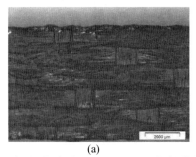 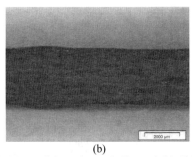

(a)   (b)

Fig. 1. Typical macrostructure of SiC$_f$/C/SiC (a) Parallel to plane of plies and (b) Side lateral view.

2.2 Thermal Soak Treatment

Each set of flexural specimens were subjected to thermal soak for 3 h at a preset temperature after which the specimens were cooled to room temperature by removing from the furnace. The thermal soaking temperature (Ts) of 1073 and 1273 K was preset and the heating rate for each temperature level was 100 K/h. An electric furnace of make M/s Therelek Engineers Pvt. Ltd., Bangalore was used for thermal soaking. After thermal soaking at each temperature, one set of specimen was air cooled (AC) and the other was water quenched (WQ).

2.3 Thermal Cyclic Treatment

Flexural test specimens were subjected to thermal cycling on a burner rig (Fig. 2) with LCS-4B controller of make M/s. BECON Inc. USA. The air turbine fuel was used for combustion purpose on the rig. Special water cooled fixture assembly was fabricated to hold the specimens in vertical up position. The specimens were positioned such that the hot flame from the combustor impinges on the mid of the specimen. This positioning was achieved by a servo hydraulic actuator controlled by control console of make M/s MTS 458.20. Each set comprising of three numbers of specimens was subjected to thermal cycling between the maximum temperature Tf of 1073 and the cooling air exhaust temperature (Approx 303 K). The specimens were exposed for 30 s each under the hot zone and the cold zone. The surface temperature of the specimen was measured with optical pyrometer. This was carried out with a timer switch coupled with a servo hydraulic control flow valve. This thermal cycle was purposely selected to simulate the intended application of an aeroengine component and a total of 11 cycles were conducted after which the specimens were unclamped for testing and further characterization. This thermal fatigue exposure was repeated on different set of specimens for Tf of 1273 and 1473 K.

(a)   (b)

Fig. 2. Burner Rig (a) Front View & (b) Right Side Angle View

2.4 Room temperature flexural strength of the composites

The three point bend flexural test was conducted on all i.e. as-fabricated, thermally soaked and thermally cycled set of specimens. These tests were carried out at room temperature on a servo hydraulically controlled 50 KN universal testing machine at M/s. BiSS, Bangalore. Usually the tests are conducted with a stroke of 1 mm per minute or less but considering the surge loads expected on the component the stroke of 10 mm per minute were selected. The specimens are positioned such that the flame hit surfaces of the thermally cycled specimens experiences the tensile forces. The ratio of span length 'L' (30 mm) to thickness 't', L/t was 8.1 as against 15 that was reported to prevent shear fracture prior to tensile fracture on the surface experiencing the tensile forces.[5,17] For all the flexure tested specimens, the failure occurred from the surface experiencing the tensile forces, hence, the microstructural analysis has been carried out on this surface. In addition the fracture surfaces are observed under stereo zoom microscope of make M/s Olympus, Japan and scanning electron microscope of make M/s Carl Zeiss, UK to observe the features of fracture process such as micro cracks, fiber debonding, fiber pull out, oxidation, fiber fracture, etc.

3. RESULTS AND DISCUSSIONS

3.1 Strength of the composite

Fig. 3-5 shows the typical three-point flexural load-displacement curves of the composites with thermal soak temperatures (Ts) and thermal fatigue temperatures (Tf). The trend of these curves are typical for SiC$_f$/C/SiC composite and similar to reported elswhere.[5,18] As all of these tests were conducted under a constant stroke control the load response observed is related to the type of the thermal effect the specimen has undergone and the associated fracture mechanism. The initial load response is attributed to the matrix loading. The onset of the non-linearity before the maximum suggests the matrix microcracking and fiber bridging behavior.[16] The displacement continues to increase in this phase. This continues to a critical point where the majority of the microcracks get connected and majorly the fibers are the load carrying members and this is observed as a localized peak in the curve near the maximum load range. Subsequent to this, two types of load response is observed. These tests being the bend tests, the onset of the macrocrack on the surface experiencing the tensile forces (on the opposite side of the plunger-specimen contact), can propagate in both ways i.e. along the laminar plies and in the direction of the applied load. For the later the crack has to come across the fiber tow which has higher strength than the matrix or the interface causing the crack to have higher propensity to propagate along the laminar plies. The two responses are; first, a sudden load drop, which is the result of the crack bridging but mainly due to propagation of these cracks along the laminar plies. Second, the not-so-sudden drop in load which is the crack arresting-diverting mechanism. In this type of fracture the propagating crack gets arrested as they approach the transverse tow of fibers and on imparting additional energy in the form of load the crack gets diverted causing a gracious fracture. These two responses are similar and are termed as cumulative and non-cumulative failure respectively.[5] The bending mechanical behavior and the Young's modulus for all the specimens are similar except for the specimens exposed at Ts of 1273 K. The bending mechanical behavior in the loading stage is more representative of the damage accumulation stage whereas the unloading stage is more representative of the nature of the critical flaw progress leading to fracture.[12,16] The work of fracture was measured as per Single Edge Notch Bend (SENB) method which is an indicative method.[19] The as-fabricated specimen exhibited work of fracture 8.98 KJ/m^2 which is lesser than that of the specimen exposed to 1273K/3h/WQ (11.79 KJ/m^2). This may be attributed to the not-so-sudden drop in the load for the specimen exposed to 1273K/3h/WQ.

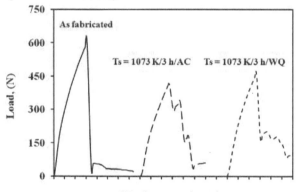

Fig. 3. Three-point flexural load-displacement curve for as-fabricated and test specimens exposed at Ts of 1073 K (Each unit on x-axis is 0.5 mm)

The flexural strength of the specimens under different conditions are plotted as bar-graph and shown in Fig. 6. The maximum flexural strength is exhibited by the as fabricated specimen. The significant drop in strength is observed for the specimens that were thermally soaked at Ts of 1273 K irrespective of the quenching media. Similar results have been reported for Ts of 1373

K for soaking period of 20 h.[14] This suggests that the temperature is playing a predominant role as compared with the quenching media. Also, no significant reduction in the flexural strength properties are observed for the specimens exposed to thermal fatigue at Tf of 1073, 1273 and 1473 K as reported elsewhere for the 3D braided composite laminates under different number of thermal shock cycles.[12] In our experiments this result is likely because of the limited specimen exposure time on a burner rig. As such the tests were carried out in the temperature range of 1073 to 1473 K with time cycles which addresses the temperature requirement for a typical aero engine component the test results of the test specimens exposed to thermal fatigue are found to be better than that of the test specimens exposed to thermal soak conditions for the same temperature levels.

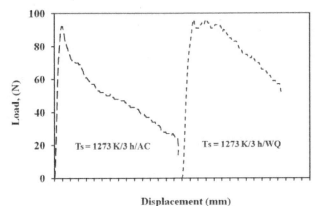

Fig. 4. Three-point flexural load-displacement curve for test specimen exposed at Ts of 1273 K (Each unit on x-axis is 0.5 mm)

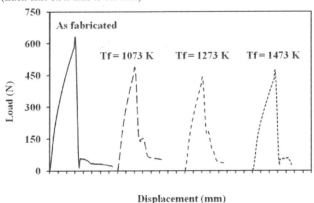

Fig. 5. Three-point flexural load-displacement curve for as-fabricated and test specimens exposed at Tf of 1073, 1273 and 1473 K (Each unit on x-axis is 0.5 mm)

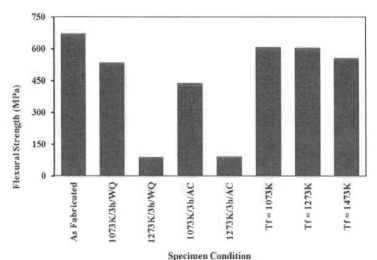

Specimen Condition

Fig. 6. The flexural strength for as fabricated specimen and the reduction in the strength levels after specimen exposure to different thermal effects.

3.2 Microscopy and Correlations

Further correlations between the observed load-response, flexural strength reduction and the fractographs are carried out. The fracture surfaces of the composites as seen under the stereo zoom microscopy are shown in Fig. 7-12. The inherent flaws (interlaminar voids) of the order of 2000 µm in length and 150 µm thickness in the as fabricated material are evident from Fig. 1 which are on the higher side. These specimens exhibited maximum flexural strength of the order of 673 MPa and minimal fiber pull out (Fig 7a) on the plane experiencing the tensile forces. Though it was expected for the as fabricated specimen to yield a gracious fracture, in actual, the peak load was followed by sudden crack propagation along the laminar plies (Fig. 7b) and further mitigation of the inherent flaws. The thermal soaked specimens at Ts of 1073 K under both AC and WQ conditions exhibited gracious fracture with extensive fiber pull out (Fig. 8 and Fig. 9 respectively). The reduction in the flexural strength for these specimens is accounted to the onset of the early matrix microcracking as compared with as fabricated specimens resulting in building up of inefficiency towards the load transfer from the matrix to the fibers. This phenomenon continues further and is witnessed by exposing the specimens to thermal soaking temperature Ts of 1273 K under both AC and WQ conditions. Under these conditions the test specimens exhibited significant drop in the flexural strength to the extent that the strength may be attributed only due to bending of fibers in the presence of extensive matrix microcracks. The authors would like to draw attention towards the non existence of the instantaneous peak in the load displacement curves and the increase in the Young's modulus only for the test specimens exposed at Ts of 1273 K. At the same time the gracious fracture seen by the load response curve as well as the extensive fiber pull-out seen is due to the fact that the matrixes no longer carries the load and under bending mode the fibers are free to bend resulting in very low loads and large displacements. Unlike other specimen fracture surfaces wherein majority of the cracks are propagating along the planar plies the specimens exposed at Ts of 1273 K shows transverse crack propagation across the thickness (Fig. 10 & 11). A decrease in the flexural strength was observed on thermal fatigue exposed specimens with increase in the Tf (Fig. 6) but no major difference in the fracture surfaces (Fig. 12) of these specimens was observed.

(a)   (b)

Fig. 7. As fabricated and flexure tested specimen (a) Surface experiencing tensile forces and (b) Side lateral view

(a)   (b)

Fig. 8. Flexure tested specimens exposed at 1073K/3H/AC (a) Surface experiencing tensile forces and (b) Side lateral view

(a)   (b)

Fig. 9. 1073 K / 3 H / WQ flexure tested specimens (a) Surface experiencing tensile forces and (b) Side lateral view

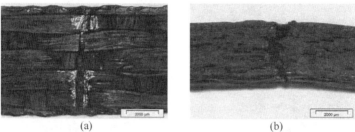

(a)   (b)

Fig. 10. 1273 K / 3 H / AC flexure tested specimens (a) Surface experiencing tensile forces and (b) Side lateral view

(a)   (b)

Fig. 11. 1273 K / 3 H / WQ flexure tested specimens (a) Surface experiencing tensile forces and (b) Side lateral view

(a)   (b)

(c)

Fig. 12 Thermal fatigue exposed and flexure tested specimen (side lateral view)
(a) Tf = 1073 K, (b) Tf = 1273 K, (c) Tf = 1473 K

The specimens were further investigated by scanning electron microscopy and the fractographs are shown in Fig. 13-17. Though some amount of fiber pull out is observed in the as fabricated flexure tested specimen (Fig. 13) the voids between the interlaminar plies are sufficiently large for the cracks to join together resulting in the fracture of the specimen. The matrix was favorably transferring the load to the fibers before interlaminar crack propagation took over causing sudden load drop response. The specimens thermally soaked at Ts of 1073 K exhibits distinct features of fiber pull-out and the gracious failure (Fig. 14).

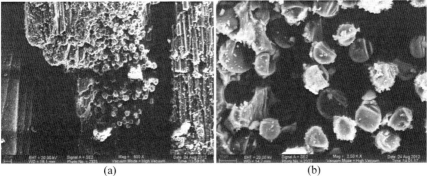

(a)   (b)

Fig. 13. Fractographs of the as fabricated flexure tested specimens (a) The voids between the tows gets bridged (b) Few fibers seen pulled out from matrix

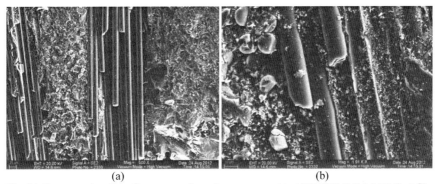

(a)   (b)

Fig. 14. Fractographs of the flexure tested specimens exposed to 1073 K/3h/AC (a) Distinct fiber pull out seen (b) Fiber pull out and sheared fibers

For the specimens thermally soaked at Ts of 1273 K and irrespective of the quenching media i.e. AC or WQ, the crack has traversed across the thickness and the fibers have also sheared on the same crack plane. Though distinct fibers are seen under SEM (Fig. 15 and 16) and these are more on the plane experiencing the tensile forces, it is not clear if this is fiber-pull out or just matrix loosening and falling apart. But as the maximum load levels are just 10% of the as fabricated specimen and no non-linear behavior in the load response, it suggests that there is no load transfer from matrix to the interface.

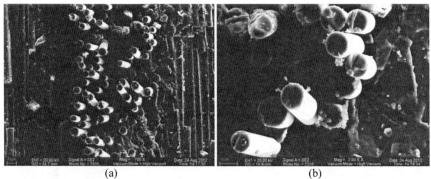

(a)   (b)

Fig. 15. Fractographs of the flexure tested specimens exposed to 1273 K/3 H/AC (a) Extensive fiber pull out (b) Matrix microcracks and loss of matrix

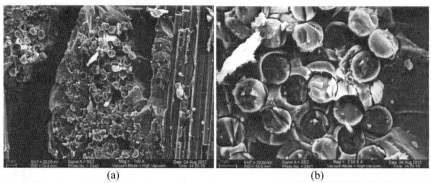

(a)   (b)

Fig.16. Fractographs of the flexure tested specimens exposed to 1273 K/3 H/WQ (a) Shearing of fibers on same plane (b) Interface oxidation and matrix microcracks

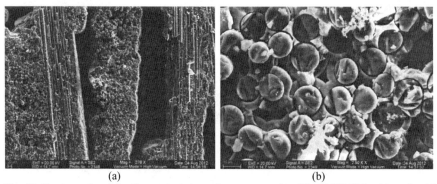

(a)   (b)

Fig. 17. Fractographs of the flexure tested specimens exposed to Tf of 1473 K for 11 cycles (a) Interlaminar crack but no sign of fiber pull-out (b) Interface oxidation

The porosities have acted as the pathway for the oxygen to diffuse and react with the interface and then with the fibers causing degradation of the fiber strength.[5,15] Extensive oxidation around the fiber and microcracks observed in the matrix is observed. Oxidation of the carbon interface acts as a weak link process causing debonding and additional pathways for oxidation of fibers at elevated temperatures.[20] The thermal fatigue exposed specimens have not shown any significant reduction in the flexural strength as compared to the as fabricated specimen. Though the oxidation of the interface (Fig. 17) is observed for the specimens with Tf of 1473 K the matrix has not fallen apart as for the thermal soaked specimens at Ts of 1273 K. The set of specimens thermally soaked at Ts of 1073 K and thermally fatigue exposed at Tf in the range of 1073 to 1473 K have not exhibited any change in the modulus or the non-linear phase of the load response which could have been attributed to the quenching cracks.[21] Thus, the severe oxidation coupled with inherent interlaminar voids results in the degradation of the fiber matrix bond and the interply integrity causing loss of the flexural strength.

4. CONCLUSIONS

The effect of the short term thermal soak and thermal fatigue exposure on the degradation behavior and the residual flexural strength of the SiC$_f$/C/SiC composite have been studied. The as fabricated specimen exhibits the maximum flexural strength and any amount of thermal exposure either by thermal soak or thermal fatigue of 1073 K and above shows degradation in the properties. The strength degradation behavior has been correlated with the microscopy of the fracture surfaces. The instantaneous peak in the load response is analogous to the strength of the fibers. This peak has not been found for highly oxidized specimens suggesting degradation of the fibers. The propensity of the crack to propagate along the laminar plies is more in presence of higher in-plane voids and the progress of these cracks across the fiber tows in the direction of applied load causes crack diversion leading to gracious failure. The temperature plays a major role for the oxidation kinetics and significant strength reduction (approx 85%) is observed for the thermally soaked specimens at Ts of 1273 K irrespective of the quenching rate. The gradual decrease in the flexural strength for the thermal fatigue exposed specimens with increase in the Tf from 1073 to 1473 K is in correlation with the higher amount of interface oxidation. For the intended application, the absence of the interface coating and seal coat, it was found that the residual flexural strength is better for the thermal fatigue exposed specimens as compared to the thermal soaked specimens.

5. ACKNOWLEDGEMENTS

The authors thank Director, Gas Turbine Research Establishment (GTRE) for permitting to publish the work. Shri. A. Udayakumar, National Aerospace Laboratories is gratefully acknowledged for many useful suggestions and the fruitful discussions during the execution of the work.

6. REFERENCES

1. James A. DiCarlo, Mark van Roode, Ceramic Composite Development for Gas Turbine Engine Hot Section Components, Proceedings of GT2006 ASME Turbo Expo 2006: Power for Land, Sea and Air, May 8-11, 2006, Barcelona, Spain
2. John, Dominy, Structural Composites in Civil Gas Turbine Aero Engines, Composites Manufacturing, Vol 5, No 2, 1994
3. H. Ohnabe et al., Potential Application of Ceramic Matrix Composites to Aero-Engine Components, Composites: Part A 30 (1999) 489–496

4. E. Bouillon, C. Louchet, P. Spriet, G. Ojard, D. Feindel, C. Logan, K. Rogers, T. Arnold, Post Engine Test Characterization of Self Sealing Ceramic Matrix Composites for Nozzle Seals in Gas Turbine Engines, Ceramic Engineering and Science Proceedings, Vol. 26, (8), 207-214

5. M. Takeda et al., High-Temperature Thermal Stability of Hi-Nicalon^{TM} SiC fiber/SiC Matrix Composites Under Long Term Cyclic Heating, Materials Science and Engineering A286 (2000) 312-323

6. L. Filipuzzini, G. Camus, R. Naslain, J. Thebault, Oxidation Mechanisms and Kinetics of 1D-SiC/C/SiC Composite Materials: 1. An Experimental Approach, Journal of the American Ceramic Society, 77 (2) (1994) 459-466

7. L. Filipuzzini, R. Naslain, Oxidation Mechanisms and Kinetics of 1D-SiC/C-SiC Composite Materials: 2. Modeling. Journal of the American Ceramic Society, 77 (2) (1994) 467-480

8. C. F. Windish Jr, C. H. Henager Jr, G. D. Springer, R. H. Jones, Oxidation of the Carbon Interface in Nicalon-Fiber-Reinforced Ceramic Matrix Composite, J. Am. Ceram. Soc. 80 (3) (1997) 569-574

9. Shijie Zhu et al., Time-Dependent Deformation in an Enhanced SiC$_f$/SiC Composite, Metallurgical and Material Transaction A, Vol 35A, September 2004-2853

10. G. Camus et al., Development of Damage in a 2D Woven C/SiC Composite Under Mechanical Loading: I. Mechanical Characterization, Composite Science and Technology 56 (1996) 1363-1372

11. T. K. Jacobsen, P. Brondsted, Mechanical Properties of Two Plain-Woven Chemical Vapor Infiltrated Silicon Carbide-Matrix Composites, J. Am. Ceram. Soc, 84 (5) 1043-51 (2001)

12. Shoujun Wu et al., Thermal Shock Behavior of a Three-Dimensional SiC/SiC Composite, Metallurgical and Material Transactions A, Volume 37A, December 2006, 3587-3592

13. James A. DiCarlo, Progress in SiC/SiC Ceramic Composite Development for Gas Turbine Hot Section Components under NASA EPM and UEET Programs, Proceedings of ASME Turbo Expo 2002, June 3-6, 2002, Amsterdam, The Netherlands

14. J. M. Agullo, F. Maury and J. M. Jouin, Mechanical Properties of SiC/SiC Composites with a Treatment of the Fiber/Matrix Interfaces by Metal-Organic Chemical Vapor Co-Deposition of C and Si$_x$C$_{1-x}$, Journal De Physique IV, Colloque C3, supplement au Journal de Physique II, Volume 3, 1993

15. A. Hahnel et al., Formation and Structure of Reaction Layers in SiC/Glass and SiC/SiC Composites, Composites Part A 27A (1996) 685-690

16. J. Mukherji, Ceramic Matrix Composites, Defence Science Journal, Vol 43, No. 4, October 1993, pp 385-395

17. M. Takeda et al., Strength of a Hi-Nicalon™/Silicon-Carbide-Matrix Composite Fabricated by the Multiple Polymer Infiltration–Pyrolysis Process, J. Am. Ceram. Soc. 82 (6) (1999) 1579-1581

18. Gregory N. Morscher, Intermediate Temperature Stress Rupture of Woven SiC Fiber, BN Interphase, SiC Matrix Composites in Air, Case Western Reserve University, Cleveland, Ohio, NASA / CR--2000-209927 April 2000

19. Yongdong Xu et.al., Microstructure and Mechanical Properties of Three-Dimensional Textile Hi-Nicalon SiC/SiC Composites by Chemical Vapor Infiltration, J. Am. Ceram. Soc, 85 (5) 1217-21 (2002)

20. W. H. Glime, J. D. Cawley, Oxidation of Carbon Fibers and Films in Ceramic Matrix Composites: A Weak Link Process, Carbon Vol. 33, No. 8, 1995, pp. 1053-1060

21. R.T. Bhatt, R.E. Phillips, Thermal Effects on the Mechanical Properties of SiC Fiber Reinforced Reaction Bonded Silicon Nitride Matrix (SiC/RBSN) Composites, NASA Technical Memorandum 101348, AVSCOM Technical Report 88-C-028

EFFECT OF PHASE ARCHITECTURE ON THE THERMAL EXPANSION BEHAVIOR OF INTERPENETRATING METAL/CERAMIC COMPOSITES

Siddhartha Roy, Pascal Albrecht, Lars Przybilla, Kay André Weidenmann, Martin Heilmaier, and Alexander Wanner

Institute for Applied Materials, Karlsruhe Institute of Technology
Kaiserstraße 12, 76131 Karlsruhe, Baden Württemberg, Germany

ABSTRACT

Interpenetrating metal/ceramic composites (IPC) consisting of percolating metallic and ceramic phases offer a good combination of properties. Thermal expansion behavior of two IPCs with different phase architectures is studied in this work during thermal cycling between RT and 500 °C. The composites were fabricated by melt infiltration in open porous ceramic preforms. One of the preform types was made by freeze-casting and had a lamellar structure; while the other type was fabricated by burning out cellulose place holders and had a highly open porous structure. Results show that the phase architecture strongly influences the thermal expansion of the composites. Freeze-cast samples display pronounced anisotropy – the thermal expansion coefficient (CTE) being low parallel to the freezing direction and high transverse to this direction. Only slight anisotropy is displayed by the highly open porous alumina based composite, with marginally higher CTE along the preform press direction. Thermal strains and CTEs of the highly open porous alumina based composite lie within the extremal values of the composite based on the freeze-cast preform. Thermal expansions of the composites correlate well with preform stiffness, with highest thermal expansion being observed along the most compliant directions in both composites.

INTRODUCTION

Metal/ceramic composites (MCC) are attractive because of the manifold advantages they offer, such as high specific stiffness and strength, better creep, fatigue and wear resistance, low thermal expansion coefficient (CTE) etc. [1, 2] Depending upon the architecture of the ceramic phase, MCCs can be classified into several classes like long fiber reinforced, short fiber reinforced, particle reinforced, interpenetrating in 3D composites etc. [3, 4] Among these, the interpenetrating composites (IPC) possess a better combination of strength, toughness and wear resistance than the other composite types. [5, 6] Moreover, the ability to easily fabricate these composites with tailor-made phase morphology have resulted in much recent research work on IPCs. [1]

In the last few years the present authors have carried out intensive research on two Al-alloy/alumina IPCs. [7, 8, 9, 10, 11, 12] One of the composites is based on freeze-cast alumina preforms and it has a lamellar domain structure. The second composite is based on a highly open porous alumina structure. Main emphases of these studies were to analyse the room temperature mechanical properties such as stiffness, compressive strength, damage mechanism and internal load transfer mechanism under external loading. Although MCCs display better high temperature properties than the unreinforced metallic alloy, they typically show a complex behavior when heated at high temperatures due to the presence of multiple phases with different CTEs. [13] Hence, an understanding of the influence of the phase architecture on the composite thermal expansion behavior is necessary. Although the thermal expansion behavior of particle reinforced [14, 15] and long fiber reinforced [16, 17] MCCs are well studied, relatively few studies are available for IPCs. The present study is aimed in this direction.

A preliminary study of the thermal expansion behavior of two IPCs with phase architectures similar to the composites mentioned above is carried out in this work. The alumina volume content in both the MCC types was similar – in the range of 38-43 vol. %. The composite samples were thermally cycled between room temperature and 500 °C and thermal expansions were measured along different directions. This allows investigating the influence of the phase morphology on the thermal expansion and the CTE of the two composites.

EXPERIMENTAL PROCEDURE
Specimen material

A detailed description of the processing route for the composite based on freeze-cast ceramic preform (hereafter named as MCC Type A) is given in earlier publications [7, 18] and will not be described here further. Alumina preforms were fabricated at the Institute for Applied Materials – Ceramics in Mechanical Engineering, Karlsruhe Institute of Technology, Germany via freeze-casting of an alumina suspension in water and subsequent freeze drying. Freeze-casting was carried out at -10 °C and the suspension had 22 vol. % alumina powder. The preforms were infiltrated by the eutectic alloy AlSi12 using squeeze-casting at the Casting Technology Centre, Aalen University of Applied Sciences, Germany.

Both the highly open porous alumina preform and the corresponding MCC (hereafter named as MCC Type B) were fabricated at CeramTec GmbH, Plochingen, Germany [19]. Alumina particles (diameter 1-3 μm) and approx. 25 vol. % cellulose fibres were first dispersed in a fluid suspension. This was then spray dried and the resulting powder mixture was uni-axially pressed into plates with nominal dimensions of $140 \times 120 \times 20$ mm³. The green plates were then sintered in air at 1400 °C. During the sintering process the cellulose fibres were burnt out and that generated the open porosity in the preforms. The porous preforms were subsequently infiltrated using the cast aluminum alloy AlSi9MgMn (trade name Silafont-36) using die casting. Infiltration was carried out at a temperature of 700 °C and under a pressure of 1250 bars.

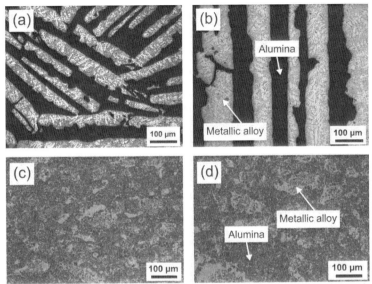

Figure 1: Optical microstructures of the two MCC types: (a) MCC Type A for the plane orthogonal to the preform freezing direction, (b) MCC Type B for the plane parallel to the preform freezing direction (= vertical direction in the image), (c) MCC Type B for the plane normal to the uni-axial press direction of the preform and (d) MCC Type B for the plane parallel to uni-axial press direction of the preform (= vertical direction in the image)

Typical optical micrographs of the two MCCs are shown in Figure 1. The MCC Type A consists of a lamellar domain structure in the plane orthogonal to the freezing direction (Figure 1a). Within each domain the alternating ceramic and the metallic lamellae lie parallel to each other. The lamellae size and spacing strongly depend on the freeze-casting parameters such as the freezing temperature, the ceramic content etc. Numerous ceramic bridges are present between the adjacent ceramic lamellae. They enhance the structural stability of the preforms and make them self-supporting. Along the freezing direction, the lamellae lie mostly parallel to each other and they are predominantly oriented along the freezing direction (Figure 1b).

The MCC Type B consists of clusters of both alumina (dark gray regions in Figure 1c-d) and aluminum based alloy (lighter regions in Figure 1b). The alumina clusters in the uninfiltrated preform had numerous micropores remaining after sintering. These micropores were also filled up by the metallic alloy during infiltration.

Measurement of the thermal strain and the CTE

Measurement of the thermal strain was carried out in a dilatometer of type DIL 805A/D from Bähr-Thermoanalyse GmbH, Hüllhorst, Germany by measuring the sample length change with temperature. For MCC Type A the samples had rectangular parallelepiped shapes with nominal dimensions of $7\times4\times4$ mm³, with length change being measured along the long axis of each sample. For

MCC Type B the samples were cylindrical with 4 mm nominal diameters and 7 mm nominal lengths. Four thermal cycles were carried out for each sample between room temperature and 500 °C. Thermal strain measurements were carried out during both heating and cooling at a constant rate of 5 °C/min in an inert helium atmosphere. Altogether 7 samples were studied – two samples of MCC Type A (one along the freezing direction and one normal to the freezing direction), three samples of MCC Type B (one each along the three orthogonal directions) and one sample of each unreinforced metallic alloy. Before measurement, the two opposite faces of each sample along the direction of strain measurement were polished to ensure plane parallelism when in contact with the measurement rods inside the dilatometer. In what follows, the direction along the preform freezing direction in MCC Type A and the direction along the preform press direction in MCC Type B will be referred to as the longitudinal direction and the directions orthogonal to these directions will be named as the transverse directions, respectively.

To determine the CTE evolution during both heating and cooling, the plot of length change vs. temperature was divided into discrete intervals of 30 °C each. A best fit straight line was fitted to the plot in each interval. Subsequently, the CTE at the mean temperature of each interval was determined following the relation

$$\alpha = \frac{1}{L_0} \cdot \left(\frac{dL}{dT} \right) \tag{1}$$

where L_0 is the sample length at the start of each interval and (dL/dT) is the slope of length change vs. temperature plot. For example, to calculate the CTE at 435 °C, the slope was determined between 420 °C and 450 °C and the sample length at 420 °C was used as L_0.

RESULTS AND DISCUSSIONS

Thermal expansion of MCC strongly depends upon the behavior of the constituent phases. Two different aluminum alloys were used as the metallic phase in the two composites and hence a study of their thermal expansion behavior is necessary for a thorough understanding. Figure 2 shows the thermal strain evolution with temperature during the four thermal cycles in the two metallic phases. Qualitatively similar behavior is observed in the two plots. Both the aluminum alloys display significant thermal hysteresis and the samples have residual compressive strain after thermal cycling. The extent of thermal hysteresis is highest in the first heating-cooling cycle.

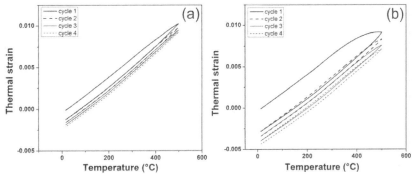

Figure 2: Evolution of the thermal strain with temperature during four thermal cycles in (a) AlSi12 (metallic phase in MCC Type A) and (b) AlSi9MgMn (metallic phase in MCC Type B)

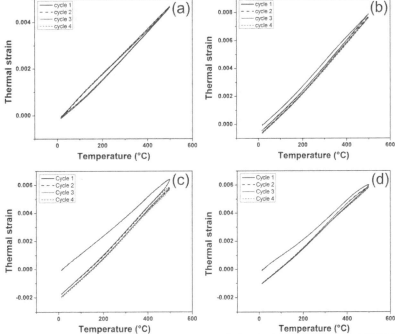

Figure 3: Evolution of the thermal strain vs. temperature during the four thermal cycles. (a) MCC Type A, along the preform freezing direction, (b) MCC Type A, transverse to preform freezing direction, (c) MCC Type B, along the preform press direction and (d) MCC Type B, transverse to the preform press direction

Thermal strain evolution in the MCC samples is shown in Figure 3. Plots (a) and (b) show the evolution in the MCC type A along and transverse to the preform freezing direction, respectively; while plots (c) and (d) show the same for the MCC Type B along and transverse to the uni-axial press direction of the preform, respectively.

The plots show that in comparison with the two unreinforced aluminum alloys, the thermal hysteresis is significantly smaller in all the MCC samples. Almost no thermal hysteresis is observed along the preform freezing direction in the MCC Type A and all four cycles show closed forms. In the remaining MCC samples, the highest thermal hysteresis is observed during the first heating cycle, while the three subsequent cycles display almost closed form shapes. The first heating cycle is typically influenced by the processing history and the previous stress state within the material. After slow heating and cooling during the first cycle the material attains a stabilised state and hence the subsequent cycles are more representative of the material behavior. Hence, for all samples the second heating-cooling cycle has been used for further discussion.

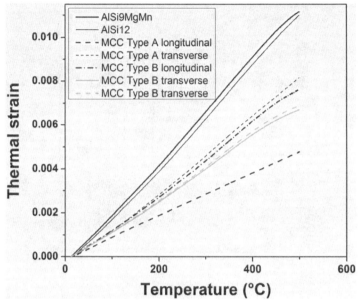

Figure 4: Evolution of the thermal strain vs. temperature during the second heating path in all studied samples

Figure 4 shows the thermal strain evolution during the second heating path in all studied samples. It is observed that the expansion is highest in the unreinforced aluminum alloy samples. In these samples, heating up to 500 °C results in almost 1.1% thermal strain. As expected, due to the presence of alumina, the thermal expansion in the MCC samples is significantly less and depending upon the direction of strain measurement and the type of ceramic architecture, it lies in the range of 0.45 – 0.8%. The MCC Type A shows a strong anisotropy, with the longitudinal thermal strain being significantly less (least among all studied samples) than the transverse thermal strain (highest among all studied samples). This can be attributed to the continuous ceramic lamellae along the freezing direction in this MCC type. This also follows our earlier studies [7, 9] where the continuous ceramic lamellae displayed maximum strength and stiffness along the freezing direction. Along the transverse direction the properties are dominated by the behavior of the metallic alloy phase. Hence, the thermal expansion is higher along this direction. In comparison, the anisotropy is significantly less pronounced in the MCC Type B and they lie within the extremities observed in the MCC Type A. However, here too a definite trend is observed with the longitudinal thermal expansion being marginally higher than the transverse thermal strains. In an earlier study the current authors carried out a thorough investigation of the elastic properties in similar MCC. [11] Even there a marginal anisotropy was reported. The longitudinal stiffness was found to be slightly less than the stiffness along the two transverse directions. This was attributed to slight orientation of the ceramic phase orthogonal to the longitudinal direction during uni-axial pressing of the powder mixture during preform fabrication. As this preferred orientation of the ceramic phase hinders the expansion of the metallic phase, marginally reduced thermal expansion is observed along the transverse direction in MCC Type B.

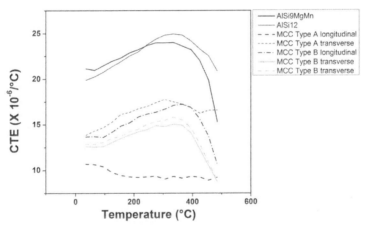

Figure 5: CTE evolution during second heating path in all studied samples

Figure 5 shows the CTE evolution during the second heating cycle in all studied samples. As expected, the highest CTE is observed in the unreinforced aluminum alloys and the two alloys show qualitatively a similar trend. CTE increases from about 20×10^{-6} / °C at room temperature to a maximum of about 24×10^{-6} / °C at about 300 °C. At still higher temperatures the CTE of the aluminum alloys decreases with increasing temperature. Huber et al. [20] also observed a similar CTE evolution in Al-Si alloys and they attributed the behavior to precipitation of pure Si from the α-solid solution up to about 300 °C and its re-dissolution at still higher temperatures.

The temperature dependence of the CTE in multiphase materials like MCCs is controlled by several parameters such as the prior stress history in individual phases, the phase architecture and the volume content of individual phases and the presence of defects such as cracks and pores etc. [13] Figure 5 shows that the levels of the CTE curves of the individual MCCs follow the same trend as with the thermal strain visible in Figure 4. Highest and lowest CTEs are observed in the MCC Type A along the transverse and the longitudinal directions, respectively. The CTE curves for the MCC Type B lie within the extremities of the MCC Type A.

The longitudinal CTE evolution curve for the MCC Type A can be divided into three regions. At low temperatures up to around 100 °C (region I) the CTE remains nearly constant with a value of around 11×10^{-6} / °C. Due to the large CTE difference between the metallic and the ceramic phases, self equilibrating residual stresses generate inside the MCC during cooling from high temperature. Due to the higher CTE of the metallic phase, these stresses are normally tensile within it and compressive within the ceramic phase. As the MCC is heated up inside the dilatometer, due to the different expansion of the two phases, stresses will alter in both the phases. At low temperatures the stresses in the two phases are still low and both phases behave elastically. At temperatures in the range of about 100 - 175 °C, the CTE drops by about 2×10^{-6} / °C. This can be attributed to a change in the stress state within the MCC. With an increase in temperature, the developing thermal stresses in both phases will first cancel out the original existing residual stresses in the phases and then further increase. Additionally, the yield stress of the metallic phase decreases with increasing temperature. A combination of these two phenomena will cause the metallic phase to undergo plastic deformation at some temperature. Subsequently, the matrix can flow into the residual pores in the MCC thereby reducing the CTE. The effect of pores on the CTE of metal matrix composites has been thoroughly discussed by Balch et al. [21], and they have shown that the presence of even a very small amount of residual porosity may significantly influence the CTE of the MCC. Finally, at temperatures between 175 – 500 °C (region III) the CTE remains almost constant at a value of about 9×10^{-6} / °C. The soft metallic phase has very little influence on the thermal expansion behavior of the MCC and it is now fully controlled by the stiffer and stronger ceramic phase. Böhm et al. [22] studied the thermal expansion of uni-axial ceramic fiber reinforced metal matrix composites and showed that when the matrix is fully plastic, the CTE of the composite along the fiber direction is controlled only by the ceramic phase and it is almost independent of the volume content of the metallic phase.

The CTE evolution along the transverse direction in the MCC Type A follows a completely different trend. The shape of the curve mostly follows the shape of the unreinforced metallic alloy, suggesting that the behavior is metallic phase controlled. Furthermore, corresponding to the drop in the longitudinal CTE due to the initiation of plastic deformation, there is a corresponding increase in the CTE along the transverse direction. This can be attributed to the stiffness anisotropy in the ceramic preform as well as in the MCC. Due to the continuous ceramic lamellae along the freezing direction, stiffness along this direction is significantly higher. However, these stiff ceramic lamellae strongly hinder the flow of the metallic phase along the longitudinal direction. As a result, once the metallic phase starts to deform plastically, due to constant volume condition, it expands more along the compliant directions. As a result, CTE increases along the transverse direction. A similar phenomenon has already been reported in existing studies with long fiber [16] and short fiber [23] reinforced metal matrix composites. In ceramic preforms fabricated by freeze-casting, the lamellae thickness changes with distance from the cold plate. With increasing distance from the cold plate, the structure becomes coarser [24]. This results in a gradient structure. Correspondingly, the local ceramic content varies with the distance from the cold plate. This should have a strong influence on the thermal expansion of the composite along the transverse direction. A finer lamellae spacing is expected to pose a stronger constriction to the softer metallic alloy phase and as a result both the transverse thermal strain and the transverse CTE should be lower. Additionally, as already reported in previous studies [7, 9], the domain orientation also controls the transverse properties of the composite. To obtain a representative transverse behavior, the length scale of the domains should be much smaller than the sample dimensions and the domains should be randomly oriented. These factors were not considered here and the studied sample was selected randomly. For a more thorough analysis, samples cut at different lengths along the freezing directions and with various domain orientations and sizes need to be studied.

Figure 5 further shows that the CTE curves for all three samples of the MCC Type B have shapes similar to the unreinforced metallic alloy, suggesting that the thermal expansion behavior is metallic alloy controlled. However, here too a marginal anisotropy is observed, with the longitudinal CTE being marginally higher than the two transverse CTEs, which are similar. Measured longitudinal elastic constants of the uninfiltrated alumina preform showed that its stiffness along the uni-axial press direction (= longitudinal direction) was significantly lower than the stiffness along the two transverse directions. Following the previous discussion for the MCC Type A, this explains the marginally higher CTE along the longitudinal direction.

CONCLUSIONS

A study of the effect of the phase architecture on the thermal expansion behavior of two interpenetrating Al-alloy/alumina metal/ceramic composites has been carried out in this work. Both composites had similar ceramic content, enabling meaningful comparison of only the influence of the phase architecture. The samples were thermally cycled between room temperature and 500 °C and the thermal expansion was measured along different directions to investigate the extent of anisotropy. Additional measurements were carried out in unreinforced Al alloys. The following main conclusions are drawn:

i) The composite based on the freeze-cast alumina preform displays significant thermal anisotropy with both the thermal strain and the thermal expansion coefficient being significantly lower along the freezing direction than along the transverse direction.

ii) Only minor thermal anisotropy is observed in the composite based on the highly open porous alumina based preform. Marginally higher thermal strain and thermal expansion coefficient are observed along the direction of uni-axial pressing employed during preform fabrication. Both the thermal strain and the CTE for this composite lie within the extremities displayed by the composite based on the freeze-cast ceramic preform.

iii) In the composite based on the freeze-cast preforms, the thermal expansion along the freezing direction is controlled by the stiff and continuous ceramic lamellae. Along the transverse direction in this composite type and along all three directions in the highly open porous alumina based composite, the thermal expansion behavior is more metallic alloy controlled.

iv) There is a good correlation between the extent of thermal expansion and the stiffness of the ceramic preform. Highest thermal expansion is observed along the most compliant direction in both composite types.

The current study shows that for similar ceramic volume content, it is possible to tailor the thermal expansion behavior of interpenetrating metal/ceramic composites by varying the morphology of the constituent phases. Effects of different ceramic volume content in the composite as well as different metal/ceramic interfaces will be investigated in our future studies.

ACKNOWLEDGEMENTS

The authors would like to thank the German research Foundation (DFG) for financial support through project RO 4164/1-1; Dr. T. Waschkies, Dr. R. Oberacker and Prof. M. J. Hoffmann (IAM-KM, KIT) for fabrication of porous freeze-cast preforms; Dr. A. Nagel and co-workers (FH Aalen) for squeeze-casting of freeze-cast preforms as well as CeramTec GmbH, Plochingen, Germany for the highly open porous alumina based composite samples

REFERENCES

[1] A. Mortensen and J. Llorca, Metal Matrix Composites, *Annu. Rev. Mater. Res.*, **40**, 243–70 (2010)

[2] N. Chawla, K. K. Chawla, Metal matrix composites, Springer (2006)

[3] T. W. Clyne, P. J. Withers, An introduction to metal matrix composites, Cambridge University Press (1993)

[4] J-M. Berthelot, Composite materials: mechanical behavior and structural analysis, Springer (1999)

[5] H. X. Peng, Z. Fan, J. R. G. Evans, Bi-continuous metal matrix composites, *Mater Sci Eng A*, **303**, 37–45 (2001)

[6] M. T. Tilbrook, R. J. Moon, M. Hoffman, On the mechanical properties of alumina–epoxy composites with an interpenetrating network structure, *Mater Sci Eng A*, **393**, 170–8 (2005)

[7] S. Roy, A. Wanner, Metal/ceramic composites from freeze-cast ceramic preforms: domain structure and elastic properties, *Compos Sci Technol*, **68**, 1136–1143 (2008)

[8] S. Roy, J. Gibmeier, A. Wanner, In-situ study of internal load transfer in a novel metal/ceramic composite exhibiting lamellar microstructure using energy dispersive synchrotron X-ray diffraction, *Adv Eng Mater*, **11**, 471-477 (2009)

[9] S. Roy, B. Butz, A. Wanner, Damage evolution and domain-level anisotropy in metal/ceramic composites exhibiting lamellar microstructures, *Acta Mater*, **58**, 2300-2312 (2010)

[10] S. Roy, J. Gibmeier, V. Kostov, K. A. Weidenmann, A. Nagel, A. Wanner, Internal load transfer in a metal matrix composite with a three dimensional interpenetrating structure, *Acta Mater*, **59**, 1424-1435 (2011)

[11] S. Roy, O. Stoll, K. A. Weidenmann, A. Nagel, A. Wanner, Analysis of the elastic properties of an interpenetrating AlSi12-Al$_2$O$_3$ composite using ultrasound phase spectroscopy, *Compos Sci Technol*, **71**, 962-968 (2011)

[12] S. Roy, J. Gibmeier, V. Kostov, K. A. Weidenmann, A. Nagel, A. Wanner, Internal load transfer and damage evolution in a 3D interpenetrating metal/ceramic composite, *Mater Sci Eng A*, **551**, 272-279 (2012)

[13] F. Delannay, Thermal stresses and thermal expansion in MMCs, Comprehensive Composite materials edited by A. Kelly and C. Zweben, Vol. 3 (2000)

[14] K-M, Shu, G. C. Tu, The microstructure and the thermal expansion characteristics of Cu/SiCp composites, *Mater Sci Eng A*, **349**, 236-247 (2003)

[15] Y. W. Yan, L. Geng, Effects of particle size on the thermal expansion behavior of SiCp/Al composites, *J Mater Sci*, **42**, 6433-6438 (2007)

[16] X. Luo, Y. Yang, C. Liu, T. Xu, M. Yuan, B. Huang, The thermal expansion behavior of unidirectional SiC fiber-reinforced Cu–matrix composites, *Scripta Mater*, **58**, 401-404 (2008)

[17] Z. H. Karadeniz, D. Kumlutas, A numerical study on the coefficients of thermal expansion of fiber reinforced composite materials, *Composite Structures*, **78**, 1-10 (2007)

[18] T. Waschkies, R. Oberacker. M. J. Hoffmann, Control of lamellae spacing during freezecasting of ceramics using double-side cooling as a novel processing route. *J Am Ceram Soc*, **92**, S79 – S84 (2009)

[19] I. T. Lenke, D. Rogowski, Design of metal ceramic composites. Int. J. Mat. Res., **97**, 676-680 (2006)

[20] T. Huber, H. P. Degischer, G. Lefranc, T. Schmitt, Thermal expansion studies on aluminium-matrix composites with different reinforcement architecture of SiC particles. *Compos Sci Technol,* **66**, 2206-2217 (2006)

[21] D. K. Balch, T. J. Fitzgerald, V. J. Michaud, A. Mortensen, Y. –L. Shen, S. Suresh, Thermal Expansion of Metals Reinforced with Ceramic Particles and Microcellular Foams. *Met Mater Trans A,* **27A**, 3700-3717 (1996)

[22] H.J. Böhm, H.P. Degischer, W. Lacom und J. Qu, Experimental and theoretical study of the thermal expansion behavior of aluminium reinforced by continuous ceramic fibers. *Composites Engineering,* **5**, 37-49 (1995)

[23] S. Kúdela Jr., A. Rudajevová, S. Kúdela, Anisotropy of thermal expansion in Mg- and Mg4Li-matrix composites reinforced by short alumina fibers. *Mater Sci Eng A,* **462**, 239-242 (2007)

[24] S. Deville, E. Saiz, A. P. Tomsia, Ice-templated porous alumina structures. Acta Mater, **55**, 1965-1974 (2007)

HIGH TEMPERATURE INTERACTIONS IN PLATINUM/ALUMINA SYSTEM

Ali Karbasi[1], Ali Hadjikhani[1,2], Rostislav Hrubiak[2], Andriy Durygin[2], Kinzy Jones[1]

1) Department of Mechanical and Materials Engineering, Advanced Materials Engineering Research Institute, Florida International University, Miami 33199, USA
2) Department of Mechanical and Materials Engineering, Center for the Study of Matter at Extreme Conditions, Florida International University, Miami 33199, USA

ABSTRACT

There is increasing demand for hermetic metal/ceramic bonds for application in biomedical engineering, in particular for use in neurostimulating prosthetic devices such as, cochlear implants, muscular stimulators and retinal prosthesis. Platinum/Alumina bonds are particularly interesting because of the proven biocompatibility of the two materials and their strong bonding. Yet, the true nature of their bonding is not clear. Platinum/alumina interactions in different atmosphere (i.e. air and hydrogen) and different temperatures were studied by means of high temperature X-ray diffraction, SEM and EDS analyses, to better understand the interfacial reactions and bonding mechanism. It was observed that upon heating the platinum/alumina system in the reduced atmosphere tetragonal Pt_3Al formed in low temperature and transformed to cubic structure at higher temperatures. In addition to that, at temperatures above 1500 °C alumina could migrate and encapsulate the platinum particles, with particle migration mechanism.

INTRODUCTION

Alumina ceramics brazed to a platinum wire pin has a long and high reliability history in implantable devices, due to biocompatibility of both platinum and alumina. Strong bonding between metal and ceramic is the most important factor to develop a hermetic device. The bonding mechanism between platinum and alumina is mostly belived to be solid-state ceramic-metal diffusion bonding. However, despite the different reaction mechanisms and products that have already been reported for the Pt/Al_2O_3 system, the exact bonding mechanism between them is not clear [1-13]. This ambiguity could be due to the fact that $Pt-Al_2O_3$ or $Pt-Al-O$ phase digrams are unknown or from the complicated behavior of metal and oxide at high temperatures.

Schulz et al. [1] first studied the reaction of refractory metals with several oxides in a reduced atmosphere by the X-ray diffraction technique. They showed that the Pt/Al_2O_3 system could go through several alloy formation reactions in temperatures above 1100 °C. It has been shown [8, 14] that platinum in the prescence of stable oxide (i.e. SiO_2 and Al_2O_3) in reduced atmposhere can go through alloy formation reaction and create Pt_3M (M:Al and Si) with the cubic Cu_3Au (Pm3m space group) structure. Tetragonal $GaPt_3$ crytal structure was also used to anlyse the same stocimetry by Bronger et al. [7]. Zhong et al. [10] showed that the mixtuire of hydrogen/Argon instead of pure hydrogen can change the resultant product. Reaction of Pt/Al_2O_3 in pure hydrogen in 700 °C formed the Pt_3Al compound [8], while the same composition at the same temperature in the H_2-Ar (10-90) will produce Pt_8Al_{21} [10].

The first solid bonding of the platinum/alumina system in a reduced atmosphere was reporetd by Allen and Borbidge [3]. They showed that bond strength increases by increasing time and temperature. They suggested that the best bonding was achieved just below the melting point of platinum and in the time of 10 hours or more. In addition, they showed that the application of force could enhance the bond strength.

Panfilov [9] used a plasma evaporation technique for joining platinum and alumina and reported the formation of the transition layer between the platinum and the alumina coating (30 μm in depth). The transition layer contained small (~ 1 μm) and large (~ 20-50 μm) alumina inclusions and possessed a rough surface on the ceramic side. Additionally, the alumina particles penetrated into the platinum matrix during the sanding stage, but this treatment has not changed the mechanical behavior of the platinum either at room or elevated temperatures. However, the true nature of the transition layer and the bonding is not clear yet [9].

De Graef et al. [15] used the platinum/alumina system with and without an SiO_2 intermediate layer, for bonding in an air atmosphere. They showed that SiO_2 could work as the liquid diffusion phase and enhanced the bonding between platinum and alumina by the assumption of the limited solubility of one of them or both. However, several studies [13, 16, 17] showed that the strength of direct bonding Pt/Al_2O_3 depends basically on the crystal orientation of platinum and alumina. When the orientation is suitable for semi-coherency, the interface has a strong driving force to adopt direct metal-ceramic bonding; when the orientation is not suitable for coherency, then a weaker metal-glass-ceramic bond is favored since the covalent bonds in the glass and alumina are more conducive to mutual bonding.

EXPERIMENTS
Sample Preparation

The spray dried platinum black (Heraeus Inc. PM-100-10) mixed with the nano alumina particle (MKnano Co. MKN-Al2O3-A040 MKN40) for X-ray diffraction analyses. Different ratios of platinum and alumina were used in this paper to study the effect of the composition on the reaction product. Each ratio was chosen based on the Pt-Al intermetallic compounds of the Pt-Al phase diagram. Each sample was rapidly heated to 1400 °C and 1600 °C and then maintained that temperature for 8 hours. Experiments were carried out in two different atmospheres (i.e. air and H_2-Ar (10-90)) to see the effect of atmosphere on the reaction.

A 10 μm thick film of platinum paste was prepared from the same platinum powder and applied on the alumina substrate to study the effect of platinum on the microstructure of alumina at high temperatures. The thick film heated up to 1550 °C in the H_2-Ar (10-90) atmosphere mixture and maintained that temperature for 3 hours.

Experimental Techniques

JEOL JSM-6330F FE/SEM and CM 200 TEM were used to capture the electron microscope image of the samples. EDS analyses was used to evaluate the diffusion profiles of elements at the interface. Measurements were taken in an arbitrary imaginary line through the interface. The results are the average measurements of the several lines. JIB 4500 Multi–beam FIB/SEM was used to section the samples for TEM analyses.

Siemens x-ray system was used to measure the crystal structure of the samples. MoKα radiation (tube voltage 50 kV, tube current 24 mA, cathode gun 0.131 mm) monochromated using an incident beam graphite monochromator was passed through a collimator of diameter 200 μm to the sample with the wavelength of 0.71073 nm. The diffracted x-ray was collected on a 512×512 pixels imaging plate area. Data were acquired for 2θ range of 5 and 40°.

High temperature *in-situ* X-ray diffraction experiments were conducted on the beam line B-2 (λ = 0.485946Å) of the Cornell High Energy Synchrotron Source (CHESS). The sample was resistively heated in a specially designed high-temperature cell. The description of the cell can be found elsewhere [18]. The sample was loaded and heated in a 0.15mm hole, which was drilled through the 1mm thick graphite-heating element. The temperature of the sample was measured using a d-type thermocouple and by measuring the thermal expansion of platinum [19] by *in-situ* x-ray diffraction from (111), (200), and (220) lattice planes. X-ray diffraction measurements in the temperature range between room temperature and 1500 °C. Diffracted X-rays were collected between Bragg angles of $2\theta=5°$ and $2\theta=25°$ using a MAR3450 imaging detector. Sample to detector distance and other diffraction geometry parameters were calibrated using a CeO_2 standard. 2D angle-dispersive diffraction images were processed using the software FIT2D to generate the intensity versus 2θ diffraction patterns. Each diffraction peak was indexed and fitted with a pseudo-Voigt function to determine its d-spacing.

RESULTS

Platinum/Alumina Reaction
        As expected, the X-ray results of the platinum/alumina powder mixture heated in air do not show any sign of reaction. Although different ratios of platinum/alumina were used, XRD results indicated that, in all cases, a series of new peaks appeared upon heating the sample in hydrogen. The first peak was at $2\theta=10.40$Å, which indicates the formation of cubic Pt_3Al. At temperatures below 1290 °C, Pt_3Al transforms to tetragonal the $GaPt_3$ structure [20], and could also goes through two other tetragonal transformations in lower temperatures [21]. While Bronger et al. [20] adopted a $GaPt_3$ tetragonal structure to analyze their experimental data, the cubic structure agrees more with our results. The observed Pt_3Al cubic has a lattice parameter of 3.88 Å, which is consistent with the literature, 3.876 Å[20]. This suggests that cubic structure was formed at high temperatures, but slow transformation kinetics limited the tetragonal structure formation in the cool-down process.

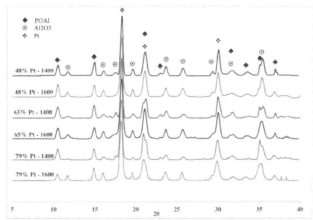

Figure 1 X-ray diffraction results of different platinum/alumina ratios fired in hydrogen at 1400 °C and 1600 °C for 8 hours

        The high temperature XRD results at different temperatures are shown in figure 2. It should be noted that graphite was used as the reducing agent to study the effect of temperature on the platinum-alumina reaction. The sample maintained at each temperature for 5 minutes and then X-ray data was recorded. Due to the short amount of reaction time, the intensity of the observed diffraction peaks in comparison to platinum is negligible. Because of that, only the portion of the diffraction pattern with more emphasis on the reaction product is shown.

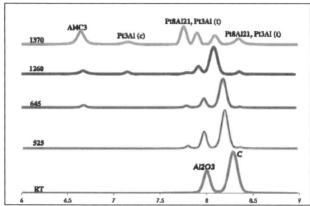

Figure 2 In situ X-ray diffraction results of 65% Pt sample in different temperatures (Tempertures shown in °C)

The first two peaks in $2\theta=7.77°$ and $8.35°$ appeared at the temperatures of above 500 °C, which are related to the tetragonal Ir_3Si structure, the other tetragonal structure of Pt_3Al [22]. By increasing the temperature up to 645 °C, another peak appeared at $2\theta=6.65°$. That peak is attributed to the formation of rombohedral structure of aluminum carbide [24]. At temperatures above 1200 °C, a new peak appeared at $2\theta=7.15°$, that corresponds to the cubic Pt_3Al structure. The Ir_3Si structure transforms to cubic Cu_3Al at high temperatures. The Pt_3Al was formed with the tetragonal structure at low temperatures and then it transformed to the cubic structure at high temperatures. With a sufficient amount of time and hydrogen concentration, the tetragonal structure transforms completely to the cubic structure. On the other hand, Zhong et al. [10] report the formation of Pt_8Al_{21} in the H_2-Ar (10-90) atmosphere. Pt_8Al_{21} also has the same crystal structure [23] and it is possible that the tetragonal Pt_8Al_{21} was formed at low temperatures and then transferred to the cubic Pt_3Al at higher temperatures. However, Zhong et al. [10] used TEM diffraction to confirm the formation of Pt_8Al_{21}, which could not rule out the formation of the tetragonal Pt_3Al instead of Pt_8Al_{21}. Phase transformation of the tetragonal Pt_3Al to cubic Pt_3Al at high temperatures is a thermodynamically favorable process [22], while Pt_8Al_{21} to Pt_3Al transformation is not. It could be concluded that the formation of tetragonal Pt_3Al at low temperatures and then phase transformation at higher temperatures is more reasonable explanation for the reaction of platinum and aluminum oxide.

Platinum/Alumina Interface Structure

The thick platinum film applied to the alumina substrate was heated up to 1550°C in the H_2-Ar (10-90) atmosphere and maintained that temperature for 3 hours. The microstructure of the fired sample is shown in figure 3. Platinum powder started to coalesce and form particles with random size distribution at high temperatures. In addition to the coalescence, upon heating, a layer of ceramic covers some of the platinum particles. This happens mostly for the larger particles, which are, by average, larger than 10 μm in diameter. SEM images of the two different platinum particles (i.e. particles (a) and (b) are 90 and 17 μm) are shown in figure 3. It is clear that increasing the particle size increases the order structure of the alumina coating. The coating on the surface of particle (b), is very rough, while for the particle (a), the coating has a very fine structure.

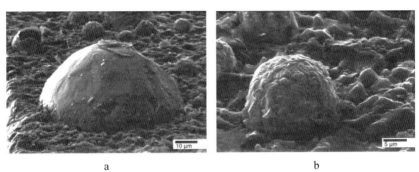

a   b

Figure 3 SEM image of thick film of platinum on alumina substrate fired at 1550 °C in the H_2-Ar (10-90) atmosphere for 3 hours

Fu et al. [25] showed that due to the difference in the surface tension and work function of metal and oxide, TiO_2 could migrate on the surface of the platinum particle and encapsulate it. Yet, this is not possible for aluminum oxide to cover the platinum under normal conditions [25]. In a reduced atmosphere at high temperatures, alumina migrates with either a mechanism of atomic or particulate migration over the surface of the platinum and covers it. Migration mechanism is atomic migration when the interaction is strong but changes to the particle migration when the interaction is weak [26, 27]. Straguzzi et al. [28] showed that the interaction force depends mostly on the surface area, which determines the binding points. For small surface area particles, the interaction is probably weak and the resistance upon migration is low, so they can migrate on the surface of oxide and sinter upon collision. For high surface area particles, the growth can occur only through the atomic migration. In the case of the particles shown in figure 3, alumina migrated with the particle migration mechanism. Those particles are formed from the agglomeration of pre-sintered submicron platinum particles. They have a low surface area, which decreases the interaction force and makes it possible for the alumina particles to migrate.

The alteration of the microstructure of alumina at temperatures above 1500 °C needs to be explained in another way and it is similar to the encapsulation process of platinum with titanium dioxide. Particles shown in figure 3 are formed from the coalescence of small particles. This enables the alumina particles to move through the platinum particle. Increasing the platinum particle size increases the distance that alumina needs to travel to reach the top surface. Hence, the larger portion of alumina dissolved in the migration process through the platinum particle and the developed coating has a finer structure. It was shown in the Pt-Al phase diagram [23, 29] that the Pt/Pt_3Al mixture becomes liquid at temperatures above 1500 °C. This liquid creates an appropriate medium for the ceramic particles' migration and increases the possibility of alumina absorption and migration. The proposed mechanism is very similar to the glass migration mechanism [30].

The results of heating the platinum/alumina compound at 1400 °C and 1600 °C are shown in figure 4. The microstructure of the alumina completely changed by increasing the temperature from 1400 °C to 1600 °C. As is shown, the agglomerated alumina particles grow and transform to a large grained structure. This could be explained with the same mechanism, in which above 1500 °C, the Pt/Pt$_3$Al phase helps alumina to grow, and because of that the surface of platinum particles is covered with alumina.

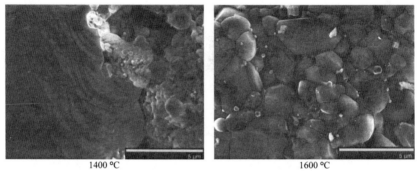

1400 °C   1600 °C

Figure 4 SEM image of platinum/alumina compound fired at 1400 and 1600°C in the H$_2$- Ar (10-90) atmosphere for 8 hours

The platinum/alumina interface in particle 3 (b) sectioned with the Focus Ion Beam and TEM image of that interface is shown in figure 5. The alumina particle in the interface still maintains its original form, however, the interface shows strong bonding between platinum and alumina, which is due to the reduction reaction of alumina and the formation of Pt$_3$Al.

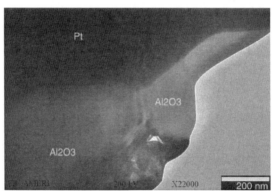

Figure 5 TEM image of the interface between platinum via and alumina particle

Diffusion profiles of the aluminum and oxygen in platinum were determined by EDS analyses and shown in figure 6. It is clear that the concentration of the aluminum gradually decreases, however the final concentration is about 5 Wt%, which is more than the solubility limit of the aluminum in platinum (~2%). This behavior promotes the formation of the Pt$_3$Al phase. The concentration of oxygen shown in figure 6 is almost zero in all cases, which is due to the reduction of alumina in the reduced atmosphere. It should be noted that EDS does not show any trace of platinum in alumina, which indicates that diffusion only happened from the alumina side to the platinum.

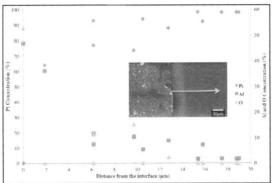

Figure 6 Concentration gradient of different element in platinum for the Pt-5 platinum powder and 99.9% HTCC ceramic tape. Feedthrough fired in the $H_2 - Ar$ (25-75) atmosphere at 1550 °C

CONCLUSION

The results of heating the platinum/alumina mixture in different conditions showed that the mixture could go through several intermetallic compound formations. The cubic structure of Pt_3Al compound is the most stable product at high temperatures. However, the tetragonal structure of Pt_3Al was identified as the reaction product at low temperatures. It then transforms to cubic Pt_3Al at high temperatures. It has been shown that despite the fact that platinum and alumina could go through alloy formation reaction at low temperatures, yet, the recrystallization and the change of the alumina structure can only be achieved at temperatures above 1500 °C. In reduced atmosphere, at high temperatures, platinum absorbs alumina particles. At temperatures above 1500 °C, these absorbed particles could be melted completely and be mixed with platinum to create a very dense structure, or they can migrate to the surface of the platinum and encapsulate the platinum particles.

REFERENCES

1.   Schulz H, Ritapal K, Bronger W, Klemm W. Über die Reaktion von Elementen der achten Nebengruppe mit Oxiden unedler Metalle im Wasserstoffstrom (About the reaction of elements of the eighth group with oxides of base metals in a hydrogen stream). Zeitschrift für anorganische und allgemeine Chemie. 1968;357:299-313.
2.   Becher PF, Murday JS. Thick film adherence fracture energy: influence of alumina substrates. Journal of Materials Science. 1977;12(6):1088-94.
3.   Allen RV, Borbidge WE. Solid state metal-ceramic bonding of platinum to alumina. Journal of Materials Science 1983;18:2835-43.

4.   Mulder CAM, Klomp JT. On The Internal Structure of Cu- and Pt- Sapphire Interfaces. J Phys Colloques. 1985;46(C4):C4-111-C4-6.
5.   Sushumna I, Ruckenstein E. Redispersion of Pt/alumina via film formation. Journal of Catalysis. 1987;108(1):77-96.
6.   Hwang CP, Yeh CT. Platinum-oxide species formed by oxidation of platinum crystallites supported on alumina. Journal of Molecular Catalysis A: Chemical. 1996;112(2):295-302.
7.   Bronger W, Wrzesien K, Müller P. High temperature phase transitions in Al1+xPt3-x. Solid State Ionics. 1997;101-103(Part 1):633-40.
8.   Penner S, Wang D, Su DS, Rupprechter G, Podloucky R, Schlögl R, et al. Platinum nanocrystals supported by silica, alumina and ceria: metal-support interaction due to high-temperature reduction in hydrogen. Surface Science. 2003;532-535:276-80.
9.   Panfilov P, Bochegov A, Yermakov A. The Transition Layer in Platinum-Alumina. Platinum Metals Review. 2004;48(2):47-55.
10.   Zhong X, Zhu J, Liu J. Study of the interfacial structure of a Pt/α-Al2O3 model catalyst under high-temperature hydrogen reduction. Journal of Catalysis. 2005;236(1):9-13.
11.   Luo M-F, Ten M-H, Wang C-C, Lin W-R, Ho C-Y, Chang B-W, et al. Temperature-Dependent Oxidation of Pt Nanoclusters on a Thin Film of Al2O3 on NiAl(100). The Journal of Physical Chemistry C. 2009;113(28):12419-26.
12.   Ivanova AS, Slavinskaya EM, Gulyaev RV, Zaikovskii VI, Stonkus ûê, Danilova IG, et al. Metal-support interactions in Pt/Al2O3 and Pd/Al2O3 catalysts for CO oxidation. Applied Catalysis B: Environmental. 2010;97(1-2):57-71.
13.   Santala MK, Radmilovic V, Giulian R, Ridgway MC, Gronsky R, Glaeser AM. The orientation and morphology of platinum precipitates in sapphire. Acta Materialia. 2011;59(12):4761-74.
14.   Klomp J. Ceramic-metal reactions and their effect on the interface microstructure. Ceramic microstructures '86: Role of interfaces; Proceedings of the International Materials Symposium; 28-31 July 1986; Berkeley, CA; UNITED STATES1987. p. 307-17.
15.   De Graef M, Dalgleish BJ, Turner MR, Evans AG. Interfaces between alumina and platinum: Structure, bonding and fracture resistance. Acta Metallurgica et Materialia. 1992;40(Supplement 1):S333-S44.
16.   Suppel KP, Forrester JS, Suaning GJ, Kisi EH. A Study of the Platinum/Alumina Interface. Advances in Science and Technology. 2006;45:1417-22.
17.   Lu H, Svehla MJ, Skalsky M, Kong C, Sorrell CC. Pt-Al2O3 interfacial bonding in implantable hermetic feedthroughs: Morphology and orientation. Journal of Biomedical Materials Research Part B: Applied Biomaterials. 2012;100B(3):817 - 24.
18.   Dubrovinsky LS, Saxena SK. Emissivity measurements on some metals and oxides using multiwavelength spectral radiometry. High Temperatures - High Pressures. 1999;31(4):393-9.
19.   Arblaster JW. Crystallographic properties of platinum. Platinum Metals Review. 1997;41:12-20.
20.   Aaltonen T, Ritala M, Sajavaara T, Keinonen J, Leskel√É¬§ M. Atomic Layer Deposition of Platinum Thin Films. Chemistry of Materials. 2003;15(9):1924-8.
21.   Alemany P, Boorse RS, Burlitch JM, Hoffmann R. Metal-ceramic adhesion: quantum mechanical modeling of transition metal-alumina interfaces. The Journal of physical chemistry. 1993;97(32):8464-75.
22.   Chauke HR, Minisini B, Drautz R, Nguyen-Manh D, Ngoepe PE, Pettifor DG. Theoretical investigation of the Pt3Al ground state. Intermetallics. 2010;18(4):417-21.
23.   McAlister A, Kahan D. The Al-Pt (Aluminum-Platinum) system. Journal of Phase Equilibria. 1986;7(1):47-51.

24.   Foster LM, Long G, Hunter MS. Reactions Between Aluminum Oxide and Carbon The Al2O3—Al4C3 Phase Diagram. Journal of the American Ceramic Society. 1956;39(1):1-11.
25.   Fu Q, Wagner T, Olliges S, Carstanjen HD. Metal-Oxide Interfacial Reactions: Encapsulation of Pd on TiO2 (110). The Journal of Physical Chemistry B. 2005;109(2):944-51.
26.   Baker RTK, Prestridge EB, Garten RL. Electron microscopy of supported metal particles : I. Behavior of Pt on titanium oxide, aluminum oxide, silicon oxide, and carbon. Journal of Catalysis. 1979;56(3):390-406.
27.   Baker RTK, Prestridge EB, Garten RL. Electron microscopy of supported metal particles II. Further studies of Pt/TiO2. Journal of Catalysis. 1979;59(2):293-302.
28.   Straguzzi GI, Aduriz HR, Gigola CE. Redispersion of platinum on alumina support. Journal of Catalysis. 1980;66(1):171-83.
29.   Kim DE, Manga VR, Prins SN, Liu ZK. First-principles calculations and thermodynamic modeling of the Al-Pt binary system. Calphad. 2011;35(1):20 - 9.
30.   Twentyman ME. High-temperature metallizing, Part 1. The mechanism of glass migration in the production of metal-ceramic seals. Journal of Materials Science. 1975;10(5):765-76.

FRACTURE MECHANICS OF RECYCLED PET-BASED COMPOSITE MATERIALS REINFORCED WITH ZINC PARTICLES

Jessica J. Osorio-Ramos, Elizabeth Refugio- García, Víctor Cortés-Suarez
Departamento de Materiales, Universidad Autónoma Metropolitana
Av. San Pablo # 180, Col. Reynosa-Tamaulipas, México, D. F. 2200

Enrique Rocha-Rangel
Universidad Politécnica de Victoria, Av. Nuevas Tecnologías # 5902, Parque Científico y
Tecnológico de Tamaulipas, Ciudad Victoria, Tamaulipas, México, 87138

ABSTRACT

The present research paper analyzes the behavior related with the impact toughness and the fracture mechanics displayed by new composite materials. These materials were manufactured by powder technology, based on recycled PET, as a polymer matrix, reinforced with 10 and 30 wt. % zinc particles respectively. The necessary test specimens for isothermal sintering were prepared and exposed at 256°C for 15 minutes periods. These preliminary specimens permitted to determine the strain intensity factor around a crack, using Peterson´s and Irwin´s models, and using the Griffith criterion in order to find the value of released energy. Therefore, the best results were obtained when the content of metallic particles were increased because the material improve its mechanical properties such as: resistance and tenacity.

INTRODUCTION

Currently polyethylene terephthalate, better known as PET, despite their high level of production is recycled in very small quantities, as an example in the 2006 year, according to data from the annual report issued by Ecoce (organization that join the major generators of PET containers in Mexico), only 20% of plastic sent to market was recycled in Mexico, while in the U.S.A. it was recycled 23%. Both country recycled very low quantities of PET. So that the management of this material has become an environmental problem, which many efforts have pledge[1].

In addition, talking about processing techniques, they mainly rely on processes such as PIM (Powder Injection Molding), which combines thermoplastic injection molding and conventional powder metallurgy. Technique used to reinforce ceramic and metallic materials[2]. However, the technique used in this project differs from the conventional, as it applies a technique very similar to powder metallurgy based on four basic stages: 1) Obtaining the powder by different techniques such as extraction, deposition and atomization[3]. 2) Grinding and homogenization of the powders[4]. 3) Forming the powder mixtures in metal matrices to reduce the gaps between the particles and increase the density of the compact, resulting in adhesion and cold welding of the particles[5]. 4) Sintering of raw or green compact by heating in an oven for a certain time and temperature to achieve union between the particles[4].

Likewise, the components to be employed, PET, is characterized by being a solid at room temperature, that are subjected to some hundred degrees becomes a viscous material easy to handle[6], while zinc alloys exhibit a low melting point, with excellent corrosion resistance, good flow and sufficient strength for structural applications[4]. So that in the polymer-metal composite, the polymeric matrix provides continuity on the structure, while the metal increases the density, improves formability, durability and mechanical resistance[7]. Therefore one of this research objective is to determine the fracture strength of these materials by increasing the metal content there in.

On the other hand, in relation to resistance and fracture, the fracture mechanics is the portion of solid mechanics to relate the size and shape of a crack and forces or loads which lead to fracture. The purpose of testing is to determine the stress intensity factor, K_{IC}, which can be known by different methods, one of that is theoretical and is based on the stress concentration.

The stress intensity factor K_{IC} is the most significant parameter of fracture mechanics linear-elastic, it defines the magnitude of stresses around a crack. Moreover K_{IC} just depends on the geometry of the piece, the crack size and the way of load application, but is independent of the type of material[8].

EXPERIMENTAL
Coding samples
To identify study samples it was used the following coding; **PEZI XZ**

Where:
PEZI: Refers to the mixture of PET with zinc.
X: Corresponds to the zinc content of the mixture (0, 10 or 30% respectively)
Z: Is a progressive serial letter.

Test specimens preparation
Using a powders technique, three test pieces were produced for each studied material and two for the blank material, to find the fracture toughness of the material. And another equal amount of samples to find the IZOD impact strength. For the preparation of the specimens they were used zinc powders with particle size between 1 and 5 microns, purity of 99.7%. Recycled PET powder mechanically processed were used in amounts of 10 and 30 wt. %, the maximum particle size of PET was 420 microns. PET and Zn powders were mixed and ground during 3 h at 375 rpm in a roller mill (Colepalmer, USA) using stabilized zirconia grinding element in a ratio of 15:1.

Subsequently, the ground and mixed powders were compacted in a uniaxial press in two different dimensions: 4.8 X 10 X 70 mm for the fracture toughness test and 10 X 10 X 70 mm for the impact test, with a subsequent isothermal sintering process in a muffle Lindberg, mod 51894 applying 256°C during 15 minutes. The term isothermal refers about the compacts were introduced in the muffle when it reaches the process temperature. After that, a notch was made in the central part of all specimens, with a 45° angle, and a depth of 2 mm. The specimen dimensions were verified with a digital vernier caliper Starrett mod. 721.

Once dimensioned the specimens, the notches were also dimensioned in relation to the aperture, depth and radius as seen in Figure 1 using a Mitutoyo profile projector mod PH-3500, which operates on the basis of coordinates.

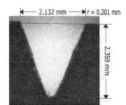

Figure 1. Profile of notch sample PEZI 10c.

All samples dimensions are shown in Table I, they were obtained in order to apply the theoretical equations of resistance, with the objective of determine the stress intensity factor K_{IC}, based on the stress concentration.

Samples	Aperture [mm]	Depth [a] [mm]	Angle [α] [degrees]	Radius [r] [mm]
PEZI 0a	1.742	1.869	42 45	0.214
PEZI 0b	3.117	3.587	39 46	0.224
PEZI 10a	2.575	2.806	45 22	0.209
PEZI 10b	2.226	2.322	44 31	0.295
PEZI 10c	2.132	2.359	41 38	0.201
PEZI 30a	3.127	3.703	43 00	0.240
PEZI 30b	3.336	3.830	42 39	0.295
PEZI 30c	2.639	2.742	46 26	0.197

Table I. Summary of the dimensions of the notches in the specimen under study.

Bending test.
    To apply the theoretical models of Peterson, Irwin and Griffith, it was first necessary to test the 8 specimens, previously prepared and tested in three-point bending to find that maximum force could with stand the test pieces before breaking. This test was conducted in the Instron universal machine, mod. 5500R, using a rate of 0.5 mm / min.

    Thereafter, defining the encoding of the specimens based on their geometry, which corresponds to a flexure beam as shown in Figure 2[8].

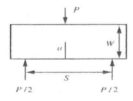

Figure 2. Geometry of a beam in bending.

Where:
W: the height of the sample [m]
P: is the applied force [N]
S: is the spam formed between the support [m]
a: is the depth of the crack [m]
Besides using other values are also useful.
B: thickness of the sample [m]
r: the radius [m]
d = W - a [m]

Thus using this coding will be able to obtain other parameters in order to find the value of K_{IC}.

    Using the model of Peterson it is possible to get K_{tn}, σ_{max} and σ_{nom},

Where:
K_{tn}: The factor of elastic stress concentration
σ_{max}: The effort at the root hub of efforts
σ_{nom}: The nominal stress

    In the concentration stress charts developed by Peterson[9] it can be found the value of K_{tn}, relating r with d and W with d. However, also it can be obtained applying the equations of the curves and formulas mentioned below[9].

$K_{tn} = C_1 + C_2(a/W) + C_3(a/W)^2 + C_4(a/W)^3$ -------------------- (1)

In the case where 2.0 < a/r < 20, the following equations are applied.

$$C_1 = 2.966 + 0.5021a/r - 0.009(a/r)^2 \quad\text{------------------------} \quad (2)$$

$$C_2 = -6.475 - 1.126a/r + 0.019(a/r)^2 \quad\text{------------------------} \quad (3)$$

$$C_3 = 8.023 + 1.2531a/r - 0.020(a/r)^2 \quad\text{------------------------} \quad (4)$$

$$C_4 = -3.572 - 0.634a/r + 0.010(a/r)^2 \quad\text{------------------------} \quad (5)$$

Where a, is the crack depth, r the radius and W the specimen height.

On the other hand to find the σ_{nom}, it may be applied to any of the following equations, with the same results.

$$\sigma_{nom} = 6M/2Bd^2 \text{ or } \sigma_{nom} = 3PS/2Bd^2 \quad\text{------------------------} \quad (6)$$

Where M refers to the bending moment, B is the sample thickness and d is the specimen resulting from the subtraction between the height of the specimen and the depth of the crack.

Finally to find $\sigma_{máx,}$, the following relationship is applied:

$$K_{tn} = \sigma_{máx} / \sigma_{nom}$$

Where: $\sigma_{máx} = K_{tn} \times \sigma_{nom}$ ------------ (7)

Obtaining the results shown in Table II, for all the specimens tested.

Samples	K_{tn}	σ_{nom} [Mpa]	σ_{max} [Mpa]
PEZI 0a	4.239	5.089	21.575
PEZI 0b	3.722	4.350	16.193
PEZI 10a	4.100	4.585	18.798
PEZI 10b	3.698	3.351	12.392
PEZI 10c	4.371	3.481	15.776
PEZI 30a	3.983	4.141	14.658
PEZI 30b	3.735	3.680	14.694
PEZI 30c	4.483	3.934	19.094

TableII. Results of K_{tn}, $\sigma_{máx}$ and σ_{nom}, for all studied samples.

Obtaining K_{IC} with the model of Irwin

Later, to obtain the K_{IC} which is the stress intensity factor around the crack, seen in the displacement mode I fracture surface, it was considered the following equation[8].

$$K_{IC} = \beta\sigma(\pi a)^{1/2} \quad\text{------------------} \quad (8)$$

Where:
β: The geometric correction factor, which for small lateral cracks corresponds to 1.12 according to the results of David Broek[10].

σ: Refers to the maximum stress calculated
a: The depth of the crack.

So that, by applying the equation (8), the values in Table III are obtained.

Table III. K_{IC} values for all studied samples			
Samples	K_{IC}	Samples	K_{IC}
PEZI 0a	1.851	PEZI 10c	1.521
PEZI 0b	1.925	PEZI 30a	1.770
PEZI 10a	1.976	PEZI 30b	1.805
PEZI 10b	1.185	PEZI 30c	1.985

Likewise, with the results of Table 3 was plotted graph of Figure 3, in which it can see that there is a little variation in the behavior of the samples with a tendency to increase the value of K_{IC} as grows the metal content of the sample.

However, the values are generally low and no remarkable difference was observed between the control sample and the others with the exception perhaps of the sample with the highest content of metal, which probably means that the internal porosity present in all samples exerts an unfavorable factor in behavior, since at present a sample with a notch or crack prefabricated, it tends to grow in a sudden and catastrophic way, characteristics of a brittle behavior.

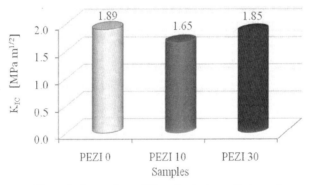

Figure 3. Comparison of K_{IC} value in the studied samples.

Energy G criterion, postulated by Griffith
        This energy criterion states that the fracture in a cracked body will come when the rate of conversion of available energy is greater than a critical value, as a property of the material.

Such that, the G value is related to the stress intensity factor K_{IC} for a linear – elastic solid. So G relates the stress intensity with the elasticity modulus, as shown below[8]:

$$G = K_{IC}^2/E \text{ ------------------ (9)}$$

Where: K_{IC} is the stress intensity factor, and E is the elasticity modulus.

In this way, the obtained G values are plotted in Figure 4.

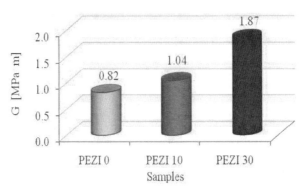

Figure 4. Comparison of G values in the studied samples.

In the graph of Figure 4 it can be noticed that at least within the fracture toughness of the metal reinforcement, in samples with low zinc content (10%), exerts no favorable influence on the material, which should be related to the porosity present in the samples, which reduces the final strength. So that, the best results by relating the value of stress intensity and the energy released upon the occurrence of fracture occur in the material with higher metal content (30%). Likewise, it can be said that only the material with 30% metal improves resistance to fracture toughness.

Impact test (IZOD)

For impact testing, it was employed a Tinius Olsen machine, model 892. And the specimens previously prepared with dimensions of 10 X 10 X 70 mm, using the results for ploting the graph presented in the Figure 5.

Observing the graph of Figure 5, it can be noted that the trend in the impact resistance is clearly upward by increasing the metal content in the samples, so that in this test is evident the function exerted by the metal, strengthening the material and increasing its resistance, which is probably because the zinc is absorbing many of the energy generated during the impact.

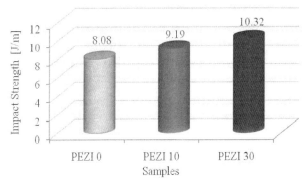

Figure 5: Impact strength of materials.

Comparing the behavior of the two samples in relation to the blank sample (PEZI 0) is perceived advantage of increasing the metal content, at least to make the material more resistant as a function of energy that can absorb before breaking, so that the behavior of samples is better than the blank sample, at least 12.0%.

Stereoscopic microscope micrographs
In the micrographs of Figure 6 it can be noted that the sample with higher content of metal is less porous, besides its can say that the images show a brittle behavior of the material, since in all the fracture surfaces can notice some surface brightness that is generated with this kind of fracture.

So that the brittle behavior of the material meets the microstructural behavior, it means that between more uniform is the microstructure and porosity less occurs, the resistance of the material results to be greater.

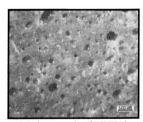

a) Micrograph of PEZI10.   b) Micrograph of PEZI30.

Figure 6: Stereoscopic microscope micrographs.

CONCLUSIONS

- o Based on the results obtained it can be said that it is evident that the metal exerts a favorable influence on the polymeric matrix, which shows a better behavior in a matter of impact resistance and fracture toughness.
- o On the other hand, to observe the material from the perspective of fracture mechanics, it is possible to see that the values obtained for K_{IC} are low, which is evidence of brittle behavior of the material, where the cracks that occur will be propagate quickly in a sudden and catastrophic way. This behavior limits their reliability for applications where components are subjected to stress.
- o Also, according to the stereoscopic microscopy micrographs it can observe certain porosity that decreases with increasing content of metal particles in the samples due to the higher density of metal, due to the difference in melting point, and especially to the higher coefficient of thermal conductivity in the metal.
- o Therefore, according to the results of fracture toughness and stereoscopy micrographs, it is possible to say that these materials are rigid and can withstand the impact stresses with the aid of metal particles being more ductile than the polymeric matrix. Because the metal particles absorb better the energy generated during application of the load.

REFERENCES

1) http://www.aprepet.org.mx/index2.htm Accessed 14/04/09.
2) A. F. Avila, M. V. Duarte. A mechanical analysis on recycled PET / HDPE composites, Journal of Polymer Degradation and Stability, vol. 80, 2003, p: 373-382.
3) J. A. Schey, Manufacturing Processes, McGraw Hill, third edition, 2000, p: 456-457
4) S. Kalpakjian, S.R. Schmid. Manufacturing, engineering and technology, Prentice Hall, Fifth Edition, 2008, p: 335 489 499
5) B. W. Niebel, A. B. Draper, R. A. Wysk, Modern Manufacturing Process Engineering, McGraw Hill, second edition, 1989, p: 265
6) http://www.brittmfg.com/technology.html Accessed 08/05/09
7) J. Delmonte. Polymer metal composites, Library of Congress Catalog, 1990, p: 39-40.
8) J. Gonzalez. Fracture mechanics, Ed Limusa, second edition, 2010, p: 46, 49 41 66
9) W. Pilkey, Stress Concentration Factors (Peterson's), Ed Wiley Interscience, second Edition, 1997, p: 110
10) D. Broek, Elementary engineering fracture mechanics; Martinus Nijhoff Publishers; 1984, p: 77

Innovative Processing

FABRICATION OF GaSb OPTICAL FIBERS

Brian L. Scott and Gary R. Pickrell
Department of Materials Science and Engineering
Virginia Tech
Blacksburg, VA

ABSTRACT

A glass-clad optical fiber with a core of gallium antimonide is presented in this paper. Fabrication of these fibers was accomplished using a draw casting method wherein the core is molten during the drawing operation. This casts the core into its final form as it freezes toward the end of the drawing process. Scanning electron microscopy, energy dispersive spectroscopy and electron backscatter diffraction were performed on the fiber samples to characterize the core integrity, composition and crystallinity respectively. A polycrystalline core was produced containing grain with lengths of 300μm and diameters encompassing the full width of the core region. Oxidation of the core is present in the core region at an 8 atomic percent concentration. This demonstrates the potential for fabrication of GaSb fibers and the possible use in- fiber opto-electronic devices.

INTRODUCTION

There has been a recent development in the fabrication and characterization of optical fibers that incorporate semiconductors as the core material.[1-5] The development of these fibers has been limited to silicon and germanium as the core material. While silicon and germanium have the advantage of being well studied platforms for electronic devices, long wavelength transmission and integration into existing silicon photonic structures, they are not suitable for efficient light generation. This is primarily due to both materials having an indirect band gap structure. While low efficiency LED's have been made with these materials it is more common for LEDs and diode lasers to be made from III-V semiconductor structures. More recently, the fabrication of a III-V semiconducting core has been accomplished using InSb.[6] Incorporation of these semiconducting materials into the core region of optical fibers will allow for the fabrication of the structures and devices for these types of opto-electronics within the fiber. It is envisioned that a series of post fiber fabrication processes will build the structures necessary for these devices. Selected regions of the fiber could be manipulated and the devices built onto an exposed end face of the fiber along with electrical contacts and then reincorporated back into the fiber network. Many challenges exist in the creating these devices in the fiber and finding a suitable semiconductor core material. However, being able to fabricate the fibers is the first step.

Results of the fabrication of a glass clad gallium antimonide optical fiber is presented in this paper. The focus on the fabrication is the establishment of a suitable platform for the future fabrication of opto-electronic devices within the core region for use in the mid infrared region. The transmission window for GaSb is from approximately 2-25μm.[7] At a stoichiometric mixture of 50% gallium and 50% antimony, the system forms a congruently melting compound upon freezing, making the composition a good choice for this type of III-V core material using a fiber draw casting method. [8] Gallium antimonide is also a good candidate for this type of semiconducting optical core material due to its melting point of 712°C, low vapor pressure, previous device fabrication and the proven integration of other III-V material components.[9]

GaSb has been used to fabricate diode lasers and LEDs with emission in the mid infrared wavelength range where GaSb has been used as the substrate for the device fabrication.[10-12]

MATERIALS AND METHODS

Starting materials for the core region of the fiber were obtained from University Wafer as a 2" Te doped GaSb wafer with a doping level of 10^{17}. Wafers with similar doping levels show a peak transmission of approximately 30-35% in the 2-5μm range.[13] The wafer was ground in a fused silica mortar and pestle until the wafer was reduced to a course powder particle size of approximately 1mm. The powder was then inserted and packed into a low softening point glass tube with a composition in the sodium borosilicate glass family of approximately 60 wt. % SiO_2 and rest made up Na_2O, B_2O_3, and Al_2O_3. Tubing inside diameter was approximately 3mm. The tube was then connected to a vacuum source to reduce potential oxidation of the GaSb and to assist in the partial consolidation of the powder and collapsing of the tubing diameters. While the tube was being evacuated, the outer part of the tube was heated with a hydrogen oxygen torch to around the softening point of the glass. The temperature of the glass was approximated visually. Visual approximation was possible due to the softening point of the glass being slightly higher than the melting point of GaSb. During heating of the tube the GaSb powder did not melt, while the tube experienced a contraction of the outside diameter.

Fiber drawing was done on a horizontal drawing system which consists of two opposing rotatable chucks. One of the chucks holds the preform, while the second chuck translates in the horizontal direction and holds a pulling rod that is fused to the preform end. The preform was inserted into the primary chuck, hooked up to a vacuum source and then a rod was fused to the end using a hydrogen-oxygen torch. Once the rod was fused to the end, a portion of the preform was heated until the GaSb was molten and the glass viscosity was sufficiently low. Fibers were then drawn from the preform end by translating the second chuck away from the preform end. Fibers were pulled from the preform at 5.5m/min, producing fibers with an outside diameter of 300 microns and core diameter of 100 microns. Lengths of approximately 0.8 meter were produced with lengths of around 0.4 meter having a consistent inner and outer diameter. The drawn fibers exhibited a uniform diameter over a substantial length of the fiber. Visual inspection of the fibers showed no indication of differences between the fiber segments. Core cracking and discontinuities were not apparent in any of the drawn fibers. Based on the homogeneity of the fibers, a sample was taken at random for SEM and EBSD analysis. While the draw lengths in these experimental runs are short in comparison to conventional drawing, the consistency of the drawn fibers indicates that a steady draw condition will produce good fiber properties, although further analysis of the fabricated fibers will be needed to confirm.

Fibers were characterized optically and with SEM, EBSD, and EDS. SEM and EDS was done with the same sample. A 1cm sample was cleaved from a longer length of fiber and mounted onto a sample stub with conductive silver paint. The end of the fiber was then polished down to 1200 grit. EBSD samples were mounted onto a copper wire scaffold, which was then mounted in Bakelite. The entire mount was then polished down to 1200 grit, followed by a vibratory polish using 0.06μm colloidal silica. The scan was conducted with a step size of 10 μm over the core region including sections of the cladding. The SEM sample is shown in Fig. 1 where the core is the lighter grey color in the center of the fiber. The EDS point scan is in Fig. 2 and shows the presence of the gallium, antinomy, oxygen, sodium, silicon and aluminum in the core. Most of the components of the glass composition in the cladding region have a low concentration in the core region with the exception of oxygen. Fig. 3 is an optical micrograph of the fiber when polished along the fiber axis. The grain structure can be seen is the micrograph as the modulations in the light reflected off of the core region. Inset into Fig. 3 is the EBSD scan of a portion of the core along with an inverse pole figure showing the grain orientations. The same grain structure can be seen in both the optical micrograph and the EBSD image. Grain size was confirmed with the use of EBSD and crystallographic data was also determined using EBSD as detailed in the inset in Fig. 3 with a section along the core being analyzed. Grain size is shown along with orientation of each grain within the scanned region. The core consists of several grains with an average diameter spanning the width of the core and an average length of approximately 300μm. Grains consists of several different crystallographic orientations between the (001) and (111) planes normal to the fiber axis.

A fiber section was optically tested but no transmission was detected. This is most likely due to the low external transmittance of the material of ~35% in conjunction with the multiple grain boundaries present in the fiber core. These grain boundaries can act as segregation centers for materials that have diffused into the core and thus will act as scattering centers further reducing the transmission. The irregularity of the core cross section and the core cladding interface will also contribute to the low transmission.

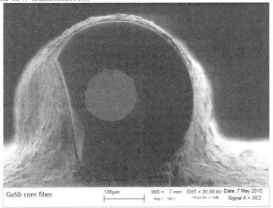

Figure 1. Scanning electron micrograph of GaSb fiber end face fracture surface

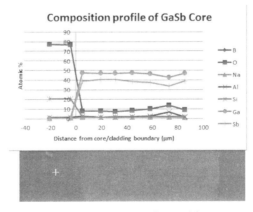

Figure 2. EDS of GaSb fiber end face

DISCUSSION

The results show a fiber that has a GaSb core with approximately 8% oxidation of the GaSb region as shown by EDS. GaSb readily oxidizes during polishing and most likely influences the percent oxygen detected. Tighter control of the fabrication process by decreasing the heating and pulling times to reduce oxygen diffusion from and through the cladding and a higher vacuum on the preform to remove the ambient environment prior to heating of the preform will reduce the oxygen content. The presence of other glass constituents in the core region show either minor diffusion into the core or possible contamination during the polishing of the fiber end face. Atomic percent of these other constituents is around 2% or less and could be attributed to either process. At present these contamination levels are too high for efficient device fabrication, so further investigation into the fabrication parameters is necessary to find the cause of and to reduce these levels or other glass types may be used to eliminate the most troublesome ones. The core region has a high degree of integrity as is seen by the fibers in Fig. 1 and Fig. 3. The images show the fiber length with a continuous core throughout the entire length with no cracking or other apparent discontinuities. This is most likely due to the close approximation of the glass softening point and the melting point of the GaSb. As the GaSb solidifies, the glass is still fluid enough to accommodate the expansion and relieve any

Figure 3. Optical micrograph of side polished GaSb fiber with EBSD inset

induced stresses due to the solidification process. Stresses related to the differential CTE between the GaSb and the glass cladding are less relevant as the core and cladding contract as it cools with the core contracting more. Stresses in the cladding and core due to high cooling rates and can be minimized by controlled the annealing of the fibers in addition to a reduction in the core size. In Fig. 1, the core region can be seen as having formed lobes or protrusions into the cladding region. This is most likely due to the softening point of the glass being close to the melting point of the GaSb. As the core freezes the glass viscosity is too low to resist the growth of the crystal into the cladding region. Alteration of the glass composition to increase the softening point will prevent the developments of these "lobes" from forming by increasing the glass viscosity at the freezing point of GaSb. While increasing the softening point of the clad glass will increase the tendency to create voids and cracks, a suitable small increase in softening point should be suitable to decrease the irregularity of core clad interface while not significantly increasing the formation of voids and cracks

The presence of grain boundaries is an important matter as these boundaries influence electronic device performance by altering the band gap near the boundary from that in the rest of the grain. Depending on the size of the grains, annealing may be necessary to minimize the presence of grain boundaries to improve device performance. Crystal orientation can also be important in the fabrication of the devices as the electrical properties can be different as well as the epitaxial growth of other materials onto the core. The grain structure in the core seen in both the optical micrograph and in the EBSD image gives an indication of the crystallization of the GaSb. In both images, the intrusion of one grain into another adjacent grain is visible. With the imaged surface being in the interior of the core and that the same crystal orientation is present on both sides of the intrusion it can be inferred that the intrusion is completely surrounded by the other grain. With this being the case, the likely explanation is that the surrounded grain grew into that region as the other grain was nucleating and growing in from the cladding/core boundary.

Crystallization of the GaSb is governed by nucleation theory which will be highly influence by the heat transfer from the core. This is expected to proceed from the outer boundary into the interior and that the nucleation of grains will begin at the glass/GaSb interface and proceed inward toward the center. During the pulling of the fiber, the amount of heat loss from the core will be greatest at the first portion of the fiber being removed from the heat zone and decrease along the fiber opposite the pulling direction. From this, one would expect that the grains will form and grow along the fiber axis until they impinge on another grain that began growing at a later time. This impingement will result in a grain boundary with the lowest energy plane facet growing the fastest, resulting in an angled boundary. While the grain and crystallinity data was collected over a small sample size, there appears that a preferential grain orientation formed in this sample. Analysis of several other samples would be needed to confirm this grain orientation may not be important optically, since GaSb is cubic and hence isotropic. However, it is important for the fabrication of devices and a preferential grain orientation would be advantageous in producing higher quality devices. The crystal data also indicates the need for the crystallinity to be interrogated at both the core surface and the core interior as surface techniques will not pick up interior grain structure.

CONCLUSION

An optical fiber which consists of a gallium antimonide core with a glass cladding has been fabricated. The core region is polycrystalline with the diameter of the grain spanning the width of the core and an average length of approximately 300 μm. A high degree of structural integrity of the core is present with little or no visible cracks or defects in the core and a distinct boundary between core and cladding. Oxidation of the core region is present at a concentration of around 8 atomic percent. The fabrication of the fiber shows proof of concept in the making this type of 3-5 semiconductor core optical fiber utilizing a draw casting approach and signifies the initial step in the fabrication of in-fiber opto-electronic devices.

REFERENCES

[1] J. Ballato, *et al.*, "Silicon optical Fiber," *Opt. Express,* vol. 16, pp. 18675-18683, 2008.
[2] J. Ballato, *et al.*, "Glass-Clad single-crystal germanium optical fiber," *Optics Express,* vol. 17, pp. 8029-8035, 2009.
[3] B. Scott, *et al.*, "Fabrication of silicon optical fiber," *Optical Engineering Letters,* vol. 48, 2009.
[4] B. Scott, *et al.*, "Fabrication of N-type Silicon Optical Fibers," *Photonics Technology Letters,* vol. 21, pp. 1798-1800, 2009.

[5] C. McMillen, *et al.*, "On crystallographic orientation in crystal core optical fibers," *Optical Materials,* vol. 32, pp. 862-867, 2010.

[6]J. Ballato, *et al.*, "Binary III-V semiconductor core optical fiber," *Opt. Express,* vol. 18, pp. 4972-4979, 2010.

[7] P. Klocek, Ed., *Handbook of infrared Optical Materials*. New York: Marcel Dekker, Inc, 1991, p.^pp. Pages.

[8] S. Adachi, "Gallium Antimonide," in *Handbook on Physical Properties of semiconductors* vol. 2- III-V compound semiconductors, ed Boston: Kluwer Academic Publishers, 2004, p. 446.

[9] A. G. Milnes and A. Y. Polyakov, "Gallium antimonide device related properties," *Solid-State Electronics,* vol. 36, pp. 803-818, 1993.

[10] A. Aardvark, *et al.*, "Devices and desires in the 2-4 micron rgion based on antimony-containing III-V heterostructurs grown by MOVPE," *Semiconductor Science and Tehcnology,* vol. 8, pp. S380-S385, 1993.

[11] H. K. Choi and G. W. Turner, "GaSb-based mid-infrared quantum well diode lasers," in *Laser Diodes and Applications*, San Jose, CA, USA, 1995, pp. 236-243.

[12] E. A. Grebenshchikova, *et al.*, "Properties of GaSb-Based Light-Emitting Diodes with Chemically Cut Substrates," in *Semiconductors* vol. 37, ed: Springer Science & Business Media B.V., 2003, pp. 1414-1420.

[13] A. Chandola, *et al.*, "Below bandgap optical absorption in tellurium-doped GaSb," *Semiconductor Science and Technology,* vol. 20, pp. 886-893, 2005.

.

CHARACTERIZATION AND SYNTHESIS OF SAMARIUM-DOPED CERIA SOLID SOLUTIONS

Aliye Arabacı
Istanbul University, Engineering Faculty, Department of Metallurgical and Materials Engineering
34320, Avcilar, Istanbul, Turkey

ABSTRACT

Doped ceria is considered as the most promising high-conducting electrolyte, alternative to the commercially used yttria-stabilized zirconia. The ceria with fluorite type structure is very tolerant to dissolution of lower valence metal ions. In this study, fully dense CeO_2 ceramics doped with 10 mol % Sm ($Sm_{0.1}Ce_{0.9}O_{1.95}$, SDC) were successfully synthesized by a simple Pechini method. X-ray diffraction analysis showed fine grained powders with single phase and cubic fluorite structure. A range of techniques including thermal gravimetric analysis (TGA/DTA), X-ray diffraction (XRD) and scanning electron microscopy (SEM) were carried out to characterize the SDC powders.

INTRODUCTION

Solid oxide fuel cells (SOFCs) are solid energy conversion devices that can directly convert chemical energy to electrical energy in an efficient and environmentally friendly way [1]. Solid oxide fuel cells are attracting widespread attention due to their high-energy conversion efficiency and low pollution. Commonly, stabilized zirconia (e.g. yttria stabilized zirconia), stabilized Bi_2O_3 and strontium/magnesium doped lanthanum gallate are used as solid electrolyte materials in SOFC. A typical SOFC uses yttria stabilized zirconia (YSZ) as the electrolyte. However, SOFC having YSZ electrolyte requires high operation temperature, 1000 °C, to have high enough ionic conductivity. Such a high temperature would cause many technological problems, such as mechanical instability, undesirable chemical reaction between cell components (electrolyte, electrode and interconnecting materials). Thermal stability and chemical compatibility of the SOFC components are key issues for designing a high performance working cell. For commercializing SOFC technology, main challenge is to lower the operating temperature while keeping high performance. Inexpensive metallic materials could then be used for the interconnect components and the diffusion between components would be restricted, thereby prolonging cell life [2, 3].

In order to eliminate technological problems of SOFC either by decreasing the electrolyte thickness or by introducing alternative electrode materials with higher ionic conductivity at lower temperatures. Doped ceria has been extensively studied with a high ionic conductivity between 500 and 800 °C as electrolytes in reduced-temperature SOFCs [4-7].

In the trivalent rare earth doped ceria, the highest conductive materials are Gadolinia-doped ceria (GDC, $Ce_{0.9}Gd_{0.1}O_{1.95}$) and samarium-doped ceria (SDC, $Ce_{0.9}Sm_{0.1}O_{1.95}$) which are considered to be most promising electrolytes for SOFCs at low operation temperature [5,8]. Therefore, the researchers have focused on the ceria based on electrolytes. To obtain these materials, various synthesis methods have been used such as homogeneous precipitation [9-11], combustion synthesis using different fuels such as urea [12,13], mechanochemical [14], hydrothermal methods [15]. polyol process [16] and sol-gel process [17].

In the present work, high purity cerium and samarium salts were used to form ceria solid solutions using Pechini method. This simple method yields high purity ultrafine powders which can form dense electrolyte at relatively low sintering temperatures. Furthermore, in this study, the effect of calcination temperature to find out the evolution of crystal structure has also been investigated. The thermal decomposition behaviors of the products were also examined by thermal gravimetric analysis. In addition, SEM technique was used for characterisation.

EXPERIMENTAL
SDC preparation
      Cerium nitrate ($Ce(NO_3)_3 \times 6H_2O$, 99%, Aldrich) and samarium nitrate ($Sm(NO_3)_3 \times 6H_2O$, 99.9%, Aldrich) were used as metal precursors and ethylene glycol (R.P. Normopur), citric acid (Boehringer Ingelheim) were selected for the polymerization treatment. Cerium nitrate and samarium nitrate salts were dissolved in de-ionized water individually and then the solutions were mixed in a beaker. Anhydrous citric acid was dissolved in de-ionized water and then was added to the cation solution. The molar ratio of total oxide (TO): citric acid (CA) and ethylene glycol: citric acid was selected as 2:1, 4:1, respectively.

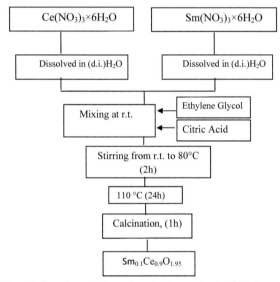

Figure 1 Experimental flow chart for the synthesis of SDC powders.

After homogenization of this solution, temperature was raised to 80°C and the solution kept for 2 h at this temperature with magnetic stirring to remove excess water. During this time, the colour of the liquid turned from white to orange then to dark brown. Then the dark brown gel was naturally cooled down to room temperature. The initial thermal decomposition of the precursor was dried 110 °C for 24 h. The details of the preparation procedure of SDC particles are given in the schematic diagram as shown in Figure 1.

Thermal analysis
     The solids were dried at 110 °C for 24 h. The conversion of the so-prepared amorphous precursors into crystalline samarium- doped CeO_2 was achieved by heating the dried solids at a

rate of 5 °C min⁻¹ for each 250, 500, 700 and 1000 °C, and keeping them at these temperatures for 1 h. The thermal behavior of the SDC particles was carried out with SII Exstar 6000 TG/DTA 6300 from room temperature to 750 °C at a heating rate of 10 °C min⁻¹ in a flow of air. Differential thermal analysis (DTA) and thermogravimetric analysis (TGA) results were plotted as a function of temperature.

X-ray analysis

X-ray diffraction (XRD) technique was used to determine the crystal structure and phase analysis. The precursors were calcined at 200, 250, 500, 700 and 1000 °C for 1 h. The X-ray spectra of samarium-doped ceria particles were obtained over the 2θ range of 10-90° by using Rigaku D/Max -2200 PC X-ray diffractometer with CuKα radiation.

Scanning electron microscopy (SEM) analysis

The particle morphology of the SDC samples calcined at 500 and 700°C and the morphology of the sintered samples at 1200, 1300 and 1350 °C were determined by scanning electron microscopy FEI Quanta FEG 450. In addition, the morphology change according to sintering temperature was studied.

RESULTS AND DISCUSSION

Thermal Analysis

Figure 2 shows TGA and DTA plots obtained for the dried gel. From Figure 2, two major thermal events were observed at 40-160 °C and at about 200 °C. As can be seen in the DTA plot, there is a weak endothermic peak at ~ 160°C. Then at 200 °C, strong exothermic peak appeared. The TG curve shows that samples lost weight rapidly at 200 °C, accompanied by the change seen in DTA. According to the weight loss from the TG curve, the first endothermic peak with 3.6 % weight loss can attributed to the dehydration of the gel.

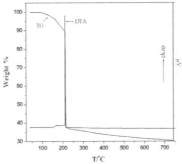

Figure2. TG–DTA curves of thermal decomposition of nanocrystalline SDC powder precursors at a heating rate of 5 °C min⁻¹ in air.

Meanwhile, the strong exothermic peak appearing in the DTA probably corresponds to the burning of the residual organic matter in the sample and to the crystallizing process of $Sm_{0.1}Ce_{0.9}O_{1.95}$. No changes were found in TG/DTA curves above 250°C which suggests the formation of crystalline $Sm_{0.1}Ce_{0.9}O_{1.95}$ as a decomposed product. The XRD results also verify this finding.

X-ray Analysis

As can be seen in Figure 3, the diffraction patterns taken after treatment at 200°C, starting with the precursor gel, showed that the material was amorphous structure. After calcination of above 250 °C, the obtained peaks are matching well with the cerium oxide JCPDS card No: 34-394. All the peaks can be assigned to the crystal planes (111) (200) (220) (311) (222) (400) (331) and (420). The XRD peak broadening in Figure 3 shows that the crystallite size of the calcined powders is small. The degree of peak broadening decreases with increasing calcination temperature which indicates the increase of the crystallite size.

The average particle size was about 11, 27 and 50.8 nm for powders calcined at 500, 700 and 1000 °C, respectively. There are no peaks detected belonging of samarium oxide phase. It indicates that the dopant ion is fully substituted in the CeO_2 lattice. The average crystallite sizes, D, were calculated from the degree of line broadening in the main (111) diffraction peak using the Scherrer formula (Eq.1) from XRD data. The calculated D_{XRD} values are presented in Table I.

$$D = 0.9\, \lambda \,/\, \beta \cos\theta \qquad\qquad\qquad \text{(Eq. 1)}$$

D = Average crystallite size
λ = Wavelength of the X-rays
β = Corrected peak at full width at half-maximum (FWHM) intensity
θ = Scattering angle of the main reflection (111)

Table I. The average crystallite size of SDC powders calcined at different temperatures for 1h.

Calcination Temperature (°C)	250	500	700	1000
D_{XRD}/ nm	9.1	11.4	27.3	50.8

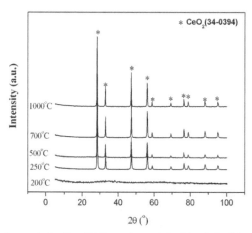

Figure 3. XRD patterns of ultrafine SDC powders calcined in air for 1 h at 200, 250, 500, 700, and 1000 °C.

SEM Results

Figure 4 a-b present typical particle morphology of the SDC samples calcined at 500 and 700°C. Figure 4 c-e give the morphology of the SDC samples sintered at different temperature for 6 h.

Sponge-like structure was observed for after calcination at 500 and 700°C for 1h. The porous morphologies of SDC samples calcined at 500 and 700 °C are likely to be caused by large volumes of gas produced in the preparation method.

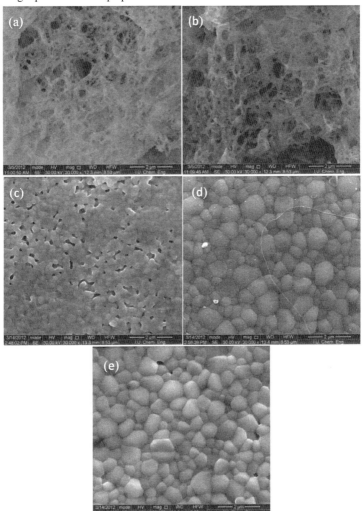

Figure 4. SEM micrographs of 10 % SDC showing morphology change of the nanoparticles annealed at (a) 500, (b) 700, (c) 1200, (d) 1300 and (e) 1350 °C.

During the combustion process, a large amount of gaseous materials is evolved and hence the combustion product is highly porous as shown in the SEM micrographs (Figure 4 a-b). The specimen sintered at 1200 °C resulted in grain growth, however no sharp grain boundary is observed. On the other hand, sintering at 1300 °C resulted in more grain growth and more densification (94 % of the theoretical value). The well-developed grain boundaries were observed in Figure 4 d. At 1350 °C, the final grain growth and densification were observed. The number of particles visible in the SEM image (Figure 4e) was counted. The average particle size was determined from approximately 140 measurements and calculated as 0,5 μm.

CONCLUSION

CeO$_2$ powders doped with 10 mol% of samarium have been synthesized via Pechni process where a relatively low calcination temperature was used in comparison with other hydrothermal treatments. By this method, the single-phase fluorite structure was obtained. The crystallization of SDC product occurred via strong an exothermic process and the final crystal structure was first formed from SDC powder precursors at 250 °C. With increasing calcination temperature, the crystal sizes increased and the crystallinity of the powders become well defined during the calcination process. With increasing sintering temperature, SDC samples are nearly fully dense with very few residual pores (Figure 4e). The calculated relative density for SDC samples sintered at 1300 °C is 94% of the theoretical value.

Acknowledgments

The present work was supported by the Research Fund of Istanbul University (Project no: 22650).

REFERENCES
[1]B.C.H. Steele and A. Heinzel, Materials For Fuel-Cell Technologies, Nature, **414,** 345-352(2001).
[2]Singh P, Minh NQ. Solid Oxide Fuel Cells: Technology Status. Int J Appl Ceram. Tech, **1,** 5-15(2004).
[3] O.Yamamoto, Solid oxide fuel cells: Fundamental Aspects And Prospects. Electrochim Acta, **45,** 2423-35(2004).
[4]S. Zha, A. Moore, H. Abernathy and M. Liu, GDC-Based Low-Temperature SOFCs Powered By Hydrocarbon Fuels, J. Electrochemical Soc., **151,** A1128-A1133(2004).
[5]B.C.H. Steele, Appraisal of Ce$_{1-y}$Gd$_y$O$_{2-y/2}$ Electrolytes For IT-SOFC Operation at 500 °C, Solid State Ionics, **129,** 95-110 (2000).
[6]M. Mogensen, N.M. Sammes and G.A. Tompett, Physical, Chemical and Electrochemical Properties Of Pure And Doped Ceria, Solid State Ionics, **129,** 63-94 (2000).
[7]M. Sahibzada, B.C.H. Steele, K. Hellgardt, D. Barth, A. Effendi, D. Mantzavinos and I.S. Metcalfe, Intermediate Temperature Solid Oxide Fuel Cells Operated With Methanol Fuels, Chem. Eng. Sci. , **55,** 3077-3083(2000).
[8]H. Yahiro, Y. Eguchi, K. Eguchi, H. Arai, Oxygen Ion Conductivity of The Ceria-Samarium Oxide System With Fluorite, J. Appl. Electrochem., **18,** 527-531 (1988).
[9]S.W. Zha, Q.X. Fu, Y. Lang, C.R. Xia and G.Y. Meng, Novel Azeotropic Distillation Process For Synthesizing Nanoscale Powders Of Yttria-Doped Ceria Electrolyte, Mater. Lett., **47,** 351-355(2001).
[10]A.I.Y. Tok, L.H. Luo and F.Y.C. Boye, Carbonate Co-Precipitation Of Gd$_2$O$_3$-Doped CeO$_2$ Solid Solution Nano-Particles, Mater. Sci. Eng. A, **383,** 229-234 (2004).

[11]H.I. Chen, H.Y. Chang, Homogeneous Precipitation Of Cerium Dioxide Nanoparticles In Alcohol/Water Mixed Solvents, Colloids Surf. A, Physicochem. Eng. Aspects, **242**, 61-68 (2004).

[12]E. Chinarro, J.R. Jurado and M.T. Colomer, Synthesis Of Ceria-Based Electrolyte Nanometric Powders By Urea-Combustion Technique, J. Euro. Ceram. Soc., **27**, 3619-3623(2007).

[13]C.C. Hwang, T.-H Huang, J.-S. Tsai, C,-S. Lin and C.-H. Peng, Combustion Synthesis Of Nanocrystalline Ceria (CeO_2) Powders By a Dry Route, Mater. Sci. Eng.B, **132**, 229-238(2006).

[14]Y.X. Li, W.F. Chen, X.Z. Zhou, Synthesis Of CeO_2 Nanoparticles By Mechanochemical Processing And The Inhibiting Action Of NaCl On Particle Agglomeration, Mater. Lett. **59**, 48-56 (2005).

[15]Y.C. Zhou, M.N. Rahman, Hydrothermal Synthesis And Sintering Of Ultrafine CeO_2 Powders, J. Mater. Res., .**8**, 1680-1687(1993).

[16] T. Karaca, T. G. Altıncekic, M. F.Oksuzomer , Synthesis of Nanocrystalline Samarium-Doped CeO_2 (SDC) Powders As a Solid Electrolyte By Using a Simple Solvothermal Route, Ceram. Int., **36**, 1101–1107(2010).

[17] G.S. Wu, T. Xie, X.Y. Yuan, B.C. Cheng, L.D. Zhang, An Improved Sol–Gel Template Synthetic Route to Large-Scale CeO_2 Nanowires, Mater. Res. Bull. , **39**, 1023. (2004)

INFLUENCE OF PRECURSORS STOICHIOMETRY ON SHS SYNTHESIS OF Ti_3AlC_2 POWDERS

L Chlubny, J. Lis

AGH - University of Science and Technology, Faculty of Material Science and Ceramics Department of Technology of Ceramics and Refractories, Al. Mickiewicza 30, 30-059 Cracow, Poland

ABSTRACT

Group of very interesting ternary compounds called MAX-phases can be found in the Ti-Al-C system. These nanolaminate materials, such as Ti_2AlC and Ti_3AlC_2, are characterised by heterodesmic layer structure consisting of covalent and metallic chemical bonds. It results with semi-ductile features locating them on the boundary between metals and ceramics. Thanks to these features they can find wide range of potential applications, for example as a part of ceramic armour. One of the most effective and efficient methods of obtaining these materials is Self-propagating High-temperature Synthesis (SHS), basing on exothermal effect of chemical reaction.

The main objective of this work was to apply SHS method to obtain fine sinterable powders of Ti_3AlC_2 and to examine influence of different stoichiometry of various precursors, particularly intermetallic powders in Ti-Al system on final product of the reaction.

INTRODUCTION

Among many covalent materials, such as carbides or nitrides, a group of ternary and quaternary compounds, referred in literature as H-phases, Hägg-phases, Novotny-phases or thermodynamically stable nanolaminates, can be found. These compounds have a $M_{n+1}AX_n$ stoichiometry, where M is an early transition metal, A is an element of A groups (mostly IIIA or IVA) and X is carbon and/or nitrogen. Heterodesmic structures of these phases are hexagonal, P63/mmc, and specifically layered. They consist of alternate near close-packed layers of M_6X octahedrons with strong covalent bonds and layers of A atoms located at the centre of trigonal prisms. The M_6X octahedral, similar to those forming respective binary carbides, are connected one to another by shared edges. Variability of chemical composition of the nanolaminate is usually labeled by the symbol describing their stoichiometry, e.g. Ti_2AlC represents 211 type phase and Ti_3AlC_2 – 312 type. Differences between the respective phases consist in the number of M layers separating the A-layers: in the 211's there are two whereas in the 321's three M-layers [1-3]. The layered, heterodesmic structure of MAX phases leads to an extraordinary set of properties. These materials combine properties of ceramics such as high stiffness, moderately low coefficient of thermal expansion and excellent thermal and chemical resistance with low hardness, good compressive strength, high fracture toughness, ductile behavior, good electrical and thermal conductivity which are characteristic for metals. They can be used to produce ceramic armor based on functionally graded materials (FGM) or as a matrix in ceramic-based composites reinforced by covalent phases.

The Self-propagating High-temperature Synthesis (SHS) is a method applied for obtaining numerous materials such as carbides, borides, nitrides, oxides, intermetallic compounds and composites. The principle of this method is utilization of exothermal effect of chemical synthesis. This synthesis can proceed in a powder bed of solid substrates or as filtration combustion where at least one of the substrates is in gaseous state. To initiate the process an external source of heat has to be used and then the self-sustaining reaction front is propagating through the bed of substrates until all of them are consumed. This process could be initiated by the local ignition or by the thermal explosion. The final form of the synthesized material may depends on kind of precursors used for synthesis and the technique applied. Low energy consumption, high temperatures obtained during the process, high efficiency and simple apparatus are the features which are characteristic for this type of synthesis. The lack of control of the process is the disadvantage of this method[4].

The main purpose of this work was application of SHS synthesis to obtain sinterable powders of Ti_3AlC_2 nanolaminate materials and investigation of influence of stoichiometry of precursors used during reaction on final products' phase composition. The final objective was to obtain powder with highest content of ternary phase (Ti_3AlC_2) and lowest content of TiC impurities, which strongly affects properties of the dense, sintered material, decreasing its pseudo-plastic behaviour.

PREPARATION

Following the experience gained during previous synthesis of ternary materials such as Ti_2AlN, Ti_2AlC and Ti_3AlC_2 [5, 6, 7], it has been decided that various materials such as elementary powders, intermetallic materials in the Ti-Al system and titanium carbide powder will be used as a precursors for synthesis of Ti_3AlC_2 powders. As a result of previous researches, intermetallic powders were mostly used in the experiments. Due to the relatively low availability of commercial powders of intermetallic materials in the Ti-Al system it was decided to synthesize them by SHS method. At the first stage of the experiment TiAl and Ti_3Al powders were synthesized by SHS method [5]. Titanium hydride powder, TiH_2, and metallic aluminium powder with grain sizes below 10 μm were used as sources of titanium and aluminium. The mixture for SHS synthesis had a molar ratio of 1:1 and 3:1 respectively (equations 1-2).

$$TiH_2 + Al \rightarrow TiAl + H_2 \qquad (1)$$
$$3TiH_2 + Al \rightarrow Ti_3Al + 3H_2 \qquad (2)$$

Powders were initially homogenized in dry isopropanol using a ball-mill. Then homogenized and dried powders were placed in a graphite crucible which was heated in a graphite furnace in argon atmosphere up to 1200°C, at this temperature SHS reaction was initiated by the thermal explosion. The obtained products were crushed in a roll crusher to the grain size ca. 1 mm and afterwards powders were ground to the grain size ca. 10 μm for 8 hours in the rotary-vibratory mill in dry isopropanol, using WC balls as a grinding medium [10]. Other powders used during synthesis of ternary compound were commercially available aluminium powder (ZM Skawina recovered from electrofilters system, grain size ca. 20 μm, +99% pure), graphite powder used as a source of carbon (Merck no. 1.04206.9050, 99,8% pure, grain size 99.5% < 50μm), titanium powder (AEE TI-109, 99,7% pure, ~100 mesh) and titanium carbide powder (AEE TI-301, 99,9% pure, ~325 mesh).

All of the Ti_3AlC_2 synthesis were conducted by SHS method with a local ignition system and with use of various precursors' stoichiometry. The mixtures of precursors for the SHS synthesis were set in assumed stoichiometric ratios and homogenized for 12 hours. The SHS reactions stoichiometries are presented in equations 6-10 respectively. The 1.1, 1.2 etc. corresponds to 10 and 20 wt.% of excess precursor respectively.

$$3Ti + Al + 2C \rightarrow Ti_3AlC_2 \qquad (3)$$
$$TiAl + 2Ti + 2C \rightarrow Ti_3AlC_2 \qquad (4)$$
$$TiAl + 2TiC \rightarrow Ti_3AlC_2 \qquad (5)$$

$$Ti_3Al + 2C \rightarrow Ti_3AlC_2 \qquad (6)$$
$$2TiC + Ti + Al \rightarrow Ti_3AlC_2 \qquad (7)$$
$$3Ti+1,1Al+2C \rightarrow Ti_3AlC_2 \qquad (8)$$
$$3Ti+1,2Al+2C \rightarrow Ti_3AlC_2 \qquad (9)$$
$$1,1TiAl+2Ti+2C \rightarrow Ti_3AlC_2 \qquad (10)$$
$$1,2TiAl+2Ti+2C \rightarrow Ti_3AlC_2 \qquad (11)$$
$$TiAl+2Ti+2C+0,1Al \rightarrow Ti_3AlC_2 \qquad (12)$$
$$TiAl+2Ti+2C+0,2Al \rightarrow Ti_3AlC_2 \qquad (13)$$
$$1,1Ti_3Al+2C \rightarrow Ti_3AlC_2 \qquad (14)$$
$$1,2Ti_3Al+2C \rightarrow Ti_3AlC_2 \qquad (15)$$

Products of synthesis were ground and the X-ray diffraction analysis method was applied to determine phase composition of the synthesised materials. The basis of quantitative and qualitative phase analysis were data from ICCD [8]. Amounts of the respective phases were calculated by the Rietveld analysis [9]. The measurements were made within an accuracy of 0.5%. Observations of the powders morphology were done by FEI Europe Company Nova Nano SEM 200 scanning electron microscope.

RESULTS AND DISCUSSION
    The X-ray diffraction analysis of SHS synthesis in the Ti-Al system proved that TiAl synthesised by SHS method was almost pure and contained only 5% of Ti_3Al impurities (Figure 1), while Ti_3Al powder did not contain any of the other phases (Fig. 2) [10].
    While synthesizing Ti_3AlC_2 by SHS method, the highest amount of ternary phase (76,4%) was achieved in the case of reaction 10. The other phases in this product were TiC (18.9%) and Al_4C_3 (4.7%). The XRD pattern of obtained powder is presented on Figure 3. The peak corresponding to the graphite that can be seen on the XRD pattern is a result of impurities introduced by the graphite combustion boat in which the synthesis was conducted and is not included in final results. This procedure was applied to all of the experimental results. Results of phase quantities analysis are presented in Table 1.

Table I. Products of SHS synthesis of Ti_3AlC_2 phase composition.

No.	Chemical reaction	Ti_2AlC[wt.%]	Ti_3AlC_2[wt.%]	TiC[wt.%]	Al_4C_3[wt.%]	Al_3Ti[wt.%]
3	$3Ti + Al + 2C \rightarrow$ Ti_3AlC_2	19.4	65.1	15.5	-	-
4	$TiAl + 2Ti + 2C \rightarrow$ Ti_3AlC_2	18.1	44.1	27.2	10.7	-
5	$TiAl + 2TiC \rightarrow$ Ti_3AlC_2	3	67.1	30	-	-
6	$Ti_3Al + 2C \rightarrow$ Ti_3AlC_2	4.6	58.6	36.8	-	-
7	$Ti + Al + 2TiC \rightarrow$	9.6	64.4	26	-	-

		Ti_3AlC_2				
8	$3Ti + 1,1Al + 2C$ $\rightarrow Ti_3AlC_2$	-	45.9	26.2	17.2	10.6
9	$3Ti + 1,2Al + 2C$ $\rightarrow Ti_3AlC_2$	-	65.0	20.1	10.3	4.6
10	$1,1TiAl + 2Ti + 2C$ $\rightarrow Ti_3AlC_2$	-	76.4	18.9	4.7	-
11	$1,2TiAl + 2Ti + 2C$ $\rightarrow Ti_3AlC_2$	41.7	17.6	27.2	13.6	-
12	$TiAl + 2Ti + 2C$ $+0,1Al \rightarrow Ti_3AlC_2$	-	49.4	27.2	23.4	-
13	$TiAl + 2Ti + 2C$ $+0,2Al \rightarrow Ti_3AlC_2$	-	60.6	16.2	19.9	3.3
14	$1,1Ti_3Al + 2C \rightarrow$ Ti_3AlC_2	18.1	31.9	32.7	17.0	0.4
15	$1,2Ti_3Al + 2C \rightarrow$ Ti_3AlC_2	37.8	19.3	31.6	11.4	-

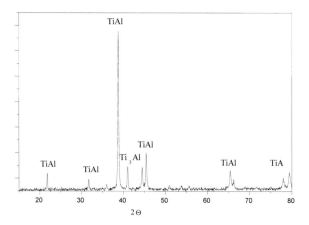

Figure 1. XRD pattern of the TiAl powders obtained by SHS

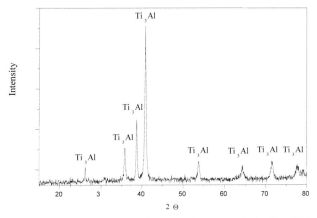

Figure 2. XRD pattern of the Ti_3Al powders obtained by SHS

It is worth to notice that further increasing of the excess TiAl precursor leads to significant decrease of the Ti_3AlC_2 content in the final product. The same situation was observed in case of reactions were Ti_3Al was used as a precursor.

Some examples of the powders morphology observations are presented on Fig.4. The elongated plate-like grains characteristic for MAX phases and regular grains of TiC can be observed in all of the obtained powders. These observations were confirmed by EDS results.

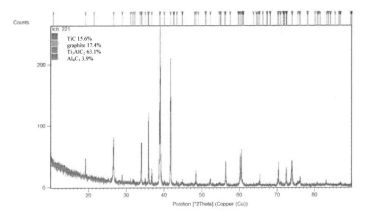

Figure 3. XRD pattern of SHS derived Ti_3AlC_2 powder
with highest content of MAX phase.

Figure 4. Morphology of SHS derived Ti_3AlC_2 powders

CONCLUSION

Obtaining of Ti_3AlC_2 powders by Self-propagating High-temperature Synthesis (SHS) is not only possible but also effective and efficient. Powders are characterized by high content of MAX phases, up to 76 wt.%, with relatively low level of TiC impurities.

The best results were achieved when excess amount (up to 10 wt.%) of intermetallic compound, namely TiAl, was used as a precursor for SHS synthesis of Ti_3AlC_2. Further increasing of intermetallic compound ratio in the reaction leads to significant decrease of desired Ti_3AlC_2 in favour of Ti_2AlC and TiC. Similar situation was observed when Ti_3Al was used as a precursor.

Careful control of the chemical reaction stoichiometry seems to play important role while obtaining Ti_3AlC_2 materials and further more detailed researches will be done.

Use of titanium carbide as one of precursors results also in quite high amount of MAX phase in the final product, but the high content of unreacted TiC may affect properties of the sintered material.

Sintering conditions of obtained powders as well as their mechanical properties will be examined.

ACKNOWLEDGMENTS

This work was supported by the National Science Centre under the grant no. 2472/B/T02/2011/40

REFERENCES
[1] W. Jeitschko, H. Nowotny, F.Benesovsky, Kohlenstoffhaltige ternare Verbindungen (H-Phase). *Monatsh. Chem.* 94, 1963, p 672-678
[2] H. Nowotny, Structurchemie Einiger Verbindungen der Ubergangsmetalle mit den Elementen C, Si, Ge, Sn. *Prog. Solid State Chem.* 2 1970, p 27
[3] M.W. Barsoum: The MN+1AXN Phases a New Class of Solids; Thermodynamically Stable Nanolaminates- *Prog Solid St. Chem.* 28, 2000, p 201-281

[4] J.Lis: Spiekalne proszki związków kowalencyjnych otrzymywane metodą Samorozwijającej się Syntezy Wysokotemperaturowej (SHS) - *Ceramics 44* : (1994) (*in Polish*)

[5] L. Chlubny, M.M. Bucko, J. Lis "Intermetalics as a precursors in SHS synthesis of the materials in Ti-Al-C-N system" *Advances in Science and Technology*, 45, 2006, p 1047-1051

[6] L. Chlubny, M.M. Bucko, J. Lis "Phase Evolution and Properties of Ti_2AlN Based Materials, Obtained by SHS Method" Mechanical Properties and Processing of Ceramic Binary, Ternary and Composite Systems, *Ceramic Engineering and Science Proceedings*, Volume 29, Issue 2, 2008, Jonathan Salem, Greg Hilmas, and William Fahrenholtz, editors; Tatsuki Ohji and Andrew Wereszczak, volume editors, 2008, p 13-20

[7] L. Chlubny, J. Lis, M.M. Bucko: Preparation of Ti_3AlC_2 and Ti_2AlC powders by SHS method MS&T Pittsburgh 09: Material Science and Technology 2009, 2009, p 2205-2213

[8] "Joint Commitee for Powder Diffraction Standards: International Center for Diffraction Data"

[9] H. M. Rietveld: "A profile refinement method for nuclear and magnetic structures." J. Appl. Cryst. **2** (1969) p. 65-71

[10] L. Chlubny: New materials in Ti-Al-C-N system. - PhD Thesis. AGH-University of Science and Technology, Kraków 2006. (*in Polish*)

CHEMICAL VAPOR DEPOSITION AND CHARACTERIZATION OF THICK SILICON CARBIDE TUBES FOR NUCLEAR APPLICATIONS

P. Drieux[1 2], G. Chollon[1], A. Allemand[2], S. Jacques[1]
[1]Laboratoire des Composites ThermoStructuraux, CNRS, Herakles, CEA, Université Bordeaux 1
[2]CEA Le Ripault
FRANCE

ABSTRACT

SiC/SiC composites have an outstanding mechanical behavior under irradiation, but their high porosity excludes their use as nuclear fuel claddings in the future nuclear power plants. A complementary thick and tight SiC sheath could be a solution to ensure the first barrier towards fissile materials.

The aim of this work is to make long, free standing and high strength SiC tubes. A few hundred micrometers thick tubular coatings were produced by chemical vapor deposition at atmospheric pressure, from $CH_3SiHCl_2/Ar/H_2$ mixtures. Their chemical compositions and microstructures were studied by electron probe microanalysis, Raman spectroscopy and scanning electron microscopy.

The deposition rate, composition and microstructure were investigated as a function of the substrate temperature and the gas flow rates. A Fourier transformed infrared spectroscopy analysis of the gas phase was carried out at the reactor outlet. The Si/C ratio, the SiC degree of crystallization and the surface morphology are strongly related to the maturation of the gas phase and the deposition regime.

INTRODUCTION

The structural components of a nuclear fusion or fission reactor must show a low induced radioactivity after neutron irradiation. Tubular SiC/SiC composites [1, 2] are materials of prime interest to be part of the nuclear fuel claddings of future power plants [3] thanks to their low neutron activation characteristics [4], but also their high strength and fracture toughness at high temperature [5].

The optimization of SiC/SiC composites for nuclear purposes requires high thermal properties as well as excellent gas impermeability [6]. The latter point has led to the idea of covering the composite with a dense and impermeable SiC_β sheath having high mechanical properties and stability at high temperature and under neutron irradiation [7].

Well known exemplary SiC coating materials, in terms of mechanical properties, are SiC monofilaments [8]. They are prepared by chemical vapor deposition at atmospheric pressure (APCVD) on a hot filament substrate, using chlorosilanes (e.g. dichloromethylsilane: DCMS) diluted in hydrogen as the SiC precursor [9]. A failure strength up to 6 GPa and a strain of about 1.5 % can indeed be reached in the high performance SCS-Ultra from Specialty Materials Inc. The aim of the present study is to prepare SiC tubes by APCVD, with a composition, microstructure and morphology compatible with high strength and strain at failure.

The present paper reports on our latest results on the elaboration and characterization of thick SiC tubular coatings. Efforts have been mainly concentrated on the adjustment of the stoichiometry and the microstructure of the SiC deposit. The influence of the deposition temperature, the initial composition of the gas mixture and the total flow rate on the physical and chemical properties and growth kinetics of the coatings have been investigated. Fourier transform infrared (FTIR) analyses of the gas phase at the reactor outlet have also been performed to follow the decomposition of the precursor.

EXPERIMENTAL APPROACH

CVD process

The laboratory-scale reactor for the APCVD of thick tubular SiC coatings is presented in Fig. 1. It consists of a horizontal silica glass tube (L = 1000 mm, $\emptyset_{int}$ = 46 mm) connected to gas inlet/outlet flanges at both ends. Tubular SiC deposits can be obtained by two ways: an outer coating on a solid graphite substrate or an inner coating inside a silica tube.

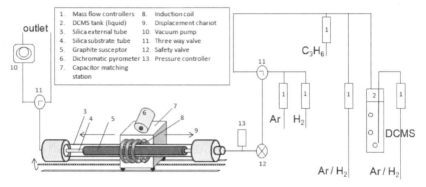

Figure 1. APCVD reactor for SiC deposition

In reference to the filaments process, a graphite solid cylinder (L = 200 mm, $\emptyset_{ext}$ = 10 mm) was first used as a substrate. In this case, the CVD reactor operated in a cold-wall configuration, since the substrate was directly heated by induction, while the external silica chamber remained relatively cold. The first experiments led to a main difficulty related to the cold-wall configuration and the horizontal position of the cylinder: a thickness gradient was indeed observed between the upper and lower parts of the rod, due to the convection of the gases. Furthermore, the removal of the graphite substrate is difficult and the final morphology of the SiC coating is strongly affected by the roughness and the porosity of the graphite substrate. The inside of a silica tube (L = 500 mm, $\emptyset_{int}$ = 8 mm) was used as an alternative substrate, the gases being forced to flow through it. The silica tube was brought to high temperature by a surrounding graphite susceptor heated by induction. This unit is placed inside the original 46 mm diameter silica tube. This new reactor was then in a hot-wall configuration, since the whole 8 mm diameter reaction chamber, including the substrate walls, was isothermal in the hot deposition area (4 cm).

Dichloromethylsilane ($H_3CSiHCl_2$, DCMS) was used as the SiC precursor [9]. The DCMS vapor saturation was obtained from a H_2 carrier gas going through a bubbler maintained at 20°C (P_{DCMS} = 48 kPa). The DCMS concentration was adjusted by further diluting the initial saturated DCMS/H_2 mixture in a secondary H_2 flow.

As the final aim of this work is the manufacturing of relatively long lengths (several tens of cm), the induction heat station was set on a motorized bench allowing the displacement of the induction coil, and therefore of the hot area, along the substrate. Although the study presented here was conducted without any displacement of the hot area, the travelling speed can range from 0.25 to 5 cm/min.

The surface temperature (T) of the graphite substrate (in the cold-wall configuration) or the graphite shell outer surface (in the hot-wall geometry) was monitored in situ during the CVD process with a high-resolution dichromatic pyrometer (Impac ISQ 5).

Deposition conditions

For all the experiments, the reactor was operated at atmospheric pressure, while the temperature and total gas flow Q_{tot} ranged from 1000 to 1100°C and from 200 to 850 sccm (standard $cm^3.min^{-1}$), respectively. A list of representative experiments is given in Table 1. Runs at higher temperatures were not carried out in the hot wall geometry due to the risk of silica substrate softening.

Table I. Deposition conditions for the CVD of SiC

N°	Q_{tot} (sccm)	$\alpha = Q_{H2}/Q_{DCMS}$	T (°C)	Si %$_{at.}$	Deposition rate (μm/h)
1	425	0	1025	54	124
2	425	1	1025	55	216
3	425	2	1025	72	296
4	425	4	1025	61	69
5	425	8	1025	67	50
6	425	12	1025	65	39
7	850	4	1025	76	237
8	213	4	1025	48	33
9	425	4	1050	57	70
10	425	4	1075	53	60
11	425	4	1100	49	72

Deposition of the pyrolytic carbon interlayer

The first coatings deposited directly on the silica tubes showed a strong chemical bonding that resulted, on cooling, to multiple cracking. A thin layer of pyrolytic carbon (PyC) was therefore deposited at the surface of the SiO_2 tube. This interlayer was deposited in the same reactor by injection of pure propene at low pressure (Q_{C3H6} = 100 sccm, T = 1000°C, P = 3 kPa). Its anisotropic structure was found to promote the SiC/SiO_2 interface delamination.

Characterization of the coatings

The morphology of the coatings was characterized with a field emission gun scanning electron microscope (SEM) (FEI Quanta 400F).

Electron probe microanalyses (EPMA) (SX 100 from CAMECA, France) were conducted on cross-sections of CVD-coated tube fragments. The Si, C and O concentrations were assessed using the wavelength dispersive spectroscopy (WDS) mode (10 kV, 10 nA). Punctual and line scan measurements were performed along the cross section of the coatings, with a spatial resolution of approximately 1 μm^3.

Raman microprobe (RM) analyses (Labram HR from Horiba Jobin Yvon) were also conducted. The excitation source was a He–Ne laser (λ=632.8 nm). The power was kept below 1 mW to avoid heating of the sample. The lateral resolution of the laser probe was close to 1 μm and the depth analyzed was less than 1 μm. As for EPMA, punctual and line scan analyses were recorded along the cross section of the deposit.

Analysis of the gas phase

A semi quantitative study of the gas phase was carried out by Fourier transform spectroscopy (FTIR) at the reactor outlet (Fig. 2). The laser beam of the spectrometer (Nicolet 550) was led through a room temperature analysis cell (with two ZnSe windows at both ends) installed at the reactor outlet (absorption configuration). While the reaction chamber was set at atmospheric pressure, the pressure in the gas cell was adjusted at 1 kPa to avoid saturation of absorption peaks on the FTIR spectra. At the outlet of the gas cell, the infrared beam was focused on the HgCdTe detector placed below. The whole infrared beam path was enclosed and constantly purged with a nitrogen flow in order to limit the absorption of atmospheric compounds. A background spectrum was first recorded with a pure H_2 flow in the reactor and subtracted to the reacting gases spectrum. The apparatus allowed the acquisition of spectra in the wave-number range 650 - 4000 cm^{-1}, with a resolution of 1 cm^{-1}.

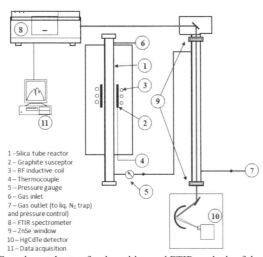

1 – Silica tube reactor
2 – Graphite susceptor
3 – RF inductive coil
4 – Thermocouple
5 – Pressure gauge
6 – Gas inlet
7 – Gas outlet (to liq. N_2 trap) and pressure control)
8 – FTIR spectrometer
9 – ZnSe window
10 – HgCdTe detector
11 – Data acquisition

Figure 2. Experimental setup for deposition and FTIR analysis of the gas phase

RESULTS AND DISCUSSION

The PyC interfacial layer deposited at low pressure before SiC (with a thickness of 350 nm) does not introduce major surface flaws in the SiC coating (Fig. 3). The EPMA and SEM measurements at opposite sides confirmed the chemical composition and thickness homogeneity of the SiC coatings deposited in the hot-wall geometry.

A number of thick SiC tubes were obtained with this APCVD process, and were successfully extracted from the silica substrate by taking advantage of the difference of thermal expansion coefficients between the silica substrate and the deposited SiC and the weak SiO_2/SiC interface provided by a PyC interlayer.

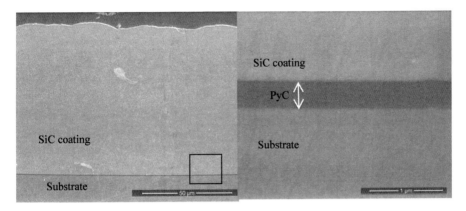

Figure 3. Cross section of the SiO_2/PyC/SiC interfacial region

It is well known [10, 11] that one of the key parameters in the deposition of SiC is the precursor dilution ratio α:

$$\alpha = Q_{H_2} \Big/ Q_{DCMS} \qquad (1)$$

The Si/C atomic ratio increases significantly with α (Table I). For α increasing from 0 to 12, the silicon atomic concentration in the coating varies from 54 %$_{at.}$ to 67 % $_{at.}$. The oxygen concentration was found to remain lower than 2 %$_{at.}$ in all specimens. Furthermore, a higher dilution of DCMS in H_2 results in a sharp decrease of the deposition rate. At 1025 °C, the deposition rate remains below 50 $\mu m/h$ when α is higher than 4. Such a deposition rate is far too low to consider a continuous deposition process for the production of thick and long tubes within a reasonable time. The Raman Spectra (Fig. 4) show no clear crystallized SiC features (expected at 750-1000 cm^{-1}) in the deposits, but a strong band at 500cm^{-1}, typical of free silicon. The intensity of the Si band varies with α in very good agreement with the Si/C ratio. However, even for $\alpha = 0$, free silicon is still present in the coating at this temperature and the SiC phase is poorly crystallized, as shown by the very broad and weak characteristic features.

A gradual improvement of the SiC crystalline state can be noticed when the deposition temperature increases. For coatings 4, 9, 10 and 11, corresponding to deposition temperatures increasing from 1025°C to 1100°C, the emergence of two bands in the 700 – 1000 cm^{-1} region, for the highest deposition temperatures confirms the coarsening of the SiC phase (Fig. 5). In parallel, the narrow peak at 520 cm^{-1} corresponding to crystallized silicon disappears as well as the surrounding bump due to amorphous Si. The Si/C atomic ratios obtained by EPMA are very consistent with the Raman analyses. It is worthy of note that a Si concentration of 49 %$_{at}$ is obtained at 1100 °C, indicating that pure SiC can be deposited in these conditions, without any additional carbon source in the gas phase.

Only few changes of the deposition rate can be noticed with the variation of temperature. The deposition process is apparently not thermally activated, suggesting a mass transfer limited regime [12]. However, since the temperature increase probably results in parallel in important changes of the gas phase composition in the hot area, the assumption of a precursor supply limitation should be considered with care.

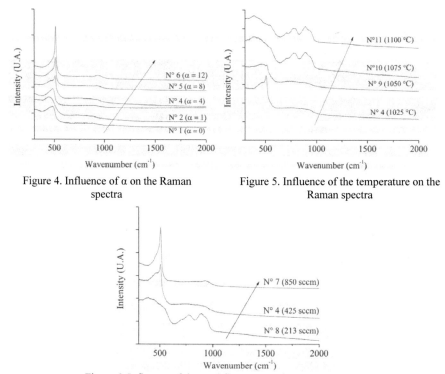

Figure 4. Influence of α on the Raman spectra

Figure 5. Influence of the temperature on the Raman spectra

Figure 6. Influence of the total flow rate on the Raman spectra

The influence of the total gas flow rate (Q_{tot}) was also examined (see deposits 4, 7 and 8 in Table I, obtained at Q_{tot} = 425 sccm, 850 sccm and 213 sccm, respectively). The gas velocity is an important parameter especially in the case of a mass transfer regime where the deposit morphology can be very sensitive to the gas flow around the substrate. In addition, Q_{tot} may affect the decomposition rate of precursors through the residence time varying along the position in the hot area. At low flow rates ($Q_8 < Q_4 < Q_7$, see Table 1), the carbon concentration, as measured by EPMA, tends to increase (from 25 %$_{at.}$ to 52 %$_{at.}$). These results are consistent with the Raman analyses (Fig. 6) showing that the peak assigned to crystalline Si tends to disappear at low Q_{tot} while simultaneously, the SiC characteristic bands gradually appear. The deposition rate increases significantly with Q_{tot} from 33 μm/h at 213 sccm to 237 μm/h at 850 sccm. This behavior is consistent with the previous study on the influence of temperature. The increase of the deposition rate with Q_{tot} is indeed indicative of a regime limited by diffusion. In parallel, the change in the composition of the deposit suggests a major effect of the maturation of gas phase, i.e. of the decomposition rate of DCMS into various intermediates, which may act as effective precursors of Si and C for the coating.

A minimal deposition temperature appears necessary to form a near stoichiometric coating with highly crystalline SiC in the reactor. Whereas there are only a few studies on the CVD from DCMS/H₂ [10, 13,14], several articles on the dichlorodimethylsilane (DCDMS)/H₂ system [14-16] and an abundant work on the MTS/H₂ system [11, 17-19] can be found in the literature. Féron *et al.* used DCMS at higher temperatures and atmospheric pressure in a cold wall reactor [10]. They obtained only Si-rich coatings (C/Si at. < 0.7) at much higher deposition rates limited by chemical reactions. This suggests the occurrence of DCMS depletion in the present case, in the hot wall configuration. On the other hand, Cagliostro *et al.* used DCDMS at similar temperatures and pressure in a hot wall reactor configuration [15, 16]. They found a dependence of the deposition rate with the total flow rate for T > 900 °C and a strong silicon excess in the coating, decreasing along the reactor axis (i.e. while the residence time increases). A similar behavior was reported in several studies of the MTS/H_2 system in relatively similar T and P conditions. Papasouliotis *et al.* and Huttinger *et al.* evidenced, for T = 1000 °C, a decrease of the deposition rate associated with a decrease of the silicon excess along the hot zone [17,18]. The decreasing of the total flow rate also reduces the deposition rate (by increasing the residence time) [17].

Since the silica substrate cannot withstand temperatures higher than 1100 °C in the hot-wall configuration, only a narrow operating temperature window is allowed. Since the deposition rate is apparently not thermally activated, the gas flow rate (or the residence time) appears as a key parameter for elaboration of long tubes.

The changes in the chemical composition of the deposit indicate an influence of the temperature on the concentration of the effective Si and C precursors brought to the deposition area. This can be explained by a change of the DCMS decomposition rate in the homogeneous (gas phase) state. To support this assumption, complementary FTIR analyses of the gas phase at the reactor outlet were performed.

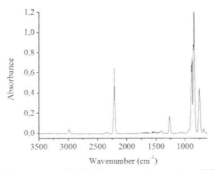

Figure 7. FTIR spectrum of DCMS at room temperature (650-3500 cm^{-1})

A room temperature spectrum of the DCMS was recorded as a reference (Fig. 7). This spectrum shows characteristic peaks of chlorosilane products, as in previous works on DCMS [20]. A table of the vibration modes of the main species from the Si-C-Cl-H system reports the peak frequencies observed experimentally and their assignments from the literature (Table II).

The infrared spectrum of DCMS at 25 °C shows the presence of a weak feature at 3010 cm^{-1} corresponding to the C-H stretching in the methyl group, a very sharp and intense peak at 2260 cm^{-1}, due to the stretching of Si-H bond, a low intensity signal corresponding to C-H wagging at 1270 cm^{-1}, and a conglomerate of intense peaks in the 900 – 650 cm^{-1} region, due to the stretching of Si-C and Si-Cl bonds (Table II). In a second step, an investigation of the temperature effect on the gas phase composition was carried out (Fig. 8).

Table II. Peak frequencies observed experimentally and assignments fro literature: Si-H-C-Cl

Observed peak position (cm^{-1})	Vibration mode	Chemical compound	Literature values
590	Si-Cl stretching	SiCl$_3$ or Si$_2$Cl$_6$	585, 592 [23]
620		SiCl$_4$	619 [23], 620 [21]
642		SiCl$_4$	641 [23]
756		DCMS	752, 762 [23]
807		HSiCl$_3$	805　　　　[24]
848	Si-C stretching	DCMS	[20]
860			[21]
884			[25]
893			
949	Si-Cl stretching	SiH$_2$Cl$_2$	960　　　　[22]
1270	C-H wagging	DCMS	1270　　　　[23] 1272　　[21]
1307		CH$_4$	1305　　　　[21] 1306　　　　[23] 1310　　[22]
2217	Si-H stretching	DCMS	2220　　[24]
2260		HSiCl$_3$	SiH$_n$Cl$_{4-n}$: 2248, 2260 [21] 2270 [22]
2970	C-H stretching	DCMS	-
3010		CH$_4$	3018　　[22]
2700-300	H-Cl stretching	HCl	2500-3095 [22]

Whereas the spectra remains unchanged for T ranging from 25 °C to 700 °C, major chemical modifications appear at higher temperatures. As the analysis is carried out ex situ at room temperature, detected molecules are necessarily stable and some of them may result from the recombination of other unstable species (e.g. free radicals). Some part of the DCMS decomposition products can also be consumed during the SiC deposition. However, the apparition of new species at the reactor outlet gives useful information for a better understanding of the APCVD process.

As shown in Fig. 8, the decomposition of dichloromethylsilane in the reactor is accompanied by the formation of HCl, HSiCl$_3$, HSi$_2$Cl$_2$ and CH$_4$. SiCl$_4$, another stable compound, is also expected (642 cm^{-1}). However, due to the poor sensitivity of the detector at low frequencies and to other peaks overlapping, this specie could not be followed as accurately as the others. To follow the evolution of the gas phase with temperature, the main peak areas were subsequently plotted versus T (Fig. 9)

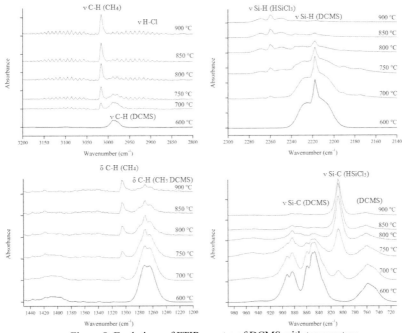

Figure 8. Evolutions of FTIR spectra of DCMS with temperature

The gradual decomposition of DCMS above 700°C (b, d, e, f, g, i) corresponds to the formation of various products in the gas phase. The apparition of methane (j, l), is progressive and tends to a plateau for the highest temperatures. The formation of trichlorosilane ($HSiCl_3$, curves c and k) shows a maximum at 900°C. Its concentration decreases at higher temperatures, indicating consumption, either due to the deposition on the reactor walls, or to the formation of other reactive intermediates. SiH_2Cl_2 also shows a similar behavior, its concentration decreasing strongly down to nearly zero for temperatures above 900 °C (h). Hence, the formation and consumption of chlorosilanes appears to be strongly related to the silicon incorporation in the coating.

A decomposition route of the DCMS during the APCVD process can be tentatively proposed by the analysis of the above results and the previous conclusions from the literature. Zhang and Huttinger [18] proposed a mechanism for the deposition of SiC at near atmospheric pressure from MTS/H_2 mixture. As for MTS, the Si-C bond is probably quickly dissociated into CH_3· and $HSiCl_2$· in the hot zone [17-19]. This assessment was confirmed by the ab initio calculations of Allendorf et al. [26]. The Si-H bond of DCMS or $HSiCl_2$· is also probably easily broken to form eventually the reactive $SiCl_2$, likely the major silica source for the coating [18]. In parallel, CH_3· readily recombines into CH_4, which has a low surface reactivity. The chlorosilanes and $SiCl_4$ detected by FTIR result from the recombination of the $HSiCl_2$· and $SiCl_2$ radicals, in the hot zone or during the cooling. The higher reactivity of $SiCl_2$ compared to CH_4 is responsible for the presence of excess silicon, especially at low temperatures and high α ratios. At high temperature (T > 1050 °C), more reactive unsatured hydrocarbons might be formed

(FTIR analyses were not carried out at these temperatures), resulting in an increase of the deposition rate and of the carbon ratio in the coating. On the other hand, higher H_2 concentrations (higher α) promote the reduction of $SiCl_2$ and therefore the deposition of free silicon, while they stabilize CH_4 and prevent the deposition of C.

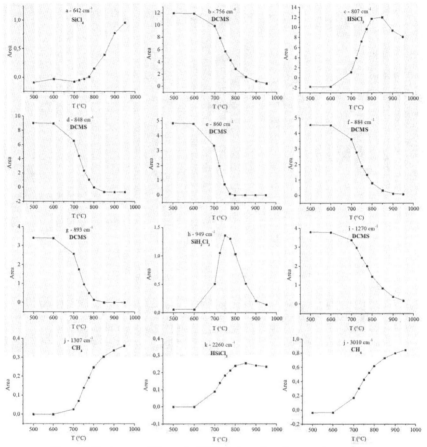

Figure 9. Peak areas of the main FTIR features (a...l) versus temperature

CONCLUSION

The present paper introduced a method to prepare long and dense monolithic SiC tubes with thick walls by APCVD. Free standing SiC tubes were obtained by deposition inside a SiO_2 tube covered with pyrocarbon. An optimization of the process was carried out by analyzing the influence of different parameters such as the precursor dilution in the injected gas mixture, the total flow rate and the deposition temperature. The increase of deposition rate with the gas flow rate, and the low thermal activation of the deposition process indicate that the process is limited by mass transfer. The modification of the gas phase itself with the different parameters leads to significant differences in the final composition of the SiC deposit.

A better understanding of the process was obtained by analyzing the gas phase at the reactor outlet by FTIR spectroscopy. The production of chlorosilanes ($SiCl_4$, $HSiCl_3$ and HSi_2Cl_2) was highlighted, along with a production of methane (CH_4) and HCl. The increase of temperature promotes the reactivity of the carbon precursors, while an important dilution inhibits them and favors the reduction of chlorosilanes, hence resulting in a Si rich deposit. The increase of the deposition rate with Q_{tot} is indicative of a regime limited by diffusion.

The elaboration of longer and thicker SiC tubes is already in progress, and the mechanical testing of the samples, for instance by C-ring tests [28] is planned.

REFERENCES

[1]C. Lorrette, C. Sauder and L. Chaffron, "Progress in developing SiC/SiC composite materials for advanced nuclear reactors" (Paper presented at the 18th International Conference on Composite Materials, Jeju Island, Korea, 21-26 August 2011), 1-4.
[2]C. Ayranci, J. Carey, "2D braided composites: A review for stiffness critical applications", Composite Structures 85 (2008) 43–5
[3]R.H. Jones et al., "Recent advances in the development of SiC/SiC as a fusion structural material", Fusion Engineering and Design, 41 (1-4) (1998), 15-24
[4]E.V Dyomina et al., "Low-activation characteristics of V-alloys and SiC composites", Journal of Nuclear Materials, 258–263 (1998), 1784–179
[5]A.R Raffray et al., "Design and material issues for high performance SiC_f/SiC-based fusion power cores", Fusion Engineering and Design, 55 (2001), 55–95
[6]B. Riccardi et al., "Issues and advances in SiC_f/SiC composites development for fusion reactors", Journal of Nuclear Materials, 329–333 (2004), 56–65
[7]H. Feinroth et al., "Multi-Layered Ceramic Tube for Fuel Containment Barrier And Other Applications In Nuclear And Fossil Power Plants", Patent WO 2006/076039, (2006)
[8]G. Chollon, R. Naslain et al., "High temperature properties of SiC and diamond CVD-monofilaments", Journal of the European Ceramic Society 25 (2005), 1929–1942
[9]T.T. Cheng et al., "The microstructure of sigma 1140+SiC fibres", Materials Science and Engineering : A 260 (1999), 139–145
[10]O. Féron, "In situ kinetic analysis of SiC filaments CVD", Diamond and Related Materials 11 (2002), 1234–1238
[11]G. Chollon, M. Placide et al., "Transient stages during the chemical vapour deposition of silicon carbide from CH3SiCl3/H2: impact on the physicochemical and interfacial properties of the coatings", Thins Solid Films 520 (2012), 6075-6087
[12]G. Astarita, "Regimes of mass transfer with chemical reaction", Ind. Eng. Chem. 58, (1966), 18–26
[13]H. Vincent, "Chemically vapour-deposited coatings of silicon carbide on planar alumina substrates", J. Mater. Chem. (1992), 567 – 574
[14]J.J. Brennan, "Interfacial studies of chemical-vapor-infiltrated ceramic matrix composites", Mat. Science and Eng. A126 (1990), 203 – 223

[15]D.E. Cagliostro and S.R. Riccitiello, "Model for the formation of silicon carbide from the pyrolysis of dichlorodimethylsilane in hydrogen", J. Am. Ceram. Soc. 76 (1993), 39 – 53

[16]D.E. Cagliostro and S.R. Riccitiello, "Comparison of the pyrolysis products of dichlorodimethylsilane in the chemical vapor deposition of silicon carbide on silica in hydrogen or argon", J. Am. Ceram. Soc. 77 (1994), 2721 – 2726

[17]Papasouliotis et al., "Experimental study of atmospheric pressure chemical vapor deposition of silicon carbide from methyltrichlorosilane", J. of Mat. Res. 14 (1999), 3397 – 3409

[18]W. Zhang and K. Hüttinger, "CVD of SiC from Methyltrichlorosilane", Chem. Vap. Deposition 7 (2001), 167 – 181

[19]F. Loumagne, F. Langlais and R. Naslain, "Reaction mechanisms of the chemical vapor deposition of SiC-based ceramics from CH3SiCl3/H2 gas precursor", J. Crystal Growth 155 (1995), 205 – 213

[20]O. Féron, "Filament CVD de SiC de composition et microstructure adaptées au procédé Snecma d'enduction par le titane liquide", (Internal report presented during postdoctoral position, Bordeaux, December 2000)

[21]S. Jonas et al., "FTIR In Situ Studies of the Gas Phase Reactions in Chemical Vapor Deposition of SiC", Journal of Electrochemical Society, 142 (1995), 2357-2362

[22]J. N. Burgess and T. J. Lewis, "Kinetics of the Reduction of Methyltrichlorosilane by Hydrogen," Chem. Ind., 976–77, 1974

[23]V. Hopfe et al., "In-Situ FTIR Emission Spectroscopy in a Technological Environment : Chemical Vapour Deposition (CVI) of SiC Composites", Journal of Molecular Structure 347 (1995), 331-342

[24]G. Socrates, "Infrared Characteristic Group Frequencies" (1980)

[25]Brennfleck et al., "In-Situ spectroscopic monitoring for SiC-CVD process control", Phys. IV 9, (2009), 1041 – 1048

[26]W. M.D. Allendorf and C.F. Melius, "Theoretical study of the thermochemistry of molecules in the Si-C-Cl-H system", J. Phys. Chem. 97 (1993), 720 – 728

[27]T.D. Gulden, "Deposition and microstructure of vapor-deposited silicon carbide", Journal of the American Ceramic Society 51, Issue 8 (1968) 424 – 428

[28]ASTM G38 - 01(2007) Standard Practice for Making and Using C-Ring Stress-Corrosion Test Specimens

UNIFORM MICROWAVE PLASMA PYROLYSIS FOR THE PRODUCTION OF METASTABLE NANOMATERIALS

Kamal Hadidi[1]†, Makhlouf Redjdal[1], Eric H. Jordan[2], Olivia A. Graeve[3], and Colby M. Brunet[3]

[1]Amastan LLC, 270 Middle Turnpike, Storrs, CT 06269

[2]Department of Mechanical Engineering, University of Connecticut, Storrs, CT 06269

[3]Kazuo Inamori School of Engineering, Alfred University, Alfred, NY 14802

†Corresponding author: khadidi@amastan.com

ABSTRACT

Uniformity, homogeneity, phase structure and size of nanostructured materials are important characteristics for a variety of applications. For a composite particle to carry uniformity in both composition and microstructure, it must be generated from molecularly mixed and homogeneous precursors and must follow a uniform thermal path during its processing. Amastan's Uniform Melt State Process (UniMelt™) achieves both requirements by processing molecularly mixed, uniform diameter precursor droplets through an axisymmetric microwave thermal plasma process with uniform high temperature and laminar entrainment flows, followed with very high quenching rates. Tuning of the UniMelt™ process allows the synthesis of uniform metastable nanopowders that can be either fully dense or porous with a very narrow particle size distribution. These characteristics are very important for the consolidation step. Amastan's amorphous nanopowders have been found to sinter under pressure at lower temperatures than usual. Furthermore, the current UniMelt™ process is easily scalable to achieve high throughput. The UniMelt™ can be exploited as a synthesis step towards producing uniform and homogenous amorphous materials that would consolidate at lower temperature and lower pressure for a variety of applications.

INTRODUCTION

Thermal processing technology for the synthesis of materials suffers from non-homogeneity in material composition of the feedstock, and non-uniformity in thermal processing paths. The compositional non homogeneity is introduced when precursors are prepared using synthesized powders such as in solid-state-reaction methods for material synthesis[1,2,3,4]. The compositional problem can be remedied by using solution-based precursors to provide a homogenous molecular mixing of constituents such as in traditional spray pyrolysis of precursor aerosols[2], Liquid-Feed Flame-Spray-Pyrolysis[5], or Solution Precursor Plasma Spray (SPPS) methods[6,7,8]. In combustion techniques using solution sources, injection of non uniform size of droplets from atomizers combined with an inherent gradient temperature in the combustion flame, impose non uniform thermal paths for particles often resulting in heterogeneity of microstructure. In plasma methods such as arc plasma, voltage source fluctuations, side injection of non uniform droplets and difference in plasma depth penetration of particles lead to similar results. This subsequently leads to difficulty in obtaining highly pure, uniform in size powders, with small grain sizes having a homogeneous microstructure. Such high end powders are needed in particular for IR domes and windows[9], and other applications where improved sinterability and moderate grain growth with temperature are desired.

In this present work, we describe the Uniform Melt State Proces (UniMelt™) as a novel thermal processing technology that incorporates a piezo driven droplet maker for the production of uniform precursor droplets, a uniform temperature profile of microwave generated plasma, and an axisymmetric geometry embodiment of a plasma hot zone. The UniMelt™ provides a uniform thermal processing path to axially inject uniform diameter droplets that leads to spherical product particles, uniform in size that can be tuned to be either nano crystalline or amorphous. In particular, we present the results of producing nanocomposite ceramics of magnesium and yttrium oxides for Infra-Red (IR) optical applications. For such composites, a homogenous distribution of MgO and Y_2O_3 grains in the matrix composite is highly desired to suppress or reduce grain growth during consolidation techniques, e.g., Spark Plasma Sintering (SPS).

EXPERIMENTAL SETUP

The experimental setup consists of a microwave generated plasma running at 2.45 GHz with a maximum power of 6 kW, coupled to a plasma chamber through a rectangular waveguide. The reflected power is minimized by using a tuner to establish a coupling efficiency of the forward microwave power to the plasma of more than 98%. This makes the entire system very efficient as more than 85% of the electricity from the plug is transformed into plasma enthalpy. The plasma is created and maintained inside a dielectric tube that is transparent to microwave radiation at 2.45 GHz. Specially designed (laminar) flows keep the plasma from attaching to the walls of the dielectric tube and maintain it centered with minimum turbulence. Plasma temperature has been measured to be around 6000 K, and is uniform across the entire plasma diameter[10].

Precursor droplets are generated by a droplet maker using a piezo electric element that generates small droplet sizes with narrow diameter distribution to within 2% of the mean diameter[11,12]. A voltage pulse to the piezo element coupled to a liquid precursor creates a disturbance in the liquid which when exiting an orifice of a capillary causes a stream breakup into uniform droplets due to Rayleigh's capillary stream breakdown principle[13]. These droplets can be as small as 20 microns.

The quality, purity, composition, and homogeneity for ceramic powder preparation depend drastically on the structure of molecular species in solution, which includes metal nitrates[14] [Saito 1988]. High miscibility of liquid phases and low melting points of precursor solutions are very important to obtain high level of precursor homogeneity. Yttrium and magnesium nitrates are ideal as they have low melting points (<100 °C) and high miscibility of the liquid phases[2]. The nitrate route is relatively cheap as the reactants sources are low cost and readily available. To produce MgO-Y_2O_3 nanocomposite powders, we mixed magnesium and yttrium nitrates with citric acid and ethylene glycol. With appropriate concentration of ethylene glycol and citric acid, this mixture allows an excellent dispersion of Mg^{2+} and Y^{3+} ions in the solution through thorough dissolution of nitrates along with increased solution stability[15]. The resulting solution precursor is thoroughly stirred using a magnetic stirrer for at least 20 minutes and filtered prior to injection. For sufficient drive voltage, the frequency of the exciting signal of the piezoelectric element is tuned to generate a stable and uniform droplet stream, with a droplet diameter of 130 micrometers. Thermal processing is achieved according to optimized experimental parameters for microwave power (< 6 KW), solution precursor injection flow rate (2 to 5 ml/mn), and carrier gas flow rate inside the plasma sheath necessary to maintain a stable and laminar microwave generated plasma. Quenching is done into atmospheric environment and particle products are collected in a custom made box with nylon filter with 10 micron pore diameter.

Amorphous powder of MgO-Y_2O_3 nanocomposites synthesized using the Uniform Melt State Process was calcined at 600 °C for one hour to remove any carbon content before consolidation. About 2.4 grams of the calcined powder was used for sintering, using Spark Plasma Sintering (SPS) in a D-25 FTC Système unit at Alfred University. SPS consolidation was conducted at die temperature of 1000 °C and sample temperature was estimated at 1100 °C. The sample was pressed at 100 MPa pressure in a 19 mm diameter die. Heating rate was 200 °C/min and a dwell time of 3 minutes was used with the total sintering cycle lasting 20 minutes.

Subsequent SEM, TEM and XRD analyses of the composite powder particles were conducted to determine particle size, size distribution, morphology, and phase microstructures.

RESULTS AND DISCUSSION

Scanning Electron Microscopy (SEM) analysis of MgO-Y$_2$O$_3$ nanocomposite powder was done on a Philips ESEM 2020 to determine particle, morphology, and texture. Figure 1 shows SEM micrographs of a single and several identical MgO-Y$_2$O$_3$ particles that are porous and spherical with a honeycomb-like morphology. The uniformity in morphology and diameter size of the particle product is due to the combination of feeding equal size precursor droplets that are processed homogenously and identically by the UniMelt™ process. Energy dispersive X-ray (EDX) analysis on the same machine revealed that the powder consists of main elemental components oxygen, magnesium, and yttrium. All microsphere particle products have approximately the same uniform diameter of 100 μm resulting from the thermal processing of 130 μm diameter precursor droplets. The particles are very volatile and easily breakable as evidenced by the wall fragments near particle in Figure 1a. However, it should be noted that the particle is already consolidated into micron sized agglomerate that requires no further consolidation for subsequent sintering at high temperature and high pressure.

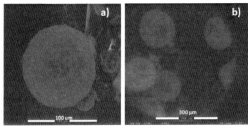

Figure 1. SEM photograph of a) single and b) many uniform and spherical MgO-Y$_2$O$_3$ nanocomposite powder particles.

X-Ray Diffraction (XRD) analysis was done on a Bruker D5005 & D8 Advance X-Ray Diffractometer to determine the phase microstructure of "as-is" MgO-Y$_2$O$_3$ powder. The results are shown in Figure 2a. It can be seen that the MgO-Y$_2$O$_3$ powders produced are totally amorphous in microstructure. Selected area electron diffraction (SAED) analysis of the same powder exhibited a diffuse pattern, as shown in Figure 2b, confirming the total amorphous nature of the magnesia-yttria nanocomposite product following thermal processing using UniMelt™.

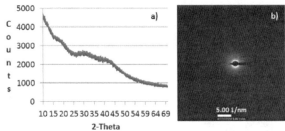

Figure 2. X-Ray Diffraction (XRD) analysis indicating amorphous MgO-Y_2O_3 powder microstructure confirmed by, b) a selected area electron diffraction (SAED) pattern analysis showing a diffuse pattern.

To investigate grain growth versus temperature and determine phase distribution of MgO and Y_2O_3 components, the "as-is" powder was respectively calcined at 1100 °C and 1200 °C for 1 hour at atmospheric pressure. The resulting microstructures were observed using TEM analysis shown in Figure 3. First, a wall fragment of the amorphous "as-is" spherical particles of Figure 1 is shown in Figure 3a. It can be seen that the microstructure is composed of undistinguishable grains with grain size not exceeding 5 nm (magnified and measured using the scale). TEM analysis of calcined powder at 1100 °C for 1 hour is shown in Figure 3b. It can be seen that the microstructure is a two-phase nanocomposite Magnesia-Yttria with average grain size smaller than 100 nm. The TEM image contrast is such that the MgO grains are light (grey) and the Yttria grains are dark with grains that are strongly-diffracting and these appear very dark. The Y_2O_3 grains appear to be homogeneously distributed in the remaining MgO grain matrix. The microstructure was found to be fully crystalline as evidenced by SADP rings of magnesia and Yttria (not shown). The 'as-is' powder calcined at 1200 °C for 1 hour is shown in Figure 3c. Grain growth can be observed, however grain sizes still remain below 200 nm. Furthermore, the homogenous distribution of Y_2O_3 grains in the MgO grain matrix is maintained. No apparent agglomeration of alike-elements of either MgO or Y_2O_3 can be observed.

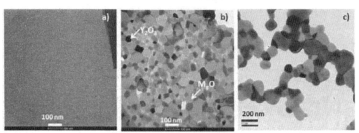

Figure 3. TEM micrographs of a) a wall fragment from a spherical "as-is" MgO-Y_2O_3 powder particle, b) heat treated at atmospheric pressure at 1100 °C, and c) at 1200 °C.

The preliminary consolidation work was performed at Alfred University. It was observed that sintering appeared to be completed at effective sample temperature close to 900 °C, well below the scheduled estimated sample temperature of 1100 °C. A fragment of the resulting consolidated disk was analyzed using a dual beam focused ion beam/scanning electron microscope (FIB/SEM). An SEM micrograph of a cross section of the consolidated MgO-Y_2O_3 powder is shown in Figure 4.   It can be seen that the microstructure of the composite is a continuous background matrix of a single phase (dark) of MgO containing well dispersed grains of Y_2O_3 oxide (gray) along with some porosity of equal size. Elongated Y_2O_3 white grains especially around pores can be seen, which might imply that these pores were Y_2O_3 grains that were chipped off the surface cross section during Fib-cutting. Further analysis of sintering parameters is required to explain the new continuous matrix of MgO as all corresponding grains seem to have fused together to form the continuous matrix. Furthermore, the size of Y_2O_3 minority grains needs to be lowered as they were found to be as large as those obtained using solid-state process using 50-100 nm raw powders.

Figure 4. SEM from a FIB-cut analysis of an MgO-Y_2O_3 sample consolidated using spark plasma sintering up to 1000 °C and 100 MPa pressure.

CONCLUSION

The Uniform Melt State Process (UniMelt™) provides a novel pyrolysis method for uniform thermal processing of materials. It combines compositional homogeneity through the use of solution precursors with uniform precursor droplet injection and uniform thermal paths. It can produce particle powders with uniform size and morphology, and homogeneous nanostructure. Heating and cooling rates can be tuned within certain limits to produce amorphous microstructure, which has been demonstrated by the production of amorphous nanocomposite MgO-Y_2O_3 powders. Spark plasma sintering of MgO-Y_2O_3 amorphous powders was achieved at lower-than-expected temperature. It was also found that the nanocomposite transformed into a continuous matrix of the higher concentration MgO constituent with well dispersed grains of Y_2O_3 oxide minority constituents. However, further work needs to be done to optimize the sintering process in order to lower grain size and improve microstructure uniformity of the sintered body.

ACKNOWLEDGMENT

This work was supported by a subcontract from the University of Connecticut, under the DARPA Nano composite materials program.

REFERENCES

[1] J. S. Abell, I. R. Harris, B. Cockayne, and B. Lent, An Investigation of Phase Stability in the Al_2O_3-Y_2O_3 System, *J. Mater. Sci.*, **9**, 527-37 (1974).

[2] M. Nyman, J. Caruso, M. J. Hamden-Smith, and T.T. Kodas, Comparison of Solid-State and Spray-Pyrolysis Synthesis of Yttrium Aluminate Powders", *J. Am. Ceram. Soc.*, **80**, 1231-38 (1997).

[3] A. Rosenflanz, M. Frey, B. Endres, T. Anderson, E. Richards, and C. Schardt, Bulk Glasses and Ultra-Hard NanoCeramics, *Letters to Nature*, **430**, 761-64 (2004).

[4] M. Medraj, R. Hammond, M.A. Parvez, R.A.L. Drew, and W.T. Thompson, High Temperature Neutron Diffraction Study of the Al_2O_3-Y_2O_3 System, *J. Europ. Ceram. Soc.*, **26**, 3515-24 (2006).

[5] J. Marchal, T. John, R. Baranwal, T. Hinklin, and R. M. Laine, Yttrium Aluminum Garnet Nanopowders Produced by Liquid-Feed Flame Spray Pyrolysis (LF-FSP) of Metallorganic Precursors, *Chem. Mater.*, **16**, 822-31 (2004).

[6] N. P. Padture, K.W. Schlichting, T. Bhatia, A. Ozturk, B. Cetegen, E.H. Jordan, M. Gell, S. Jiang, T.D. Xiao, P.R. Strutt, E. Garcia, P. Miranzo, and M.I. Osendi, Towards Durable Thermal Barrier Coatings with Novel Microstructures Deposited by Solution-Precursor Plasma Spray", *Acta Materialia*, **49**, 2251-57 (2001).

[7] L. Xie, X. Ma, E.H. Jordan, N. P. Padture, T.D. Xiao, and M. Gell, Identification of Coating Deposition Mechanisms in the Solution-Precursor Plasma-Spray Process using Model Spray Experiments, *Mat. Sci. Eng*, **A362**, 204-12 (2003).

[8] D. Chen, E. H. Jordan, M. G., and X. Ma, Dense Alumina-Zirconia Coatings Using The Solution Precursor Plasma Spray Process", *J. Amer. Ceram. Soc.*, **91**, 359-65 (2008).

[9] D.C. Harris, Materials for Infrared Windows and Domes: Properties and Performance, SPIE Optical Engineering Press, Washington (1999).

[10] P. P. Woskov, K. Hadidi, M. C. Borrás, P. Thomas, K. Green, and G. J. Flore, Spectroscopic Diagnostics of an Atmospheric Microwave Plasma for Monitoring Metals Pollution, *Rev. Sci. Inst.*, **70(1)**, 489-92 (1999).

[11] M. Orme and E. P. Muntz, The Manipulation of Capillary Stream Breakup Using Amplitude Modulated Disturbances: a Pictorial and Quantitative Representation, *Phys. of Fluids,* **2**(7), 1124-40 (1990).

[12] M. Orme, K. Willis, and V. Nguyen, Droplet Patterns from Capillary Streams, *Phys. of Fluids,* **5**, 80-90 (1993).

[13] Lord Rayleigh, On the instability of jets, *Proc. of the London Math. Soc.*, 10(1),) 4-12 (1879.

[14] S. Saito, Advanced Ceramics, Oxford Univ. Press and Ohmsha LTD, pp. 174 (1988).

[15] Gupta, S. and Katiyar, RS, Temperature-Dependent Structural Characterization of Sol–Gel Deposited Strontium Titanate ($SrTiO_3$) Thin Films Using Raman Spectroscopy, *J. Raman Spectroscopy*, **32**, 885-91 (2001).

CHARACTERIZATION OF THE CONDUCTIVE LAYER FORMED DURING μ - ELECTRIC DISCHARGE MACHINING OF NON-CONDUCTIVE CERAMICS

Nirdesh Ojha, Tim Hösel, Claas Müller, Holger Reinecke
Laboratory for Process Technology, IMTEK-Department of Microsystem Engineering, University of Freiburg, Georges-Koehler-Allee 103, 79110 Freiburg, Germany

Keywords: Electric Discharge Machining, Spark Erosion, Non-conductive Ceramic, μ-EDM

ABSTRACT

Electric discharge machining (EDM) widely used to machine hard metals requires both the tool and the work piece to be conductive. However, recently it has been shown that even non-conductive ceramics can be machined effectively with EDM by applying a conducting layer of assisting electrode on top of the insulating ceramics. Discharges are initially triggered by the assisting electrode and a conductive layer which is produced during the spark erosion process ensures that the top layer of the insulating ceramic remains conductive. After the machining process, the conductive layer can be easily removed.

This paper presents the investigations made on the conductive layer generated during the μ-EDM of three widely used non-conductive ceramics, namely ZrO_2, Si_3N_4 and SiC. Various aspects of the generated transitional layer including the thickness, the surface resistance, the surface roughness and the surface topography have been reported. Based on the surface topography of the transitional layer, the material removal mechanisms responsible for μ-EDM of different non-conductive ceramics have been discussed.

INTRODUCTION

Ceramics are increasingly becoming the ideal material choice for various novel applications like for example in the field of biomedical devices and aerospace parts. This is primarily due to the excellent material properties of ceramics such as high material strength, stiffness, bio-compatibility and temperature resistance. However, these properties also make it very difficult to structure these materials using traditional techniques such as milling, grinding or polishing with an appropriate accuracy. On the other hand, innovative application requires very precise μ-structuring of ceramic materials. Therefore novel ways to structure ceramics is in demand.

One of the non-conventional methods of structuring ceramics is by using spark erosion. The spark erosion process is commonly known as electric discharge machining (EDM) and is well-established for structuring conductive materials including hard metals. However, the spark erosion process requires that both the tool and the work piece be conductive [1]. Nevertheless, several interesting advanced ceramics such as ZrO_2, Si_3N_4 and SiC are non-conductive.

Recently, it has been shown that non-conductive ceramic such as ZrO_2 can be successfully structured using the spark erosion process by applying a conductive layer, called assisting electrode (AE) on the surface of the non-conductive ceramic [2–6]. Discharges are initially triggered by the AE and a conductive layer which is produced out of the dielectric during the spark erosion process ensures that the top layer of the insulating ceramic remains conductive. After the machining process, the conductive layer can easily be removed by simply heating the EDM processed sample.

EXPERIMENTAL
Charmilles RoboForm35® standard sinking EDM machine was used for the spark erosion process during this study. Table I show the most important parameters commonly used during the study. These parameters have been used to structure non- conductive ZrO_2 in previous studies [7].

Table I. Main parameters used for the spark erosion of non-conductive ceramics

Open voltage	Current	Pulse on time	Pulse off time	Gap Voltage
200 V	0.5 A	0.8 μs	1.6 μs	25 V

Samples of ZrO_2, Si_3N_4 and SiC non-conductive ceramics were received from CeramTec GmbH. All the samples had a polished surface and were 19.5 x 19.5 mm^2 each. A lacquer based carbon-conductive ink was screen printed on to the surface of the non-conductive ceramic. The screen printing was performed on an EKRA M2 screen printing machine. The samples were then heated in the oven at 120°C for 2 hours. Upon heating, the lacquer became conductive (resistivity of 200 Ω μm) and acted as an AE to initiate the spark erosion process. Figure 1 shows a picture of non-conductive ZrO_2 ceramic sample as received (a), after applying the AE (b) and after the complete top surface has being processed with EDM (c). The dark layer that is visible on figure 1 c) is the transitional layer formed during the EDM process. This layer can be removed by heating the sample in an oven and is not present in the final product. The top portion of this layer is made out of carbon black, formed by the degeneration of the dielectric oil during the sparks [8]. Therefore, the top portion of the transitional layer is conductive and is often referred to as the conductive layer. The transitional layer plays a vital role during the EDM of non-conductive ceramics by sustaining the electric contact.

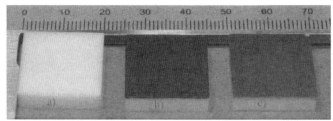

Figure 1. Sample of a non-conductive ZrO_2 a) as received, b) with assisting electrode on top and c) after being processed over the complete top surface by EDM.

RESULTS AND DISCUSSIONS

Thickness
A cross section of processed ZrO_2 sample was prepared to visualize and measure the thickness of the transitional layer and the conductive layer generated during the process. Scanning electron microscope (SEM) image of the cross section is shown in Figure 2. The transitional layer corresponding to the dark top layer in Figure 1c) can also be distinguished in the SEM image in Figure 2a). In this image, the transitional layer is distinct and has a thickness ranging from 0.9 μm to 1.5 μm. This transitional layer corresponds to the white layer observed during the EDM of hard metals and contains lot more carbon than the base material [9]. A higher

resolution SEM image (Figure 2b) of the transitional layer is presented in Figure 2b). A distinct conductive layer could not be visualized in this micrograph. In this image the underlying grains of ceramic is visible, which suggests that the conductive layer covering the transitional layer should be locally less than 1μm thick as the underlying bulk can be visualized through a thin layer of carbon.

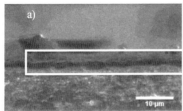 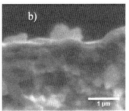

Figure 2. SEM images of the cross section of EDM processed ZrO₂. A distinct layer is visible (a) at lower magnification. The underlying ceramic grains can be seen (b) at much higher magnification

Mean surface roughness and sheet resistance

The mean surface roughness (Ra) of the top surface of the EDM processed ZrO₂, Si₃N₄, and SiC was measured using white light interferometer and is presented in Figure 3. The Ra value of the transitional layer generated on ZrO₂ (0.38 μm) is less than that of Si₃N₄ (0.56 μm) and SiC (0.54 μm). Generally, ceramic micro-structures with lower value of surface roughness exhibit higher mechanical strength. Therefore a lower value of surface roughness is usually desirable as the surface roughness of the transitional layer will directly affect the surface roughness of the final ceramic product. The standard deviation of the mean surface roughness of the transitional layer signals the stability of the μ-EDM process. The lower value of the standard deviation for ZrO₂ shows that the μ-EDM process went very smoothly in this material.

However, the lower values of the surface roughness and the standard deviation of the surface roughness obtained in case of ZrO₂ could be due to the fact that the parameters used for the spark erosion process in this study was optimized for ZrO₂ in previous studies [7]. Using another set of parameters, optimized for Si₃N₄ or SiC, could perhaps provide a lower value of the mean surface roughness for these ceramics respectively along with the decrease in standard deviation.

Also shown in Figure 3 is the sheet resistance of the surface of the transitional layer of the EDM processed non-conductive ceramics measured using standard 4-point-measurement setup. The value of the sheet resistance of the surface of the transitional layer on ZrO₂ (1.68 kΩ/□) and Si₃N₄ (1.69 kΩ/□) are comparable. The sheet resistance of the transitional layer on SiC (0.57 kΩ/□) was significantly lower than that of the other two ceramics. The resistivity (as obtained from the data sheet of the manufacturer) of the non-conductive ceramics used in this study is tabulated in Table II.

The lower value of sheet resistance of the surface of the transition layer on SiC could be due to two reasons. Firstly, the resistivity of the ceramic bulk of SiC is much smaller (less than 10000 times) than that of other two ceramics. Secondly, when the SiC ceramic is decomposed into its constituent elements, the carbon present in SiC is readily available to be used up in creating a conductive layer.

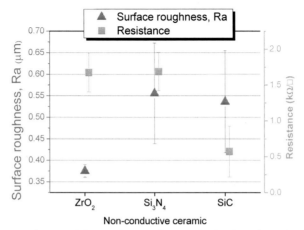

Figure 3: Surface roughness and sheet resistance of the conductive layer formed during the spark erosion process of non-conductive ceramics

Table II. Resistivity of non-conductive ceramic bulk at 20°C

ZrO_2	Si_3N_4	SiC
1×10^{12} Ω cm	1×10^{14} Ω cm	5×10^{7} Ω cm

The sheet resistance of the surface of the transitional layer formed on the ZrO_2 was also measured using Van Der Pauw method and found to be 1.58 kΩ/□. This value of sheet resistance corresponds well to the value obtained by using the 4 point method. As shown in equation (1), the sheet resistance (R_s) is simply the resistivity (ρ) divided by the thickness (t) of the layer. The resistivity of a carbon layer is 35 Ω μm [10]. Now, assuming that the conductive layer is composed entirely of carbon, the conductive layer would be approximately 20 nm thick. This would explain the inability to visualize the conductive layer at higher magnification of SEM. However, it should be noted that conductive layer produced during the EDM is not homogenously distributed and can vary in thickness.

$$R_s=\rho/t$$

SEM of EDM surface

The EDM processed surface was observed in scanning electron microscope (SEM). Figure 4 shows SEM micrograph (top view) of a) ZrO_2, b) Si_3N_4 and c) SiC samples immediately after being processed. The aim here was to observe the conductive layer generated during the process but if the layer is less than a micrometer thick as mentioned in the previous paragraphs, then what is seen on the SEM images should be the transitional layer and not the conductive layer itself. However, these SEM images allow analyzing the material removal mechanism (MRM) responsible for spark erosion process. The surface topography of the different ceramics shown is visibly different suggesting a different material removal mechanism is involved during the spark erosion process.

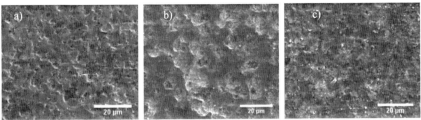

Figure 4. SEM images of the top surface of a) ZrO₂, b) Si₃N₄ and c) SiC samples immediately after being processed by EDM.

On the surface of the processed ZrO_2 it appears like several flakes of molten material has been casted over the ceramic bulk. This re-casted material does not have any particular crystal structure and could be in amorphous state. A closer observation at a higher magnification (Figure 5a), reveals the presence of many micro cracks on the flakes (in the direction parallel to the plane of the paper) and pores going through the flakes into the ceramic bulk (in the direction perpendicular to the plane of the paper). The underlying ceramic bulk also appears to be porous. These facts suggest that the most probable MRM of ZrO_2 ceramic is by melting and by decomposition due to redox reaction.

On the surface of the spark eroded Si_3N_4 (Figure 4b), it appears as if flakes of the material of various sizes have been removed from the surface of the bulk material. In contrast to ZrO_2, a higher resolution SEM image of Si_3N_4 (Figure 5b), does not show the presence of micro cracks or pores. Most likely, the scraps of the material have been removed by spalling due to repeated thermal shock induced during the spark erosion process.

The surface of the spark eroded SiC (Figure 4c) consists of micro cracks and pores. A higher magnification SEM image of spark eroded SiC did not provide additional information than that provided by the one shown in Figure 4c and therefore has not been presented in this paper. The most probable MRM of SiC ceramic is by decomposition due to redox reaction.

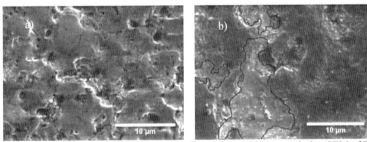

Figure 5. a) Cracks and pores are distinguishably visible at higher resolution SEM of EDM processed ZrO_2. b) Scraps of the bulk material removed during EDM of Si_3N_4 ceramic is marked

CONCLUSION

EDM process has been successfully carried out in various non-conductive ceramics such as ZrO_2, Si_3N_4 and SiC. The transitional layer formed during EDM of the non-conductive samples has been investigated. The transitional layer formed on EDM processed ZrO_2 could be visualized in a SEM micrograph. However a distinct layer on top of the transitional layer could not be visualized in SEM. At a higher magnification SEM image the underlying ceramic bulk could be seen suggesting the conductive layer to be very thin. The surface roughness (Ra) of the transitional layer of the EDM processed samples varied from 0.38 μm for ZrO_2 to 0.56 μm for Si_3N_4. The sheet resistance of the conductive layer on EDM processed ZrO_2 ceramic was measured to be as low as 1.58 kΩ/□ using Van Der Pauw method. The sheet resistance of the conductive layer as measured by 4-point method for Si_3N_4 and SiC was 1.69 kΩ/□ and 0.57 kΩ/□ respectively.

Observing the SEM micrographs of the processed samples allowed for predicting the possible material removal mechanisms. Apart from the conventional MRM, such as melting and evaporation which is common for EDM of hard metals, spalling and decomposition due to redox reaction are other possible MRM during EDM of non-conductive ceramics. A combination of several MRM is responsible for the removal of the material during the spark erosion process of non-conductive ceramics. Material removal by the decomposition of the non-conductive ceramic into its constituent element due to redox reactions was present in EDM of all three ceramics. However, in case of ZrO_2 ceramic, material removal due to melting was much more dominant MRM. In case of non-conductive Si_3N_4 ceramic, spalling due to thermal shock seems to be the most dominant MRM.

REFERENCES

[1]   W. Konig, "EDM-Future Steps towards the Machining of Ceramics," *CIRP Annals - Manufacturing Technology*, vol. 37, no. 2, pp. 623-631, 1988.

[2]   N. Mohri, Y. Fukuzawa, T. Tani, N. Saito, and K. Furutani, "Assisting Electrode Method for Machining Insulating Ceramics," *CIRP Annals - Manufacturing Technology*, vol. 45, no. 1, pp. 201-204, 1996.

[3]   T. Hösel, C. Müller, and H. Reinecke, "Spark erosive structuring of electrically nonconductive zirconia with an assisting electrode," *CIRP Journal of Manufacturing Science and Technology*, 2011.

[4]   A. Schubert, H. Zeidler, N. Wolf, and M. Hackert, "Micro Electro Discharge Machining of Electrically Nonconductive Ceramics," *Physics*, vol. 1308, pp. 1303-1308, 2011.

[5]   Y. LIU, X. LI, R. JI, L. YU, H. ZHANG, and Q. LI, "Effect of technological parameter on the process performance for electric discharge milling of insulating Al2O3 ceramic," *Journal of Materials Processing Technology*, vol. 208, no. 1–3, pp. 245-250, Nov. 2008.

[6]   A. MUTTAMARA, Y. FUKUZAWA, N. MOHRI, and T. TANI, "Probability of precision micro-machining of insulating Si3N4 ceramics by EDM," *Journal of Materials Processing Technology*, vol. 140, no. 1–3, pp. 243-247, Sep. 2003.

[7]   T. Hösel, C. Müller, and H. Reinecke, "Analysis of Surface Reaction Mechanisms on Electrically Non-Conductive Zirconia, Occurring within the Spark Erosion Process Chain," *Key Engineering Materials*, vol. 504, pp. 1171–1176, 2012.

[8]   T. Hösel, P. Cvancara, T. Ganz, C. Müller, and H. Reinecke, "Characterisation of high aspect ratio non-conductive ceramic microstructures made by spark erosion," *Microsystem Technologies*, vol. 17, no. 2, pp. 313-318, Apr. 2011.

[9]   J. Kruth, L. Stevens, and L. Froyen, "Study of the white layer of a surface machined by die-sinking electro-discharge machining," *CIRP Annals-Manufacturing*, vol. 44, no. I, pp. 169-172, 1995.

[10]   J. D. Cutnell and K. W. Johnson, *Physics*, 8th ed. John Wiley & Sons, 2009, p. 603.

FORMING MULLITE-CERAMICS REINFORCED WITH ZrO_2-t STARTING FROM MULLITE-ZrO_2-t AND KYANITE-Al_2O_3-ZrO_2-t MIXTURES

Elizabeth Refugio-García, Jessica Osorio-Ramos
Departamento de Materiales, Universidad Autónoma Metropolitana
Av. San Pablo # 180, Col Reynosa-Tamaulipas, México, D. F., 02200

José G. Miranda-Hernández
Universidad Autónoma del Estado de México (UAEM-Valle de México), IIN, Blvd.
Universitario S/N, Predio San Javier, Atizapán de Zaragoza, México, 54500

José A. Rodríguez-García, Enrique Rocha-Rangel
Universidad Politécnica de Victoria, Avenida Nuevas Tecnologías 5902
Parque Científico y Tecnológico de Tamaulipas, Tamaulipas, México, 87138

ABSTRACT

Mullite-ceramics reinforced with 8 and 10 vol. % YZT were prepared starting from two different powder mixtures. The first one was a mixture of pure mullite and YZT, and the other was a mixture of kyanite, alumina and YZT. The production of ceramics consisted of the pressureless-sintering of mixtures powders which were thoroughly mixed under high energy ball-milling. During sintering, kyanite and alumina of the second mixture react between them to form mullite. Measurements of density, microhardness and K_{IC} were carry out in all produced materials, from it was obtained that samples produced with pure mullite reach major densities, microhardness and toughness in comparison with the values displayed for samples prepared with kyanite and alumina. The reason of this has its explanation since during sintering of the kyanite and alumina samples; there is a competition between two phenomena; the reaction to form mullite and the sintering of the product. As the reaction occurs at low temperature and has high energy consumption, the activation energy necessary for diffusion during sintering is not reached; therefore this phenomenon is the controlling agent during the processing.

INTRODUCTION

Besides its traditional uses, mullite has attracted attention in recent years, as a material for high temperature structural applications, mainly because at these temperatures it may retain a significant portion of the mechanical strength that it has at room temperature, also, because it has a low thermal expansion coefficient, low dielectric constant, high melting point, high creep resistance and high chemical stability[1]. Table 1 shows some of the values of the main properties of the mullite.

Table 1. Main properties of pure mullite at room temperature[1].

Property	Value
Thermal expansión coefficient	5×10^{-6} K^{-1}
Flexural resistance	200 MPa
Fracture toghness (K_{IC})	2 $MPam^{1/2}$
Young modulus	231 MPa
Density	3.16 gcm^{-3}

As can be seen in this table, fracture toughness (K_{IC}) is a characteristic in which mullite is deficient. Furthermore, to obtain dense bodies of mullite, it is required long sintering treatment at elevated temperatures (>1700 °C)[2], this is due to the high value of the activation energy necessary for ion diffusion occurs through the network of mullite[3]. Because of these difficulties, over the past 15 years in many countries around the world, they were conducted a series of investigations in the seek for new processing methods by which they could obtain dense mullite bodies. To increase its tenacity, in some of these studies, it has proposed the use of ZrO_2 as reinforcing material[4-10], thus values of K_{IC} that have been obtained, varying between 3 and 3.2 MPa·m$^{1/2}$ for 10 to 20 wt. % ZrO_2 content.

Some researchers[6] have mentioned that large amounts of ZrO_2 on mullite-based composites cause thickening of the microstructure and thus the difficulty in the retention of the tetragonal form of ZrO_2. This retention of the ZrO_2 is important, because it has been suggested that one method by which the mullite matrix is reinforcing by ZrO_2 is the transformation of ZrO_2-tetragonal to ZrO_2-monoclinic[11]. However, other authors have suggested different possible reinforcement mechanisms, such as microcracking induced by the same transformation of the ZrO_2[9-11], strengthening grain boundaries caused by a metastable solid solution of ZrO_2[10-12] and deflection of cracks due to the presence of acicular microstructure[13].

The aim of this work is to study the formation of mullite composites reinforced with ZrO_2-t, starting from mixtures of mullite + ZrO_2-t and Kyanite + Al_2O_3 + ZrO_2-t, using conventional methods of sintering without the application of pressure in an electric furnace.

Because pure millita is a highly expensive mineral, it is suggest the use of kyanite as mullite precursor material. Kyanite mineral is cheaper than the mullite mineral, with chemical formula ($Al_2O_3 \cdot SiO_2$), this mineral when is heated at high temperatures (~ 1300 ° C) decomposes forming mullite + silica, for this reason in the mixture with kyanite is added certain amount of alumina to compensate silica excess resulting from the decomposition of kyanite in order to form more mullite.

EXPERIMENTAL PROCEDURE

Starting materials were: Mullite powder (99%, 1μm, Virginia Milling Corporation, USA), Kyanite powder (99%, 3-5 μm, Virginia Milling Corporation, USA), Al_2O_3 powder (99.9 %, 1μm, Sigma, USA) and YZT powder (99.9 %, <1 μm, Tosho, Japan). Final YZT contents in the produced composites were: 8 or 10 vol. %. Powder blends of 20 g were prepared in a ball mill with ZrO_2 media, the rotation speed of the mill was of 300 rpm, and the studied milling time was 12 h. With the milled powder mixture, green cylindrical compacts 2 cm diameter and 0.2 cm thickness were fabricated by uniaxial pressing, using 300 MPa pressure. Then pressureless sintering in an electrical furnace was performed, at 1500 °C during 2 h. Characterization of sintered samples was as follows: Densities were measured using the Archimedes' principle. The microstructure was observed by scanning electron microscopy (SEM), equipped with an energy dispersive spectroscopy analyzer (EDS). Phases present in the sintered composites were determined by X-ray diffraction (XRD). Fracture toughness was estimated by the fracture indentation method[14], (in all cases ten independent measurements per value were carrying out).

To identify study samples, it was used the following code;
C8Z: Sample prepared with Kyanite + Al_2O_3 + 8 % vol. ZrO_2
C10Z: Sample prepared with Kyanite + Al_2O_3 + 10 % vol. ZrO_2
M8Z: Sample prepared with Mullite + 8 % vol. ZrO_2
M10Z: Sample prepared with Mullite + 10 % vol. ZrO_2

Table I. Summary of the dimensions of the notches in the specimen under study.

Samples	Aperture [mm]	Depth [a] [mm]	Angle [α] [degrees]	Radius [r] [mm]
PEZI 0a	1.742	1.869	42 45	0.214
PEZI 0b	3.117	3.587	39 46	0.224
PEZI 10a	2.575	2.806	45 22	0.209
PEZI 10b	2.226	2.322	44 31	0.295
PEZI 10c	2.132	2.359	41 38	0.201
PEZI 30a	3.127	3.703	43 00	0.240
PEZI 30b	3.336	3.830	42 39	0.295
PEZI 30c	2.639	2.742	46 26	0.197

Bending test.

To apply the theoretical models of Peterson, Irwin and Griffith, it was first necessary to test the 8 specimens, previously prepared and tested in three-point bending to find that maximum force could with stand the test pieces before breaking. This test was conducted in the Instron universal machine, mod. 5500R, using a rate of 0.5 mm / min.

Thereafter, defining the encoding of the specimens based on their geometry, which corresponds to a flexure beam as shown in Figure 2[8].

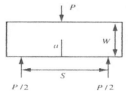

Figure 2. Geometry of a beam in bending.

Where:
W: the height of the sample [m]
P: is the applied force [N]
S: is the spam formed between the support [m]
a: is the depth of the crack [m]
Besides using other values are also useful.
B: thickness of the sample [m]
r: the radius [m]
d = W - a [m]

Thus using this coding will be able to obtain other parameters in order to find the value of K$_{IC}$.

Using the model of Peterson it is possible to get K$_{tn}$, σ$_{max}$ and σ$_{nom.}$

Where:
K$_{tn}$: The factor of elastic stress concentration
σ$_{max}$: The effort at the root hub of efforts
σ$_{nom}$: The nominal stress

In the concentration stress charts developed by Peterson[9] it can be found the value of K$_{tn}$, relating r with d and W with d. However, also it can be obtained applying the equations of the curves and formulas mentioned below[9].

$$K_{tn} = C_1 + C_2(a/W) + C_3(a/W)^2 + C_4(a/W)^3 \quad \text{-----------------} (1)$$

In the case where 2.0 < a/r < 20, the following equations are applied.

ray diffraction patterns do not indicate complete reaction, since there are several peaks correspond to Al_2O_3 and SiO_2 in both cases, however these peaks are very low in intensity, therefore reaction is almost complete. The most important observation here is the presence of ZrO_2 in its monoclinc form, and this situation could be a problem, because the reinforcement by transformation of zirconia would not operate in this situation.

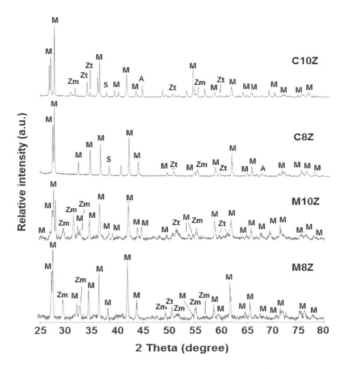

Figure 3. X-ray diffraction patterns of different prepared composites.
M-mullite, S-silica, A-alumina, Zt-tetragonal- ZrO_2, Zm-monoclinic- ZrO_2.

Microhardness

The microhardness evaluated in different prepared samples is reported in Figure 4. In this figure it is seen that those samples prepared with mullite + ZrO_2-t exhibit higher hardness values, compared to the samples prepared with the mixture of kyanite + Al_2O_3. The greater degree of densification and the situation of has a more homogeneous material in chemical composition in samples prepared with ZrO_2 and mullite, are the cause responsible for this condition. Since materials made from kyanite besides being less dense, have a more heterogeneous chemical composition, as have been indicated by XRD results, where phases such as mullite, monoclinic and tetragonal ZrO_2, Al_2O_3 and SiO_2 are present.

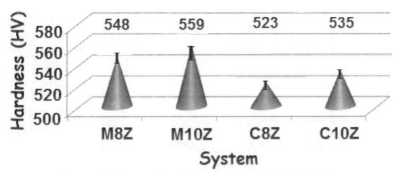

Figure 4. Hardness of samples sintered at 1500°C, during 2 h.

Fracture toughness

Figure 5 shows the values of fracture toughness determined in different samples prepared here. As expected samples prepared with mullite + ZrO$_2$-t exhibit larger values of fracture toughness because therein ZrO$_2$ is present in tetragonal form, giving the feasibility of the reinforcement by the transformation of ZrO$_2$-t to ZrO$_2$-m exist. On the contrary, the fracture toughness is less in the samples prepared with kyanite + Al$_2$O$_3$ + ZrO$_2$-t, because it was not possible the ZrO$_2$-t retention at the end of processing.The reason of these behaviors have its explanation since during sintering of the kyanite and alumina samples; there is a competition between two phenomena; the reaction to form mullite and the sintering of the product. As the reaction occurs at low temperature and has high energy consumption, the activation energy necessary for ion diffusion through the network of mullite during sintering is not reached; as a result this phenomenon is the controlling agent during the processing.

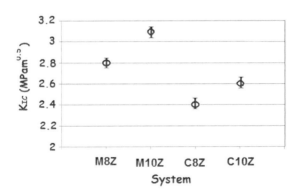

Figure 5. Fracture toughness of samples sintered at 1500°C, during 2 h.

Microstructure

Typical microstructures of the samples prepared with 10% vol. ZrO_2-t are shown in Figure 6. Figure 6a which corresponds to sample prepared with mullite + ZrO_2-t, clearly shows a good distribution of fine particles of a second phase (ZrO_2-t) in the matrix (mullite). While figure 6b corresponds to a sample prepared with the kyanite + Al_2O_3 and ZrO_2-t mixture, the ZrO_2 distribution is not homogeneous because it is appreciate colony of ZrO_2 agglomerates. ZrO_2 was added to mullite for the purpose of serve as reinforcing material of the same, situation unsuccessful, due to the transformation of ZrO_2-t to ZrO_2-m during any processing stage. The size of the ZrO_2-t in figure 6a is about 1 to 3 μm and the location thereof is in intergranular regions of the mullite matrix. The reinforcing mechanism in these kinds of samples is the transformation of ZrO_2-t to ZrO_2-m, mechanisms before widely documented[2,4,6,7,8,11].

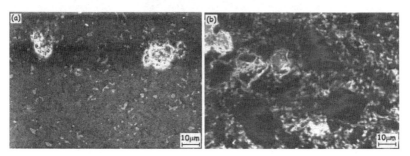

Figure 6. Microstructure observed in scanning electron microscope of samples sintered at 1500°C, during 2 h. (a) mullite + ZrO_2-t, (b) kyanite + Al_2O_3 + ZrO_2-t.

CONCLUSIONS

Samples produced with pure mullie reach major values of density, microhardness and toughness. During sintering of kyanite-alumina mixtures; there is a competition between two phenomena; the reaction to form mullite and the sintering of the product. As the reaction occurs at low temperature and has high energy consumption, the activation energy necessary for diffusion during sintering is not reached; therefore this phenomenon is the controlling agent during the processing. The reinforcing mechanism in these kinds of samples is the transformation of ZrO_2-t to ZrO_2-m.

ACKNOWLEDGMENT

Authors would thank Universidad Autónoma Metropolitana, Universidad Autónoma del Estado de México and Universidad Politécnica de Victoria for the financial and technical support given for the realization of this work.

REFERENCES

[1] Mullite Processing, Structure and Properties, Topical Issue, *J. Am. Ceram. Soc.*, **74**, (1991).
[2] P. Boch and J. P. Giry, Preparation and Properties of Reaction-Sintered Mullite-ZrO2 Ceramics, *Mater. Sci and Eng.*, **71**, 39-48 (1985).

[3] M. D. Sacks and J. A. Pask, Sintering of Mullite-Containing Materials: Y Effect of Composition, *J. Am. Ceram. Soc.* **65**, 65-70 (1982).

[4] J. S. Moya and M. I. Osendi, Effect of ZrO2 (ss) in Mullite on The Sintering and Mechanical Properties of Mullite/ZrO2 Composites, *J. Mater. Sci. Lett.,* **2**, 599-601 (1983).

[5] T. Koyama, S. Hayashi, A.Yasumori and K.Okada, Preparation and Characterization of Mullite-Zirconia Composites from Various Starting Materials" *J. Eur. Ceram. Soc.,* **14**, 295-302, (1994).

[6] J. S. Moya and M. I. Osendi, Microstructure and Mechanical Properties of Mullite-Zirconia Composites, *J. Mater. Sci,* **19**, 2909-2914 (1984).

[7] T. Koyama, S. Hayashi, A. Yasumori and K. Okada, Contibution of Microstructure to The Toughness of Mullite/Zirconia Composites, *Ceram. Trans.* **51**, 695-700 (1995).

[8] A. Leriche, Mechanical Properties and Microstructure of Mullite-Zirconia Composites, *Ceram.Trans.* **6**, *Mullite and Mullite Matrix Composites*, ed. by S. Somiya, R. F. Davis and J. A. Pask, 541-552 (1991).

[9] P. Descamps, S. Sakaguchi, M. Poorteman and F. Cambier, High-Temperature Characterization of Reaction Sintered Mullite-Zirconia Composites, *J. Am. Ceram. Soc.,* **74**, 2476-2481 (1991).

[10] T. Koyama, S. Hayashi, A. Yasumori and K. Okada, Microstructure and Properties of Mullite/Zirconia Composites Prepared form Alumina and Zircon Under Various Fairing Conditions, *J. Eur. Ceram. Soc.,* **16**, 231-237 (1996).

[11] J. S. Wallace, G. Petzow and N. Claussen, Microstructure and Property Development of In Situ Reacted Mullite-ZrO2 Composites, *Advanced Ceramics,* **12** *Science and technology of Zirconia,* ed. by N. Claussen, M. Ruhle and H. Heur, 436-442 (1984).

[12] G. Orange and G. Fantozzi, High Temperature Mechanical Properties of Reaction Sintered Mullite-Zirconia and Mullite-Alumina-Zirconia Composites, *J. Mater. Sci.,* **20**, 2533-2540 (1985).

[13] K. Srikrishna, G. Thomas and J. S. Moya, Sintering additives for Mullite/Zirconia Composites, *Advances in ceramics* **24** *Science and Technology of Zirconia III.* Ed. by S. Somiya, N. Yamamoto and H. Yanagida, 276-286 (1988).

[14] A. G. Evans and E. A. Charles, Fracture Toughness Determination by Indentation, *J. Am. Ceram. Soc.,* **59**, 371-372 (1976).

IMPACT OF NANOPARTICLE-MICROSTRUCTURE ON COSMECEUTICALS UV PROTECTION, TRANSPARENCY AND GOOD TEXTURE

Yasumasa Takao

National Institute of Advanced Industrial Science and Technology (AIST)

Nagoya, Aichi JAPAN (http://staff.aist.go.jp/yasumasa.takao/indexE.htm)

ABSTRACT

Particle processing nanometer-scale mesostructures of composite-particle and granule are of interest for recent well-controlled cosmetics with three conflicting properties, ultraviolet rays (UV) protection, visible light transparency, good texture-feel. Titania(TiO_2)-mica ordered composite-structures can be prepared by simply powder-mixing, liquid process by controlling copolymer-liquid-crystalline phases, and spray-pyrolysis. A variety of mesostructures have been providing plate-shaped particle, composite-particle, and granule. However, the degree of ordering and the range of the composite and granule mesostructures have been limited. Here we report a rapid, spray-drying and beads-milling based process for synthesizing uniform & "controlled-inhomogeneous" TiO_2-mica composite-particle, and solid & hollow TiO_2 spheres. Our method relies on a combination of (1) a precipitation-induced interfacial self-assembly confined to spherical aerosol droplets, and (2) a layered-nanostructure of hexagonal topology during the precipitation process, without aggregation of TiO_2 nanoparticles. This simple, generalizable process can show the diversity of nanoparticle-mesostructure construction and UV protection, transparency, good texture.

INTRODUCTION

With the expansion of the progress of global warming, the amount of UV radiation reaching the skin is increasing. UV causes skin disease and photo aging. It must be prevented from reaching skin directly. Cosmetic is a useful UV shielding measure without deteriorating the quality of life, QOL. The combined material with cosmetic and pharmaceutical properties, as it is called "COSMECEUTICAL", is increasingly powerful measure.

Figure 1 shows an apparent limit of making a cosmeceutical' pearl-pigment powder. It is a composite-particle consisting of mica $KAl_2AlSi_3O_{10}(OH)_2$ raw material and TiO_2[1-4]. Generally, TiO_2, the UV shielding agent, is a superfine particle (nanoparticle) having 0.01-0.1μm diameters. Mica, the scale-like (plated structure) clay-mineral particle, is 1-100μm longest-length and 0.1-1μm thickness (see the aspect-ratio in Fig.2). Superfine and plated particles are facing difficulties to reduce their own aggregation before making composite-microstructure by surface attraction (i.e., van der Waals force). Therefore, to dispersing uniformly each other, the dispersing chemical (ex. surfactant) was indispensable. Even if using the chemicals, it could not make a composite-particle completely [5-10].

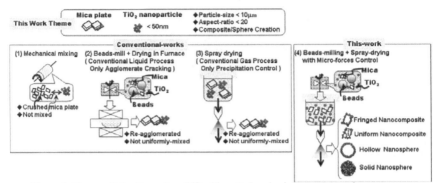

Figure 1. Nanoparticle dispersion difficulties. Ordered-microstructures via our process

Figure 2 shows the existing products map of pearl-pigment. Apparently, there is a technological limit for the prior arts. E.g., IRIODIN® (MERCK Co.,Inc.), COVERLEAF® (JGC CATALYSTS AND CHEMICALS Ltd.), SILSEEM® (NIHON KOKEN KOGYO Co.,Ltd.), and NABEYAMA Mica (HIKAWA KOGYO Inc.) [1-4], these pearl-pigment material's performances are consisting of particle-size (both mica & TiO_2) and aspect-ratio of mica. The smaller TiO_2 size is, the higher UV-B shielding property; however, considerably aggregation in the superfines. The smaller aspect-ratio of mica is, the higher powder-molding strength and smooth impression from the use of a powder-foundation cosmetic product; however, difficult to make the smaller ratio [1-4].

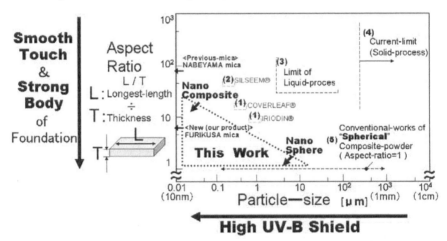

Figure 2. Existing products of pearl-pigment and their map of aspect-ratio & particle-size

We propose the (1) FURIKUSA mica having an unprecedented smaller aspect-ratio, (2) UV shielding products having 50nm particle-size & 10 aspect-ratio (Nanocomposite), and (3) 5μm granulated powder size consisting of 50nm nanoparticle (Nanosphere), subsequently [11-12].

EXPERIMENTALS

Figure 3 shows our scheme of spraying, beads-milling and numerical-calculation. Figure 4 indicates a detail of the numerical-calculation to decide a threshold of TiO_2 addition amounts. To disperse nanoparticle uniformly and prevent its aggregation, the threshold is under 10vol% in 20nm's TiO_2, and is possible over 30vol% in 50nm case [11-12].

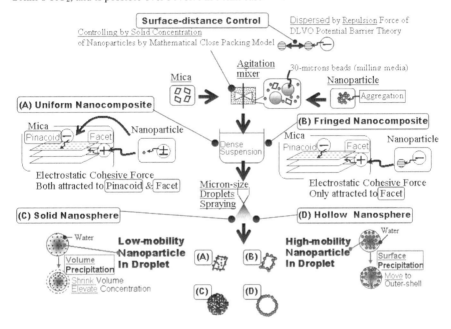

Figure 3. . Scheme of spraying, beads-milling and numerical-calculation.

Slurry Preparation (numerical-calculation and beads-milling)

As shown in Fig.3, nanoparticles are attracted and/or repulsed by van der Waals' attraction force, and electrostatic repulsion (and/or attraction) forces, simultaneously. When an excessive amount of nanoparticle is loading, the nanoparticles begin to contact each other, i.e., aggregation.

Figure 4 shows a potential barrier calculated from attractive potential by van der Waals' force, and electrostatic repulsive potential. Interparticle distances (L_{DLVO} and $L_{Woodcock}$ in Fig.4) are calculated by assuming the hexagonal close-packed structure compared with the nanoparticle loading amounts.

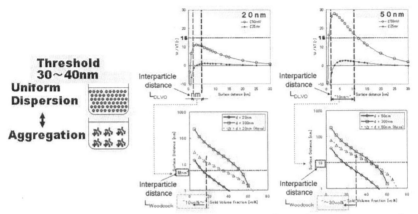

Figure 4. Numerical-calculation to disperse TiO_2 nanoparticle and prevent its aggregation.

DLVO (Professors B.Derjaguin, L.Landau, E.Verwey and T.Overbeek) theory allows dispersing particles based on the electrostatic potential barrier. The theory shows that, over 15 potential barrier value, loading particles disperse without aggregation (Fig.4). However, in case of nanoparticle, its uniform dispersing is difficult using only the DLVO interactions; because of its relatively small potential barrier (the effect of van der Waals' attraction force is relatively large). L_{DLVO}, a limit of allowable loading amount for uniform dispersing, is very short; few nanometers. So, only as for very little quantity of nanoparticles cannot fill it.

Woodcock, firstly, reports the relationship between the nanoparticles volume fraction (concentration of solid loading amounts) and their interparticle distances; i.e., the non-equilibrium molecular dynamics (NEMD) method[10-13].

E.g. Fig.4, the model shows that, the nanoparticles form an aggregate when overloading larger than the volume fraction, which calculated from the interparticle distance, i.e., $L_{Woodcock}$. The NEMD method describes that the L_{DLVO} of 20nm's TiO_2 is a few nanometer under the 10 vol% solid loading. Apparently, to disperse nanoparticle without its aggregation, the threshold is under 10vol% in 20nm's TiO_2. On the other hand, the L_{DLVO} of 50nm's TiO_2 is over 10 nm. In this case, we can fill it over 30vol%[11-12].

Further, we firstly apply an accelerated flow field theory to this nanoparticle dispersion model[11-12]. Generally, a dispersion of agglomerated particles in the accelerated flow field (i.e., beads-milling) utilizes the 2 micro dispersion forces; (1) resistance of particle to the flow, (2) inertial force[14,15]. Surface force (capillary, electrostatic) is in proportion to the diameter, resistance to flow (cross-sectional area) is to a square of the diameter and inertia (gravity) is a cube. Less than around 5μm, an influence of the surface is dominant rather than inertia. We apply the 20nm's TiO_2 of no electrostatic charge and 50nm's TiO_2 of the negative electrostatic charge (Fig.3). In conclusion, the electrostatic and resistance forces of the 50nm are larger than those of the 20nm; i.e., 50nm has a higher "MOBILITY".

We applied the NEMD method and the accelerated flow field theory to make a composite-particle and a granule of nanoparticles[11-12]. Nanocomposite shown in Fig.3(A), TiO_2 nanoparticles coated on mica-plate uniformly, is made by 20nm's TiO_2 of no electrostatic charge The solid-core sphere (Fig.3(C)) is also prepared by 20nm's TiO_2 to lower the nanoparticle mobility in slurry.

Meantime, the composite of nanoparticles coated on only around the fringe of mica, Fig.3(B), and hollow nanosphere, Fig.3(D), are made by 50nm's TiO_2 of the negative electrostatic charge. Using a high mobility of nanoparticle to move fast onto the mica facet charged on the positive, it enables to create these anisotropic composite microstructure (Figs.3(B)(D)) as described below[11-12].

Beads-mill (UAM015, KOTOBUKI KOGYO Ltd.) disperses the nanoparticle uniformly, and looses its aggregation using ZrO_2 plasma-beads (D_{50} 30μm, DAIKEN CHEMICAL Co.,Ltd.). TiO_2 (50nm, MT500H, TAYCA Co.; 20nm, TTO-51, ISHIHARA SANGYO KAISHA Ltd.) loading in distilled water is 2vol%. In this case, $L_{Woodcock} < L_{DLVO}$ is satisfied gratifyingly even in Fig.4's severe case (i.e., 20nm). It can be distributed nanoparticles uniformly and prevented their aggregation [11-12]. Subsequently, FURIKUSA mica (7μm, Sericite, FSE, SANSHIN MICA Co.) is mixed with the nanoparticle's slurry (50vol% solid loading).

Spraying

Spray-dry (MDL-050B, FUJISAKI ELECTRIC Co.Ltd.) creates a well-dispersed single-micron's droplet by triple fluids nozzle (PN3005, FUJISAKI ELECTRIC Co.Ltd.). The droplet includes just only one 7μm's mica coated with nanoparticles (Fig.3(A),(B)). This micron-sized droplet making is essential to control the nanoparticle coating condition. Also, enable to make a narrow size distribution of granules consisted with nanoparticles (Fig.3(C),(D)).

Volume precipitation[5] is dominant in case of the lower mobility of nanoparticle in slurry (Figs.3(A),(C)). This low mobility enables to shrink droplets gradually. This gradual shrinkage allows the charged-neutral 20nm's TiO_2 nanoparticles to coat onto all over the mica-plate uniformly (Fig.3(A)). Also, the gradual shrinkage makes the higher-densified solid nanosphere during its sufficiently long drying period (Fig.3(C)).

Surface precipitation[5], meantime, is in the central part for the higher mobility (Figs.3(B),(D)). The higher value shrinks droplets rapidly. This rapid shrinkage drives to move the solute to the droplet's surface. Thus, this higher mobility aggravates the 50nm's TiO_2 of the negative electrostatic charge to attach onto the mica facet charged on the positive (Fig.3(B)). Also, the rapid shrinkage causes a cave in the lower-densified hollow granule during the considerably shorter drying time (Fig.3(D)).

Evaluations

Mesostructures of composite-particle and granule are confirmed by analytical electron microscopy (AEM) and scanning electron microscopy (SEM).

Spectrophotometer (UV160A, SHIMAZU Co.) indicates the transmissivity (among 290~800nm) of silicon film representing human skin which puts our prepared powders on.

Shear properties of compacted powder (NS-S200, NANO SEEDS Co.) represents a variety of feeling textures of cosmetic powder-foundation. Granule's crushing strength (NS-A100, NANO SEEDS Co.) is also measured to collect supporting evidence of the feeling texture.

RESULTS & DISCUSSION

Mesostructures (a diversity to design cosmeceutical performances)

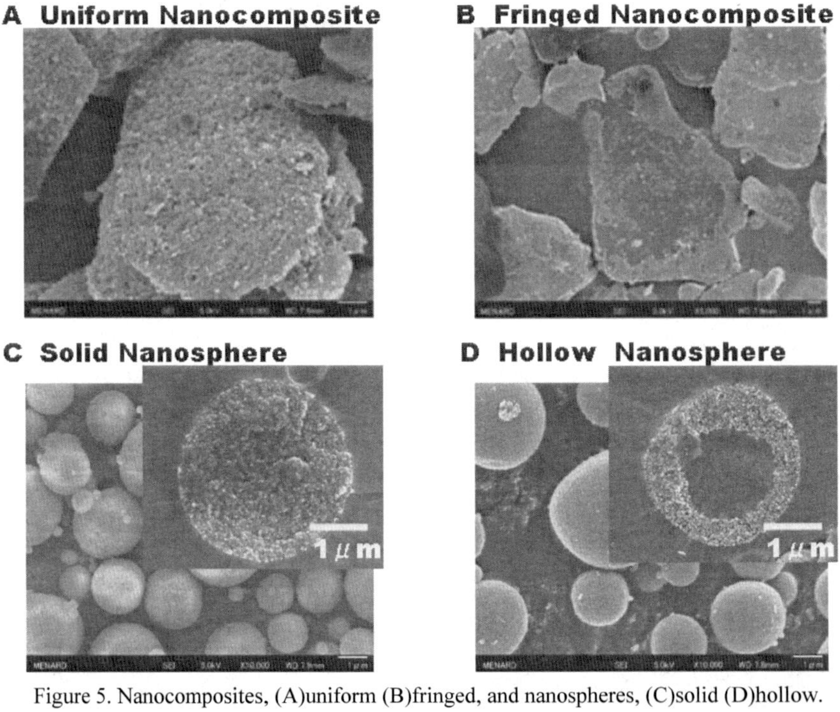

Figure 5. Nanocomposites, (A)uniform (B)fringed, and nanospheres, (C)solid (D)hollow.

Figure 5 shows the nanocomposites and nanospheres shown in Fig.3(A)-(D). "Uniform nanocomposite" in Fig5(A) indicates 50nm's TiO_2 nanoparticles coated on 7μm's mica-plate uniformly. "Fringed nanocomposite (Fig.5(B))" demonstrates the composite-microstructure of nanoparticles coated on only around the facet of mica.

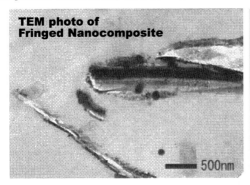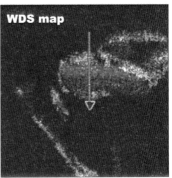

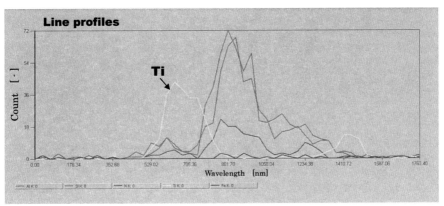

Figure 6. AEM photos and line profile of Fringed nanocomposite (Fig.5(B)).

None of the mica pinacoid on, which AEM photos and line profile in Fig.6 demonstrate that. Ti component is only detected at both ends of the other components (Si, Al, K and Fe). This line profiles reveal that the composite of nanoparticles coated on only around the fringe of mica, Fig.3(B), is created.

"Solid-core sphere" of 20nm's TiO_2 nanoparticles is Fig.5(C). "Hollow nanosphere (Fig.5(D))" having a 1μm cave in granule-microstructure is prepared by 50nm's TiO_2.

Figure 7 shows their granule's crushing strengths. The solid (Fig.5(C)) indicates the higher strength and compression distance by the crushing force. These higher values depend on consisting with the smaller (20nm) nanoparticle, and higher granule density (i.e., non-cave). We can select the nanospheres (Fig.3(C),(D)) according to the need of the feeling textures of cosmetic powder-foundation.

The mesostructures shown in Figs. 5-7 illustrate the diversity of nanocomposite and nanosphere constructions attainable by our method.

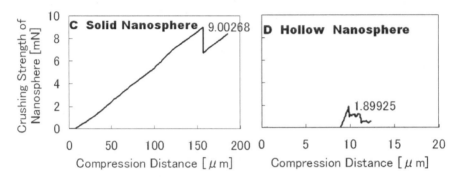

Figure 7. Granule's crushing strength of Solid (Fig.5(C)) Hollow (Fig.5(D) spheres

UV protection and transparency

Figure 8 shows a shielding ability in UV light region and transparency in visible light region. Figure 8(a) shows our human skin model prepared by nanocomposite (Figs.5(A),(B)). Figure 8(b) is a powder-foundation consisting with a commercialized powder for UV protection (COVERLEAF®, JGC CATALYSTS AND CHEMICALS Ltd.). Figure 8(c) is the result of a silicon film dispersed of mica raw powder (for a reference).

Desired ideal state is the lower light transparency under UV range, especially UV-B field of 280-315nm for sun protection factor. Simultaneously, high transparency is indispensable under visible light region of 380-780nm. Then, a drop sharply of transparency from visible to UV ranges is a performance index of UV shielding ability.

Figure 8(a), our model prepared by blending the nanocomposite (Figs.5(A),(B)), shows the sharply-drop performance. It depends on a combination of 2 effects. (1) Firstly, "Uniform nanocomposite (Fig5(A))" shields enough UV light (but also visible light). (2) Secondly, "Fringed nanocomposite (Fig.5(B))" puts visible through (but UV protection the less). The combination creates the sharply-drop performance from visible to UV ranges.

Meanwhile, the previous powder-foundation is insufficient for UV protection and visible transparency (Fig.8(b)). Because of a large amount of TiO_2 nanoparticle aggregation. This aggregation is unavoidable via a conventional preparation method as analyzed in Fig.2. The nanoparticle agglomerate deteriorates an efficiency of UV protection and opacifies a visible transparency.

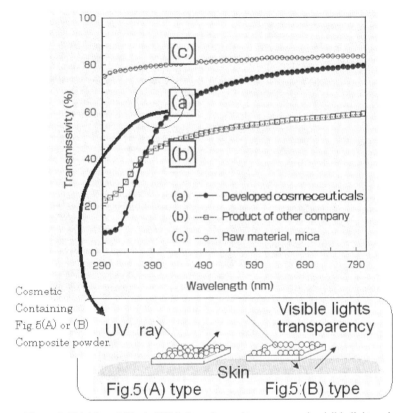

Figure 8. Shielding ability in UV light region and transparency in visible light region.

Feeling textures of cosmetic powder-foundation

Figure 9 shows a variety of feeling textures of cosmetic powder-foundation. Figure 9(a) shows our human skin model prepared by blending the nanospheres (Figs.5(C),(D)). Figure 9(b) is a powder-foundation consisting with the commercialized powder for UV protection (COVERLEAF®, JGC CATALYSTS AND CHEMICALS Ltd.). Figure 9(c) is the result of a compacted powder consisted of mica raw material only (for a reference).

One of representative index for feeling textures of cosmetic powder-foundation is a smaller shear stress value when applying the same verticality (normal) stress. Because, the smaller value causes to feel a higher smoothness on human skin.

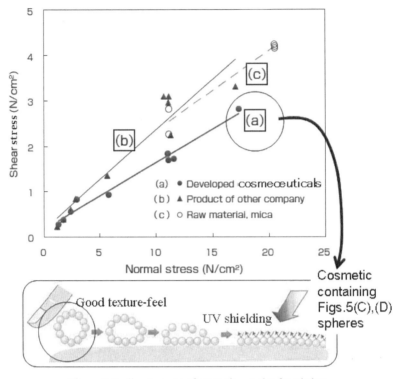

Figure 9. Feeling textures of cosmetic powder-foundation.

Figure 9(a), our model prepared by the nanosphere (Figs.5(C),(D)), shows the smallest shear force performance. It depends on a combination of 2 effects. (1) Firstly, "Hollow nanosphere (Fig.5(D))" loosens its cave by finger-pressure, and spreads over the skin. (2) Secondly, "Solid nanosphere (Fig.5(C))" puts a rolling contact touch on the skin. The combination creates the smaller shear stress value and higher smoothness on human skin.

On the other hands, the previous powder-foundation is insufficient for lowering the shear stress (Fig.9(b)). Because of a large amount of TiO_2 nanoparticle aggregation. This nanoparticle agglomerate deteriorates a scratch resistance. As a result, the shear stress overgrows rather than the compacted powder consisted of mica raw material only (Fig.9(c))

CONCLUSION

We obtained the following results.

1. "Uniform Nanocomposite" having 50nm's TiO2 nanoparticles coated on 7μm's mica-plate

uniformly is created.

2. "Fringed Nanocomposite" with the composite-microstructure of nanoparticles coated on only around the facet of mica is synthesized.

3. "Solid Nanosphere" is made by using of 20nm's TiO_2 nanoparticles.

4. "Hollow Nanosphere" having a 1μm cave in granule-microstructure is prepared by 50nm's TiO_2.

5. Transmissivity sharply-drop performance of UV protection and visible transparency appears a combination of 2 effects. (1) Uniform Nanocomposite shields enough UV light. (2) Fringed Nanocomposite puts visible through.

6. Smaller shear force performance of a compacted powder reveals a combination of the below. (1) "Hollow Nanosphere" loosens its cave by finger-pressure, and spreads over the skin. (2) "Solid Nanosphere" puts a rolling contact touch on the skin.

REFERENCES

[1] Du Pont., Micaceous Flake Pigment, *USA patent* **3087827** (1963).

[2] T. Miyazaki; "COVERLEAF®" Characteristics of Highly Functional Inorganic Particles in Cosmetic Field, *Technol. Report JGC Catalysts and Chemicals Ltd.*, **17**, 67-74 (2000)

[3] F. Suzuki, Y. Murui, M. Adachi, K. Hashimoto, Y. Toda; "SILSEEM®" Preparation of Interference Colors and High Luster Pearlescent Pigments Provided by Thin Film Laminations with Different Refractive Index, *J. Jpn. Soc. Colour Mater.*, **81**, 429-436 (2008).

[4] T. Hoshino; "IRIODIN®" Pigments using Principles of Optical Interference Colors, *J. Jpn. Soc. Colour Mater.*, **84**, 246-253 (2011).

[5] G. L. Messing, S. Zhang, G. Jayanthi; Ceramic Powder Synthesis by Spray Pyrolysis, *J. Am. Ceram. Soc.*, **76**, 2707–2726 (1993)

[6] W. Walker, J. Reed, S. Verma; Influence of Slurry Parameters on the Characteristics of Spray-dried Granules, *J. Am. Ceram. Soc.*, **82**, 1711–1719 (1999)

[7] G. Bertranda, C. Filiatreb, H. Mahdjouba, A. Foissyb, C. Coddeta; Influence of Slurry Characteristics on The Morphology of Spray-Dried Alumina Powders, *J. Eur. Ceram. Soc.*, **23**, 263-271 (2003)

[8] Y. Kamata, H. Oka; Development of Multifunctional Marimo Shaped Titanium Dioxide and Application as A Cosmetic Ingredient, *FRAGRANCE J.*, **35 (No.3)**, 45-49 (2007)

[9] M. Tanaka; Preparation of Fine Composite Particles by Suspension Polymerization, *J. Jpn. Soc. Colour Mater.*, **82**, 23-29 (2009)

[10] M.Iijima, M.Tsukada, H.Kamiya, Effect of Particle Size on Surface Modification of Silica Nanoparticles by using Silane Coupling Agents and Their Dispersion Stability in Methylethylketone, *Journal of Colloid and Interface Science*, **307**, 418-424 (2007)

[11] Y. Takao and M. Sando, Products and Evaluation Device of Cosmetics for UV Protection

(AIST Commercialization based on regional Collaboration that Combines The Current Strategic Logic, and An Intermediary'S Experience and Trial-and-Error Approach), *Synthesiology English Edition* **3(No.2),** 140-150 (2010).

[12] Y. Takao, T. Asai, H. Asano, K. Tsubata, T. Okuura and S. Nakata, Production-method and Apparatus of Composite-particle and Granules by Adhering Nanoparticles and Loosening their Aggregation, *Japan-patent* **2011-116569** (2011).

[13] L.V. Woodcock, Entropy Difference between the Face-centred Cubic and Hexagonal Close-packed Crystal Structures, *Nature,* **385**, 141-142 (1997); Molecular dynamics and relaxation phenomena in glasses, *Proc. a workshop held at the Zentrum für Interdisziplinäre*, Forschung Universität Bielefeld, **277**, 113–124 (1987).

[14] Y Kousaka, Y Endo, T Horiuchi and T Niida, Dispersion of Aggregate Particles by Acceleration in Air Stream, *J. Chem. Eng. Jpn.*, **18**, 233-239 (1992).

[15] Y. Endo, and Y. Kousaka, Dispersion mechanism of coagulated particles in liquid flow, *Colloids and Surfaces A: Physicochemical and Engineering Aspects*, **109**, 109-115 (1996).

PIEZOELECTRIC THICK-FILM STRUCTURES FOR HIGH-FREQUENCY APPLICATIONS PREPARED BY ELECTROPHORETIC DEPOSITION

Danjela Kuscer[1,2], Andre-Pierre Abelard[1,3], Marija Kosec[1,2], Franck Levassort[3]
[1]Jožef Stefan Institute, Ljubljana, Slovenia
[2]Centre of Excellence NAMASTE, Ljubljana, Slovenia
[3] Université François Rabelais, GREMAN CNRS 7347, Tours, France

ABSTRACT
    We have studied the processing of thick films based on lead-zirconuim-titanate (PZT) by electrophoretic deposition (EPD) for high-frequency ultrasound transducer applications. The PZT powder prepared by a conventional solid-state synthesis was dispersed in ethanol using a polyacrylic acid. The negatively charged PZT particles were deposited on a Au/Al_2O_3 substrate at a constant current density of 1.56 mA/cm^2. To prevent any crack formation in the PZT layer during the drying, some polyvinyl butyral (PVB) was added to the suspension. The PVB did not significantly influence zeta-potential of the particles and the conductivity of the suspension, but it decreased the mobility of the particles by increasing the viscosity of the suspension. The PZT thick films were sintered at $950^\circ C$ for 2 h in the presence of a PbO-rich liquid phase. After the sintering the thick films were homogeneous, without any cracks, and had a density of between 80 and 85 % of the theoretical value. The thicknesses of the sintered PZT thick films varied from 15 to 38 µm and depended on the deposition conditions,i.e., on the deposition time and on the properties of the suspension, namely the viscosity, the conductivity of the suspension and the zeta-potential of the particles. The PZT thick films were then electromechanically characterised. The resonant frequencies of the transducer, which are defined by the thicknesses of the piezoelectric layers, were 75 MHz and 36 MHz for PZT layers with thicknesses of 15 and 38 µm, respectively. All the sintered PZT thick films had a high coupling factor of around 50 %. The results indicate that piezoelectric PZT thick films with a tailored thickness and density can be prepared using theelectrophoretic deposition process and that these PZT thick films can be used for the fabrication of transducers operating at high-frequencies.

INTRODUCTION

    Ultrasonic systems are frequently used in medicine for imaging and therapy because of their relatively low costs and the non-ionizing character of the ultrasound. A system operating at high frequency enables non-destructive investigations of the skin, the eyes and blood vessels, as well as the imaging of small animals[1]. The performance of the system is determined by the characteristics of the transducer, i.e., the electrical input impedance, the electroacoustic response and the radiation pattern[2]. A single-element transducer consists of a piezoelectric element positioned between two electrodes. The thickness of the piezoelectric layer defines the resonant frequency of the transducer because the thickness represents half the wavelength of the ultrasound wave. In addition to the piezoelectric layer, the backing, matching layer and lenses are needed to dampen the resonance, improve the sensitivity and the axial resolution of the transducer and to focus the acoustic beam, respectively[1,2].

    For piezoelectric transducer applications the most frequently used piezoelectric material is based on lead-zirconium titanate, a solid solution of $PbZrO_3$ and $PbTiO_3$. The piezoelectric and ferroelectric properties of the solid solution are tailored by varying the Zr/Ti ratio in the $Pb(Zr,Ti)O_3$ perovskite structure. The composition at the morphotropic phase boundary $Pb(Zr_{0.53}Ti_{0.47})O_3$, which is denoted PZT, is characterized by a high dielectric constant at room temperature, i.e.,1500, and a high coupling factor of around 50 %. The values of the dielectric constant and the coupling factors are required to be high for a piezoelectric transducer. In addition, the piezoelectric material should have a low acoustical impedance. For dense PZT it is around 30 MRa; however, it is significantly lowered when the material consists a certain amount of pores[3].

PZT is processed by solid-state synthesis from its constituent oxides. In the first step TiO_2 and PbO form $PbTiO_3$ at 500 °C. Than, upon heating the ZrO_2 and PbO_2 react with the $PbTiO_3$, forming thermodynamically favourable Zr-rich solid solutions[4,5]. In the final step these solid solutions react with $PbTiO_3$ and form the $Pb(Zr,Ti)O_3$ solid solution. Fine-grained powders, high processing temperatures and long sintering times are used to avoid the presence of any intermediate products in the final PZT. However, the high processing temperature causes PbO-losses and consequently a change in the PZT's stoichiometry [6].

For high-frequency ultrasound transducer applications bulk PZT ceramic are not frequently used, because the ceramic must be cut and polished to specified dimensions. The final step is the assembling of a layer of ceramic, a few tens of micrometres thick, with the other components. This imposes limits on the minimum dimensions of the manufactured parts and it constrains the geometry of the parts to simple shapes, like discs, plates, rings, cylinders, etc. A more frequently used approach is thick-film technology, where a layer is built on the substrate. This technology enables the direct integration of layers onto the substrates and therefore eliminates the difficulties associated with handling thin, bulk ceramics [7].

In order of geometrically focus of the ultrasound wave, the thick-film piezoelectric layer must be patterned on a non-flat substrate. Electrophoretic deposition (EPD) is a method that enables the deposition of charged particles in a conductive, complex-shaped substrate when applying a DC electric field[8]. The method makes possible the control of the thickness and morphology of the as-deposited layer by varying the deposition conditions, such as the deposition time and the electrical field. For this reason we found the method to be attractive for processing ultrasound transducers[9].

In order to be able to process the thick films by EPD a stable suspension with well dispersed, charged particles is required. The anionic polyelectrolyte polyacrylic acid (PAA) dissociated in ethanol in the presence of the organic base n-butlyamine is used for charging Al_2O_3 [10] as well as the PZT particles [9] in ethanol. The challenge is in the drying and sintering of the as-deposited layer, since the layer may crack spontaneously during the drying after a critical thickness [11]. During the sintering the as-deposited layer is in constrained conditions[12]. It shrinks due to the driving force for sintering, but it is clamped on the rigid substrate and as a consequence the stresses are created in the layer . The stresses are relaxed via the formation of cracks that are highly undesirable in piezoelectric thick films.

The aim of this study was to process a piezoelectric PZT thick film on a flat electroded alumina substrate using EPD. The layers with a targeted thickness of around 30 µm and a density of around 80 % of the theoretical value were characterized and assesses for high-frequency-transducer applications.

EXPERIMENTAL

The $Pb(Zr_{0.53}Ti_{0.47})O_3$ (denoted PZT) powder was synthesized from PbO (99.9 %, Aldrich, Germany), ZrO_2 (99 %, Tosoh, Japan) and TiO_2 (99.8 %, Alfa Aesar, Germany) by solid state synthesis. The oxides were mixed in a molar ratio corresponding to a stoichiometry of $PbZr_{0.53}Ti_{0.47}O_3$. They were homogenized in a planetary mill for 2 hours. After drying, the mixture was calcined at 950 °C for 2 hours, re-milled and re-calcined at 1100°C for 1 h. After a second calcination the powder was milled for 8 hours in an attritor mill and dried.

The electrophoretic deposition experiments were performed using ethanol-based suspensions containing 1 vol. % of PZT powder that were prepared as follows. Polyacrylic acid

(PAA) from Alfa Aesar (50 wt. % of PAA in water, molar mass 2000) was mixed with n-butylamine (BA) from Alfa Aesar. The amount of PAA is given as a mol of PAA per mass of PZT powder, i.e., μmol/g. The PAA/BA molar ratio was 1/2.5. Into the PAA and BA mixture we added ethanol (Carlo Erba) and PZT powder. The suspension was homogenised in a ZrO_2 planetary mill at 150 rpm for 1 h. An identical procedure was used for processing the PbO suspension. The PZT and PbO suspensions were mixed in a stoichiometry corresponding to 98 mol. % of PZT and 2 mol. % of PbO. The suspension is denoted as PZT PbO.

In some cases polyvinyl butyral (PVB, Aldrich, molar mass 50 000-80 000) was added to the suspension. A total of 3 wt. % of PVB was added to the PZT PbO ethanol-based suspension and mixed with a magnetic stirrer for 8 hours. The suspension is denoted as PZT PbO PVB.

A standard electrophoretic deposition setup was employed. The vertically aligned electrodes were separated by a distance of 17 mm. Two identical Al_2O_3 square-plates (Kyocera, 99.9 %) with a thickness of 0.5 mm and a length of 12.5 mm, covered on one side by gold electrodes (ESL 8884-G), were used as the working and counter electrodes. The area of the electrode was 0.64 cm^2. The deposition process was performed at a constant current density of 1.56 mA/cm^2, provided by a Keithley 2400 source meter. The experiments were conducted at ambient temperature and without any mechanical stirring. The deposition times were 60, 80 and 90 seconds. After the deposition the samples were placed in an ethanol-rich atmosphere and dried at ambient temperature.

The particle size and the particle-size distribution were determined using a static light-scattering particle-size analyser (Microtrac S3500, USA). The zeta-potential of the particles was measured in diluted ethanol dispersions using a zeta-potential analyser ZetaPALS (Brookhaven Instruments Corporation, USA). The viscosity of the suspensions was determined at 25°C for shear rates between 10 and 100 s^{-1} after an equilibration period of 30 s with a CC 27 cylindrical system using a Physica MCR 301 rheometer (Anton Parr, Austria). The conductivity of the suspensions was measured using an InoLab Cond 730 (WTW, Germany).

The sintered thick films were investigated with a scanning electron microscope (JEOL 5800, Tokyo, Japan). For each sample we took two representative scanning electron microscopy images. The density of the sintered thick-film structures was calculated from a quantitative characterization of the ceramic microstructure using computerized image-analysis software ImageTools 3.0 (The University of Texas Health Science Center in San Antonio, USA). A binary image obtained from the original micrograph was used to determine the quantity of pores.[13] The density of the layer was calculated by deviding the area covered with pores with a total area. The density of the PZT thick films was given as an average value of the two measurements. The error is ± 2 %.

The dielectric, mechanical and piezoelectric parameters were deduced by measuring the complex electrical impedance around the fundamental thickness mode of the resonance. The experimental set-up was composed of an HP 4395 vector analyser and its impedance test kit. To simulate the theoretical behaviour of the electrical impedance of the samples as a function of frequency for the thickness mode, an equivalent electrical circuit model was used. The KLM scheme was retained[14]. From the complex electrical impedance and the fitting process, the thickness-mode parameters of the EPD piezoelectric thick films were deduced. The structure of the samples is composed of three inert layers (two golds electrodes and an alumina substrate) and one piezoelectric layer (PZT thick film)[15]. The parameters of those layers are fed into the KLM model and considered to be constant. The parameters values of the inert layers[16,17] are given in Table I.

Table I: Material properties of the inert layers of the thick-film structure

Layer	Material	thickness (μm)	v_L(m/s)	ρ (kg/m^3)	Z (MRa)
Front electrode	Gold	0.2	3240	19700	63.8
Bottom electrode	Gold	10	3240	19700	63.8
Substrate	Alumina	640	3900	10500	41

V_L: longitudinal wave velocity; ρ: density; Z: acoustic impedance

With the KLM model, five parameters were deduced (i.e., the effective thickness-mode coupling factor k_t, the longitudinal wave velocity v_L, the dielectric constant at constant strain ε_{33}^S, and the loss factor (mechanical δ_m and dielectric δ_e). On this basis and from additional measurements of the density and the thickness of the piezoelectric thick film, the acoustical impedance (Z) and the resonant frequency of the thick film (f_0) in free mechanical resonance conditions were deduced. The accuracy of this characterization method for a multilayer structure was discussed in earlier work[18,19] .

RESULTS AND DISCUSSION

Properties of PZT PbO and PZT PbO PVB suspensions

The PZT and PbO ethanol-based suspensions were stable and no sedimentation of the particles was observed after two hours. The zeta-potential of the PZT particles in the ethanol was 50 ± 5 mV. The conductivity of the PZT suspension was 24 μS/cm at 25°C. The zeta-potential of the PbO particles in the ethanol was 40 ± 5 mV and the conductivity of the suspension was 12 μS/cm at 25°C. The PZT PbO suspension exhibited well-dispersed particles, and also in this case we did not observe any sedimentation after two hours. The zeta-potential of the particles was -50 $\pm$ 7 mV. The conductivity of the PZT-PbO suspension was 20 μS/cm at 25°C. After the addition of 3 wt. % of PVB to the PZT PbO suspension the zeta- potential of the particles did not change significantly; it was -47 $\pm$ 7 mV. The conductivity of the PZT-PbO-PVB suspension was 22 μS/cm at 25°C.

The viscosity of the PZT PbO and PZT PbO PVB suspensions in a share-rate range from 10 to 100/s is presented in Figure 1. It is evident that the viscosity of the PZT PbO suspension that contained 1 vol. % of solid particles in the ethanol was 1.2 mPas, and it did not vary significantly in the reported share-rate range. A slightly higher viscosity of 1.4 mPas was measured for the PZT PbO PVB suspension that contained 1 vol. % of solid particles. The addition of PVB to the suspension may increase the viscosity of the suspensions, which was already reported in the literature. [20] The interactions between PZT, PAA and PVB in ethanol-based suspensions were studied and are reported elsewhere[21]

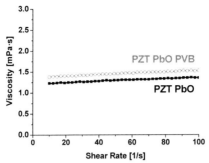

Figure 1: The viscosity of the PZT PbO and PZT PbO PVB suspensions, both containing 1 vol % of solid particles, as a function of the share rate in the share-rate range 10 to 100/s.

We have measured the mass of the deposit obtained from the PZT PbO and PZT PbO PVB suspensions for different deposition times. Masses of 9 mg, 12.1 mg and 14.4 mg were obtained from the PZT PbO suspension after 60, 80 and 90 seconds. Masses of 7.5 mg, 11.2 mg and 12.2 mg were obtained from the PZT PbO PVB suspension after 60, 80 and 90 seconds. The mass of the deposit increased with the increasing deposition time. It is evident that the mass obtained from the PZT PbO suspension was slightly higher than the one obtained from the PZT PbO PVB suspension. This phenomenon may originate from the mobility of the particles. The mobility of the particles during the electrophoretic deposition process depends on many parameters, including the viscosity of the suspension[8]. A higher viscosity of the suspension results in a lower mobility of the particles. In the PZT PbO and PZT PbO PVB suspensions the zeta-potential of the particles and the conductivity of the suspensions were similar, and therefore we consider that the lower deposition mass in the case of the PZT PbO PVB suspension was the result of the higher viscosity of this suspension.

The layers deposited for 60, 80 and 90 seconds were crack-free after drying. However, when a longer deposition time was used some cracks appeared. This was particularly evident in the deposits from the PZT PbO suspension. The addition of the PVB significantly reduced the formation of cracks in the as-deposited and dried layers. Kuscer et all [21] studied the interactions between PZT, polyacrylic acid and PVB in ethanol. They showed that carboxyl groups interacted with PZT and therefore the polyacrylic acid was bonded to PZT particles in ethanol-based suspensions. PVB was not bonded to PZT particles but it was detected in the as-deposited layers as a "free", non-bonded polymer. We observed that some cracks are formed in the as-deposited PZT PbO layer during the process of drying. However in the presence of PVB the cracks were not observed or appeared after longer time. In the presence of polymer PVB, the liquid migration rate may be reduced in the system. The evaporation of ethanol from the PZT PbO PVB as-deposited layer may be reduced and therefore the formation of cracks may be hindered.

PZT thick films on flat alumina substrates
The dried deposits obtained from the PZT PbO and PZT PbO PVB suspensions after the depositions for 60, 80 and 90 seconds were sintered at 950°C for 2 hours in a PbO-rich atmosphere. The samples deposited from the PZT PbO suspensions for 60, 80 and 90 seconds are denoted PZT PbO_60, PZT_80 and PZT PbO_90, respectively. The samples deposited from the PZT PbO PVB suspensions for 60, 80 and 90 seconds are denoted, respectively, PZT PbO

PVB_60, PZT PbO PVB_80 and PZT PbO PVB_90. The microstructures of the samples from the PZT PbO and PZT PbO PVB suspensions are shown in Figure 2 and Figure 3, respectively.

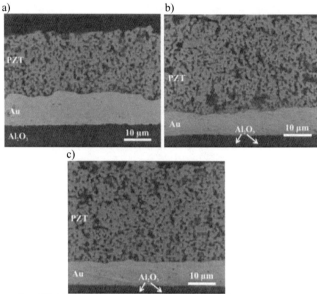

Figure 2: Cross-section SEM image of the deposits obtained from the PZT PbO suspension sintered at 950°C for 2 hours. The samples were deposited at various deposition times. a) PZT PbO_60; b) PZT PbO_80; c) PZT PbO_90.

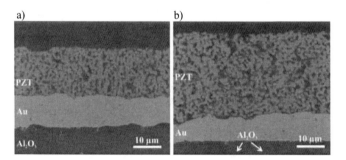

c)

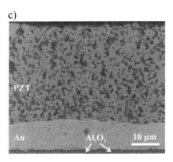

Figure 3: Cross-section SEM image of the deposits obtained from the PZT PbO PVB suspension sintered at 950°C for 2 hours. The samples were deposited at various deposition times. a) PZT PbO PVB_60; b) PZT PbO PVB_80; c) PZT PbO PVB_90

From the images it is evident that the PZT thick films are relatively homogeneous and consist of one- to two- micrometer-sized PZT grains. The PZT layer is well adhered to the Au/Al$_2$O$_3$ substrate. We did not observe any delamination of the structures. The thickness of the deposits was measured from the cross-section SEM images. The thickness of the sintered PZT thick films as a function of the deposition time obtained from the PZT PbO and PZT PbO PVB suspensions is shown in Figure 4.

The densities of the PZT thick films were also evaluated from SEM cross-section images of the sintered layers. The density values of the sintered PZT thick films are collected in Table II. The theoretical density of PZT is 8 g/cm^3.

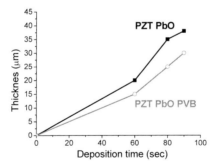

Figure 4: Thickness of PZT thick films as a function of the deposition time obtained from the PZT PbO and PZT PbO PVB suspensions. All the deposits were prepared at a constant current density of 1.56 mA/cm^2 at a distance between the electrodes of 17 mm and sintered at 950°C for 2 h.

Table II: The density of the PZT thick films sintered at 950°C for 2 hours obtained from the PZT PbO and PZT PbO PVB suspensions.

Sample	Deposition time (sec)	Density* (%)
PZT PbO_60	60	83 ± 2
PZT PbO_80	80	84 ± 2
PZT PbO_90	90	82 ± 2
PZT PbO PVB_60	60	82 ± 2
PZT PbO PVB_80	80	84 ± 2
PZT PbO PVB_90	90	83 ± 2

* % of the theoretical density of PZT

Electromechanical characterization of thick films on flat substrates

The complex electrical impedance of the PZT PbO_60, PZT PbO_80, PZT PbO_90, PZT-PbO-PVB_60, PZT PbO PVB_80 and PZT-PbO-PVB_90 were measured with the experimental set-up described previously in the experimental part. For the six samples, the five parameters were deduced (i.e., the effective thickness-coupling factor (k_t), the dielectric constant at constant strain (ε_{33}^S), the longitudinal wave velocity (v_l), and the mechanical and dielectric losses (respectively, δ_m and δ_e) measured at the anti-resonant frequency). The following two figures (Figure 5 and Figure 6) show the superposition of the theoretical and experimental electrical impedance curves after the fitting process for the samples PZT PbO_60 and PZT PbO PVB_80, respectively. The coupling of the resonances in the alumina substrate with those in the piezoelectric thick film is clearly observed.

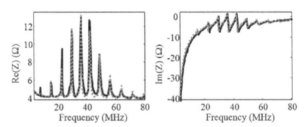

Figure 5: Complex electrical impedance measurement of PZT PbO_60 (black solid line: theoretical, grey dash points: experimental).

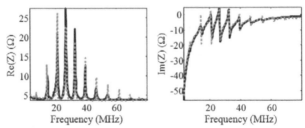

Figure 6: Complex electrical impedance measurement of PZT PbO PVB_80 (black solid line: theoretical, grey dash points: experimental).

The measured parameters are summarized in Table III. First, and as expected, the resonant frequency (f_0) of the thick films from the PZT-PbO suspension and the PZT-PbO-PVB suspensions decreased as the thickness of the thick films increased. The thickness-coupling factor has a comparable value of around 50 % for the two compositions and the six corresponding samples. For bulk standard PZT, this thickness-coupling factor can be lower than 50 % and consequently this result reveals the strong potential of this process for efficient thick films for ultrasonic transducer applications. However, the two lowest values are obtained with the thinner samples (PZT PbO_60 and PZT PbO PVB_60). This can be related to the clamping effect of the substrate, which grows in importance with a decrease of the thickness. This clamping effect decreases the piezoelectric performance, in particular the d_{33} coefficient[22] and consequently the effective thickness-coupling factor. The longitudinal wave-velocity and dielectric-constant values are lower than those of the bulk and dense ceramic with similar compositions due to the porosity content and these values are coherent with those measured in [9]. For the mechanical losses, similar values were already measured in [9].

Table III: Properties of piezoelectric PZT thick films sintered at 950°C for 2 hours. PZT thick films were prepared from PZT PbO and PZT PbO PVB suspensions using various deposition times.

Samples	Thickness (μm)	S (mm²)	v_L (m/s)	ε_{33}	k_t (%)	f_0 (MHz)	δ_m (%)	δ_e (%)
PZT PbO_60	20	7.3	2780	320	48	69	6 ± 0,5	1 ± 0,5
PZT PbO_80	35	7.7	2930	340	51	42	12 ± 0,5	1 ± 0,5
PZT PbO_90	38	7.5	2771	280	50	36	6 ± 0,5	1 ± 0,5
PZT PbO PVB_60	15	7.7	2324	270	47	75	13 ± 0,5	1 ± 0,5
PZT PbO PVB_80	25	7.6	2456	270	51	49	10 ± 0,5	1 ± 0,5
PZT PbO PVB_90	30	7.3	2690	320	52	44	12 ± 0,5	1 ± 0,5

S: area of the top electrode; V_L: longitudinal wave velocity, ε_{33}^S: dielectric constant at constant strain; k_t: effective thickness coupling factor; f_0: antiresonant frequency of the thick film (in free mechanical conditions); δ_m: mechanical losses; δ_e: dielectric losses.

CONCLUSION

Piezoelectric PZT thick films with a tailored thickness and porosity were processed from PZT PbO and PZT PbO PVB ethanol-based suspensions using an electrophoretic deposition process. The PZT and PbO particles were dispersed in ethanol using polyacrylic acid. The suspension containing 98 mol. % of PZT and 2 mol. % of PbO, having a conductivity of around 20 μS/cm and a viscosity of 1.2 mPas, was used for processing the PZT thick films on a Au/Al$_2$O$_3$ substrate at a constant current density of 1.56 mA/cm². In order to prevent the as-deposited layer from crack formation during the drying, 3 wt. % of PVB was added to the PZT PbO suspension. The deposits from the PZT PbO and PZT PbO PVB suspensions were dried and sintered at 950°C for 2 hours in the presence of a liquid phase. The sintered thick films obtained from the PZT PbO PVB suspension were thinner than the ones obtained from the PZT PbO suspension for an identical deposition time. Moreover, these deposits did not tend to crack after longer deposition times. The sintered PZT thick film had a density between of 80 and 85 % of the theoretical density. Its thickness increased with the increasing deposition time and was 15 and 38 μm for 60 and 90 seconds, respectively. The corresponding resonant frequencies were 75 and 36 MHz. All the reported PZT thick films were characterized by a high coupling factor of around 50 %. These results indicate that homogeneous PZT thick films with tailored thicknesses and porosities can be prepared by an electrophoretic deposition process from ethanol-based suspensions by controlling the deposition and the sintering conditions. The electromechanical properties indicate that PZT thick films can be used for processing ultrasound transducers for medical imaging and diagnostic applications.

ACKNOWLEDGMENT

The financial support of the Slovenian Research Agency is acknowledged (Research program P2-0105). Part of the work was performed within and Piezo Institute. Mrs. Milena Pajič and Mrs. Petra Kuzman are thanked for the preparation of the samples and Dr. Janez Holc for fruitful discussions.

REFERENCES

[1] F. S. Foster, C. J. Pavlin, K. A. Harasiewicz , D. A. Christopher, D. H. Turnbull , Advances in ultrasound biomicroscopy, *Ultrasound in Med & Biol.*, **26**, 1-27 (2000).

[2] M. Lethiecq, F. Levassort, D. Certon, L. P. Tran-Huu-Hue, Piezoelectric Transducer Design for Medical Diagnostic and NDE, in *Piezoelectric and Acoustic Materials for Transducer Applications*, Ed. A. Safari, E. K. Akdogan, Springer (2008), 191-215.

[3] F. Levassort, J. Holc, E. Ringaard, T. Bove, M. Kosec, M. Lethiecq, Fabrication, modeling and use of porous ceramic for ultrasound trandsucer applications, *J. Electroceram.*, **19**, 125-137 (2007).

[4] S. S. Chandratreya, R. M. Fulrath, J. A. Pask, Reaction Mechanisms in the Formation of PZT Solid Solutions, *J. Am. Ceram. Soc.*, **64** [7] 422–425 (1981).

[5] D. L. Hankey and J. V. Diggers, Solid-State Reactions in the System $PbO-TiO_2-ZrO_2$, *J. Am. Ceram. Soc.*, **64** [12] c172 – c173 (1981).

[6] B. V. Hiremath, A. I. Kingon, J. V. Biggers, Reaction Sequence in the Formation of Lead Zirconate-Lead Titanate Solid Solution: Role of Raw Materials, *J. Am. Ceram. Soc.*, **66** [11] 790–793 (1983).

[7] M. Kosec, D. Kuscer, J. Holc, Processing of Ferroelectric Ceramic Thick Films pp. 39-41 in *Multifunctional Polycrystalline Ferroelectric Materials: Processing and Properties*; Springer, Dordrecht, Heildelberg, London, New York (2011).

[8] B. Ferrari, R. Moreno, EPD kinetics: A review, *J. Eur. Ceram. Soc.* **30** [5], 1069-1078 (2010).

[9] D. Kuscer , F. Levassort , M Lethiecq , A-P Abellard , M Kosec, Lead-Zirconate-Titanate Thick Films by Electrophoretic Deposition for High-Frequency Ultrasound Transducers, *J. Am. Ceram. Soc.* **95**, 892–900 (2012).

[10] A. M. Popa, J. Vleugels, J. Vermant, O. Van der Biest, Influence of surfactant additon sequence on the suspension propoerties and electrophoretic deposition behaviour of alumina and zirconia, *J. Eur. Ceram. Soc.* **26** [6], 933–939 (2006).

[11] E.G. Jason, A. U. Scott J. Moon, M. J.Cima, E. M. Sachs, Controlled Cracking of Multilayer Ceramic Bodies, *J. Am. Ceram. Soc.,* **82**, 2080-2086 (1999).

[12] R. K. Bordia, R. Raj, Sintering Behaviour of Ceramic Films Constrained by a Rigid Substrate, *J. Am. Ceram. Soc.*, **68**, 287-292 (1985).

[13] D. Kuscer, J. Korzekwa, M. Kosec, R. Skulski, A- and B-compensated PLZT x/90/10: Sintering and microstructural analysis, *J. Eur. Ceram. Soc.*, **27** [6], 4499-4507 (2007).

[14] D. A. Leedom, R. Kimholtz, and G. L. Matthaei, New equivalent circuit for elementary piezoelectric transducers, *Electronic Lett.*, **6** [13], 398-399 (1971).

[15] P. Maréchal, F. Levassort, J. Holc, L.P. Tran-Huu-Hue, M. Lethiecq, High frequency transducers based on integrated piezoelectric thick films for medical imaging, IEEE *Transactions on Ultrasonics, Ferroelectrics and Frequency Control*, **53** [8] 1524-1533 (2006).

[16] A.R. Selfridge, Approximate material properties in isotropic materials, *IEEE Trans. Sonics Ultrason.*, **32** [3] 381-394 (1985).

[17] http://www.ondacorp.com/tecref_acoustictable.shtml (Last viewed 23th of May 2010)

[18] M. Lukacs, T. Olding, M. Sayer, R. Tasker, S. Sherrit, Thickness mode material constants of a supported piezoelectric film, *J. Appl. Phys.*, **85** [5] 2835-2843 (1999).

[19] A. Bardaine, P. Boy, P. Belleville, O. Acher, F. Levassort, Improvement of composite sol-gel process for manufacturing 40 μm piezoelectric thick films, *J. Eur. Ceram. Soc.*, **28** [8] 1649-1655 (2008).

[20] W. J. Tseng, C-L. Lin, Effect of polyvinyl butyral on the rheologycal properties of $BaTiO_3$ powder in ethanol-isopropanol mixtures, *Mat. Lett.*, **57**, 223-228 (2002).

[21] D. Kuscer, T. Bakaric, B. Kozlevcar, M. Kosec, Interactions between Lead–Zirconate Titanate, Polyacrylic Acid, and Polyvinyl Butyral in Ethanol and Their Influence on Electrophoretic Deposition Behavior, *J. Phys. Chem. B* (2012) dx.doi.org/10.1021/jp305289u.

[22] A. Barzegar, D. Damjanovic, N. Setter, The effect of boundary conditions and sample aspect ratio on apparent d_{33} piezoelectric coefficient determined by direct quasistatic method , *IEEE Trans. Ultrason. Ferroelectr. Freq. Control*, **51** [3] 262–270 (2004).

LOW TEMPERATURE GROWTH OF OXIDE THIN FILMS BY PHOTO-INDUCED CHEMICAL SOLUTION DEPOSITION

Tetsuo Tsuchiya*, Tomohiko Nakajima, Kentaro Shinoda
1-National Institute of Advanced Industrial Science and Technology (AIST),Tsukuba Central 5,
1-1-1 Higashi Tsukuba, Ibaraki, 305-8565, Japan

ABSTRACT

Printed electronics is creating many new products given the benefits of the technology compared to conventional electronics, such as thinness, flexibility, cost, ease of manufacture, fast production turn around, "green" technology, power efficiency and more. For this purpose, we have developed the photo reaction of nano-particle method (PRNP) for the preparation of the patterned metal oxide thin film on organic and glass substrates. In this paper, we demonstrate the preparation of transparent conducting thin film by the PRNP process, and their properties of the obtained film. By using the combination of the two procedures, the resistivity of the film was 5.94×10^{-4} Ωcm. The resulting ITO film showed mobility as high as 9.99 cm^2 V-1 s-1at the carrier density of 1.05×10^{21} cm^{-3}. The film was compact in comparison with simple thermal process. Sheet resistance of the ITO film on glass and PET by using the PRNP method was 50 and 150Ω/sq.

INTRODUCTION

The tin doped indium (ITO) thin film is widely used for optical and electronic applications, such as touch panel contacts, electrodes for LCD and electro-chromic displays, and energy conserving architectural windows because of its good electrical and optical properties. In most case, the ITO film, which has a high conductivity ($2 \times 10^{-4}\Omega$cm), for industry is produced by physical vapor deposition, such as sputtering. The chemical solution process is a good candidate method for the preparation of the ITO film because of its simplicity and low facility investment. Ota and coworkers reported that the high conductivity ($2.5 \times 10^{-4}\Omega$cm) ITO film could be prepared using the metal organic deposition (MOD) at 600 °C in a N$_2$–0.1% H$_2$ atmosphere [1]. However, this high temperature heat treatment is considered to be unfavorable when the film is produced on a polymer substrate as in the case of electronics devices. In a previous study, we developed the excimer laser-assisted metal organic deposition (ELAMOD) process for the preparation of ITO films [2, 3]. Recently, another problem that indium resource will be scarce in the next generation has become recognized more and more. Therefore, to decrease the actual use of the indium in the thin film process, the direct writing of the ITO electrode using an ink-jet process would be useful instead of the photo lithography process. So far, ITO nano-particles have been synthesized by many methods [4-11]. However, the resistivity of the ITO film prepared by a

143

simple heating method was 10^{-3} Ω cm [6]. Therefore, to lower the resistivity of the ITO films, we have developed a photo reaction of nano particle (PRNP) process [12]. In this paper, we investigated the microstructures of the ITO films prepared by the simple thermal reaction and photo reaction process of nano-particles, and clarify the effect of the laser irradiation on the electrical properties of the films.

2. EXPERIMENTAL

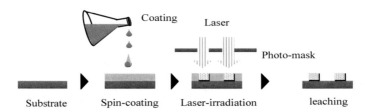

Fig. 1   Schematic drawing of patterning in PRNP.

The ITO nano-particles were synthesized by the solvothermal method. The particle size was 4-6nm. Figure 1 shows schematic drawing of patterning in PRNP. The starting solution was spin-coated onto a SiO_2 substrate, and then dried at 100 °C in air to eliminate the solvent. The nano-particle coated film was irradiated by an excimer lamp and KrF excimer laser at room temperature.

Experimental A: Optimum irradiation condition for these studies is that the coated ITO nano-particle film on an SiO_2 substrate was irradiated by 222 nm excimer lamp for 5 min, and then, irradiated by the KrF laser at a fluence of 80mJ/cm^2 at 1Hz for 5shots(process 1). After producing the two layer of ITO film by process1, then, the other three layers was prepared by the direct KrF laser irradiation of the coated nano-particle film at a fluence of 90mJ/cm^2 and 130mJ/cm^2 for 1shot.

Experimental B: Procedure of the two-step irradiation is that the film on a SiO_2 substrate was irradiated by 222 nm excimer lamp for 5 min, and then, irradiated by the KrF lase at a fluence of 80mJ/cm^2.

Experimental C: To compare with photo reaction process, the coated nano-particle film on a SiO_2 substrate was heated at 300, 400 and 500 °C.

The crystallinity of the obtained films was determined by x-ray diffraction. The morphology of the film was based on SEM and TEM measurements. Their resistivity, carrier concentration and mobility of the ITO film were measured by the Van der Pauw method.

3. RESULTS AND DISCUSSION

Fig. 2 shows a TEM image of the nano-particles prepared by the solvothermal method. As can be seen, a particle size is 4-6 nm. Also, condensed particles with 30 nm are observed. The nano-particles solution was spin-coated on a SiO_2 substrate.

Figure 2 TEM of the nano-particles.

The electrical properties(R, μ_h and N_h) of the films are summarized in Table I. The resistivity of the film (sample A) prepared by experimental A was 5.94×10^{-4} Ωcm. The resulting ITO film showed mobility as high as 9.99 cm^2 V^{-1} s^{-1} at the carrier density of 1.05×10^{21} cm^{-3}. On the other hand, in the case of using the two-step irradiation method (experimental B), the electrical resistivity, Hall mobility μ_h, and carrier concentration N_h of the ITO film (sample B) was 3.73×10^{-3} Ωcm, 1.27×10^{21} cm^{-3} and 1.32cm^2 V^{-1} s^{-1}, respectively. Also, when the film is heated at 500 °C in N_2, the electrical resistivity, Hall mobility μ_h, and carrier concentration N_h of the ITO film (sample C) was 4.10×10^{-3} Ωcm, 8.00×10^{20} cm^{-3} and 2.00cm^2 V^{-1} s^{-1}, respectively.

Table I   R, μ_h and N_h of the films prepared by photo reaction and thermal reaction[12].

	$R(\Omega cm)$	$N_h(cm^2 V^{-1} s^{-1})$	$\mu_h(cm^{-3})$
Sample A	5.94×10^{-4}	1.05×10^{21}	9.99
Sample B	3.73×10^{-3}	1.27×10^{21}	1.32
Sample C	4.10×10^{-3}	8.00×10^{20}	2.00

To investigate the effect of the sintering temperature on the film structure and electrical properties of the ITO film using the simple thermal process, we observed an SEM image of the films heated at 250, 400 and 500°C in air as shown in Fig. 3. At 250°C, the film retains the cubic particle structure. When the heating temperature increased above 400°C, sintering of the particle was observed. However, the grain size is small and pores remain in the film. Fig. 3 shows the electrical properties of the films at different temperatures. The resistivity of the film decreased with an increase in the temperature. The lowest resistivity of the film in air was 2.2×10^{-2} Ωcm. This is due to the improvement of the sintering of the film as shown in Fig. 3. To decrease the electrical properties of the film, we annealed the film in a N_2 atmosphere at 500°C. The resistivity of the film heated at 500 °C in a N_2 was 4.1×10^{-3} Ωcm (sample C). This resistivity is consistent with the result reported by Ederth et al. [6], whereas the cubic small size nano-particle prepared by solvothermal method was used.

Figure 3 SEM of the ITO films prepared by thermal process.

In order to clarify the difference in the electrical properties, we measured the microstructure of the ITO films using TEM. Fig. 4 shows the TEM image of the film of (a) Sample A and (b) sample B.

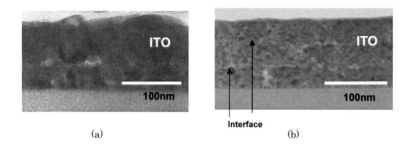

Figure 4 Cross sectional TEM of ITO films by PRNP.

As can be seen from figure, Sample A has large grain size and compactness. On the other hand, grain size of the Sample B is smaller than that of the Sample A. in addition, the interphase grain boundary was also observed. Thus, it was found that the improvement in the resistivity of the

ITO film (sample A) would be due to the improvement of the grain size of the film. According to the XRD measurements, the crystallinity of the film prepared by the two combined procedures method is also higher than that of the film (sample B) prepared by excimer lamp and laser irradiations. Therefore, a laser irradiation to nano-particles using experiment A would be found to be effective for an improvement of a grain size for thin film applications.

Finally, we tried to prepare the ITO film on PET substrate by using PRNP. We have successfully obtained the ITO film on PET substrate as shown in Fig. 5.

 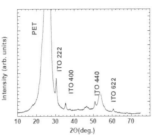

Figure 5 Photograph and XRD pattern of ITO/PET film by PRNP.

The Sheet resistance of the ITO film on PET by using the PRNP method was 150Ω/sq. Also we tried to prepared the other oxide materials, and obtained VO_2, $CaTiO_3$:Pr films on PET substrate. Thus, PRNP method is useful for the preparation of oxide film at low temperature.

CONCLUSIONS

We have successfully obtained an ITO thin film by the photo reaction of the nano-particles (PRNP) using an excimer lamp and laser at room temperature without vacuum. Using the combined two-step irradiation and one-step irradiation process(Experimental A), the electrical resistivity, carrier concentration, hall mobility of the ITO film were $5.94 \times 10^{-4}\Omega$cm, 1.05×10^{21} cm^{-3} to and 9.99 cm^{-2} V^{-1} s^{-1}, respectively. The improvement of the resistivity of the film prepared by experimental A process was found to be due to the well-sintering of the nano-particles and the improvement of the mobility. Also, by using the PRNP, we have successfully obtained the ITO film on PET substrate.

ACKNOWLEDGEMENTS

This study was supported by Industrial Technology Research Grant Program in 2012 from New Energy and Industrial Technology Development Organization (NEDO) of Japan.

REFERENCES

[1] R. Ota, S. Seki, M. Ogawa, T. Nishide, A. Shida, M. Ide, and Y. Sawada, Thin Solid Films, **411** (2002) 42.

[2]T. Tsuchiya, H. Niino, A. Yabe, I. Yamaguchi, T. Manabe, T. Kumagai and S. Mizuta, Appl. Surf. Sci. **197-198** (2002) 512.

[3]T. Tsuchiya, A. Watanabe, H. Niino, A. Yabe, I. Yamaguchi, T. Manabe, T. Kumagai and S. Mizuta, Appl. Surf. Sci. **186** (2002), p. 173.

[4]N. Nadaud, M. Nanot, and P. Boch, J. Am. Ceram. Soc. **77** (1994) 843.

[5]M. Toki and M. Aizawa, J. Sol-Gel. Technol. **8** (1997) 717.

[6]J. Ederth, P. Johnsson, G. A. Niklasson, A. Hoel, A. Hultåker, P. Heszler, C. G. Granqvist, A. R. van Doorn, M. J. Jongerius, and D. Burgard, Phys. Rev. B **68** (2003) 155410.

[7]Ki Young Kim, Seung Bin Park, Mater. Chem.and Phys., **86** (2004) 210.

[8]Jin-Seok Lee, Sung-Churl Choi, J. Eur. Ceram. Soc., **25** (2005) 3307.

[9]Hee Dong Jang , Chun Mo Seong, Han Kwon Chang and Heon Chang Kim, Curr. Appl. Phys. **6** (2006) 1044.

[10] Sung-Jei Hong and Jeong-In Han, Curr. Appl. Phys. **6** (2006) 206.

[11]M. Gross, A. Winnacker, P. J. Wellmann, Thin Solid Films, **515**(2007) 8567.

[12] T. Tsuchiya, F. Yamaguchi, I. Morimoto, T. Nakajima and T. Kumagai, Appl. Phys.A **99**, 745 (2010).

The effect of active species during TiN thin film deposition by the cathodic cage plasma process

Natália de Freitas Daudt[1], Júlio César Pereira Barbosa[2], Danilo Cavalcante Braz[1],
Marina de Oliveira Cardoso Macêdo[1], Marcelo Barbalho Pereira[3], Clodomiro
Alves Junior[1]

[1] *Universidade Federal do Rio Grande do Norte, Natal, RN, Brazil*
[2] *Universidade Federal Rural do Semi-Árido, Mossoró, RN, Brazil*
[3] *Universidade Federal do Rio Grande do Sul, Porto Alegre, RS, Brazil*

Keywords: cathodic cage, plasma deposition, TiN thin film, Optical Emission
Spectroscopy, optical properties.

Abstract: Thin solid films of titanium nitride were grown on glass using the cathodic
cage plasma deposition technique. The plasma atmosphere used was a mixture of Ar, N_2
and H_2. The H_2 amount in the mixture was 0, 12.5 and 22.5%. The deposition process
was monitored by Optical Emission Spectroscopy in order to determine the effect of
active species in plasma on deposited film characteristics. The maximum luminous
intensity of the $H\alpha$ species was observed at 12.5% of H_2 gas content, but decreased
when H_2 gas content was increased to 22.5%. This trend strongly affected the crystalline
structure, optical properties and topography of the TiN film. Grown films were
composed by nano-size TiN particles and contained various amount of nitrogen into the
crystal. The film deposited in a plasma atmosphere without H_2 gas had the highest
transmittance. The film obtained a plasma atmosphere containing 12.5% of H_2 had the
lowest transmittance, the highest homogeneity and stoichiometry.

Introduction

The cathodic cage plasma technique is an ionic nitriding adaptation, which was
developed in order to reduce defects such as edge effect, overheating obtained when
using conventional technique and allow to nitride pieces with complex geometry [1, 2].
The cathodic cage consists of a cylindrical plate with holes and a cylindrical cover with
similar holes. In this configuration, the cage acts as the cathode which the electrical
potential difference is applied in relation to the chamber walls. Thus, the sputtering
process occurs on the cage and damage on the sample surface is avoiding. Sputtered
atoms can bond with the reactive particles of plasma and be deposited on the sample
surface. Moreover, it is possible to produce a hybrid process of deposition and
diffusion, since the substrate temperature is high enough to promote diffusion [3-4].

Titanium nitride films are used in various industrial applications: as coatings for
high hardness and low friction in metallurgical industry [5], as decorative coatings
replacing gold, since different colors may be achieved by varying of the N/Ti ratio [6],
as well as coatings of solar cells and solar control windows [7], biocompatible coatings
for biomaterial alloys [8] and coatings for microelectronic semiconductors [9].

The optical properties of titanium nitride films are very sensitive to even small
variations in chemical composition and thickness. This sensibility can be used to verify
the process control. Moreover, because this technique was recently developed, few
literature reports have been regarding the effect of process parameters on characteristics

of the grown film. Therefore, film deposition using the cathodic cage plasma technique was monitored by Optical Emission Spectroscopy (OES) in order to determine the plasma active species effect on the deposition rate and the properties of deposited film.

Materials and methods

Rectangular glass substrates (each with 25 x 10 mm^2 surface area and 2 mm thickness) were used in the present study. The cage was made of titanium and had holes each 12 mm in diameter. Film deposition was conducted in an ionic nitriding reactor adapted to cathodic cage configuration in the Labplasma facilities. The cage was placed on the cathode, as illustrated in Figure 1. The substrates were in a floating potential because they were electrically isolated from the cathode trough an alumina disc.

The argon (Ar) and nitrogen (N_2) gas flow rates were fixed at 4 sccm and 3 sccm, respectively. The hydrogen (H_2) gas flow rates used were 0, 1 and 2 sccm. Thus, the $H_2/(Ar + N_2 + H_2)$ ratios used were 0, 12.5 and 22.5%. Argon gas was used to increase the titanium sputtering rate and to control the nitriding rate of cage [10]. Hydrogen was used to reduce superficial oxides and to increase the process efficiency [11]. Pressure, temperature and duration of deposition were fixed at 1.5 mbar, 450 °C and 120 minutes, respectively.

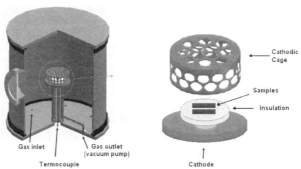

Figure 1. Schematic illustration (not scale) of a plasma reactor in cathodic cage configuration.

The film deposition process was monitored by Optical Emission Spectroscopy (OES), using an Ocean Optics USB 4000 spectrograph. X-Ray Diffraction analyses were performed with a Shimadzu XRD-6000 diffractometer using Cu radiation Kα and an accessory for Grazing Incidence X-ray Diffraction (GIXRD). The deposited films were also analyzed by Atomic Force Microscopy (AFM) in contact mode with a Shimadzu model SPM 9600 microscope. Optical properties were analyzed by transmittance, reflectance using a Varian Cary-5000 spectrophotometer with an integration sphere. Furthermore, transmittance values of laser HeNe (λ = 632.8 nm) were measured at three different points on the film surface (A = 0.5 cm, B = 1.25 cm and C = 2.0 cm). Dispersion curves of the deposited films, n(λ) and k(λ), were obtained using Sopra GES-5E Spectral Ellipsometer. These data were used in determining the deposited film thickness.

Results and Discussion

Optical Emission Spectroscopy

The results of the optical emission spectroscopy (Figure 2) provided evidence that behavior of plasma species was not linear. The lines for the Hα (2s – 3p) species in 656.3 nm and the band for the N_2^+ ($B^2\Sigma_u^+ - X^2\Sigma_g^+$) species in 391.4 nm had maximal value of luminous intensity in the plasma atmosphere that contained 12.5% of H_2. The band for the N_2 ($C^3\Pi_u - X^2\Pi_g$) species in 337.1 nm decreased with increasing content of H_2.

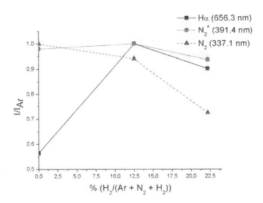

Figure 2. Effect of hydrogen content on the luminous intensity of plasma species (relative to Ar 750.4 nm) during the film deposition by cathodic cage plasma.

The N_2^+ and Hα luminous intensities had a maximum in plasma atmosphere containing 12.5 % of H_2 gas. This result is due to hydrogen action, by introducing effective collisions enhances excitation of N_2. When the H_2 content was increased above 12.5%, however, the nitrogen species amount decreased due to a decrease in the N_2/H_2 ratio [13].

X-Ray Diffraction

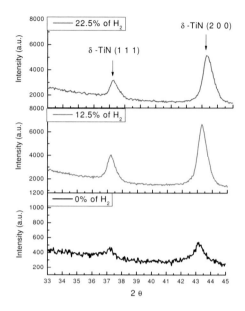

Figure 3. X-Ray pattern of films grown in plasma atmosphere containing various $H_2/(Ar + N_2 + H_2)$ ratios.

The X-ray diffraction patterns (Figure 3) exhibited two peaks at 37.3° and 43.3° corresponding to δ-TiN (111) and δ-TiN (200), respectively. TiN films deposited in the presence of 12.5% of H_2 gas in a plasma atmosphere exhibited the maximum intensity observed in the present study. It is related to higher amount of stoichiometric TiN in film. The TiN peaks for films deposited in a plasma atmosphere without H_2 gas had the lowest intensity observed in the present study intensity and they are wider.

Table I. Lattice parameter values for TiN films as a function of H_2 gas content in plasma.

H_2 content (%)	Lattice Parameter (nm)
0	0.4240
12.5	0.4133
22.5	0.4151

The film grown in a plasma atmosphere without H_2 had the highest lattice parameter (Table I); this increase was due to the highest content of nitrogen atoms in the titanium nitride compound. This result, therefore, provided evidence that there is a relation between the nitrogen content in the lattice and the content of nitrogen $(N_2/(N_2+Ar+H_2))$ in the plasma atmosphere.

The highest amount of stoichiometer TiN in films deposited in a plasma atmosphere containing 12.5% of H_2 is related to the highest luminous intensity of the $H\alpha$ species. The hydrogen enhances the deposition rate because of the chemical reduction of the superficial oxide. The sputtering rate in the cage was higher, because it was easier to sputter from the cage titanium particles than titanium nitride compound [13-14].

Atomic Force Microscopy (AFM)

The results obtained from AFM analysis provided evidence regarding with deposited film contained nano-size particles. The particle size was higher when the film was deposited in a plasma atmosphere with 12.5% of H_2 (Figure 4).

Table II shows values of the $I_{H\alpha}/I_{Ar}$ line intensity obtained from OES spectra (Figure 2), full width half maximum (FWHM) values obtained from XRD patterns (Figure 3) and roughness values were obtained from results regarding surface nanotopography (Figure 4). There was a correlation between the luminous intensity of $H\alpha$ species and the characteristics of grown films, since roughness increased when the $I_{H\alpha}/I_{Ar}$ decreased. The highest deposition rate evidenced by the intensity of TiN (200) diffraction peak and the highest crystallinity (demonstrated by the FWHM of TiN peak obtained from X-Ray Pattern) occurred when the $I_{H\alpha}/I_{Ar}$ was higher.

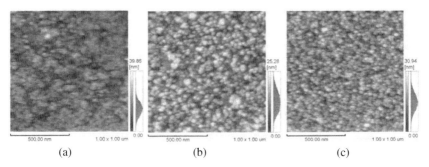

(a)   (b)   (c)

Figure 4. Representative nanotopograph using AFM analysis of the surface of films grown in different plasma atmospheres, specifically, (a) without H_2 gas, (b) containig 12.5% of H_2 gas and (c) containing 22.5% of H_2 gas.

Table II. Characteristics of the crystalline structure and topography of films deposited in plasma atmosphere containing various amount of H_2.

H_2 content (%)	$I_{H\alpha}/I_{Ar}$	Particles diameter (nm)	Intensity δ–TiN (200)	FWHM	Roughness	
					Ra	Rms
0	30267	28.6 ± 4	264	2.75	4.49	3.60
12.5	53643	33.6 ± 5	5176	0.91	4.27	3.53
22.5	48337	29.2 ± 4	3158	1.05	4.32	3.49

Transmittance spectra

The results of the transmittance spectra (Figure 5) show in the absence of H_2 in plasma atmosphere, there was increasing in the transmittance values. This result explained when it takes into consideration the higher N/Ti ratio in the titanium nitride film [7]. Furthermore, the deposition rate was lower in the absence H_2 in the plasma atmosphere, because of a more efficient nitriding in the cage that result in reduced sputtering rate. The process used in the present study is similar to that of target poising which occurs in reactive sputtering techniques [15]. Similar results were also reported by Priest [14], who obtained a thicker film after pre-treatment in a plasma atmosphere of H_2.

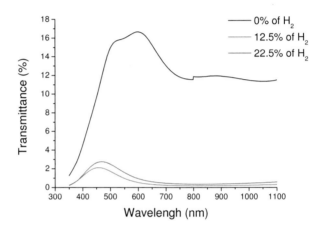

Figure 5. Transmittance spectra for films deposited in plasma atmosphere of various $H_2/(Ar + N_2 + H_2)$ ratios.

The variation of transmittance results acquired using a monochromatic laser in three different positions on sample surfaces (Table III) indicated that the film was not homogeneous when deposited in plasma atmosphere without H_2.

Table III. Monochromatic laser transmittance values obtained at different positions (designated A, B and C) on the surface.

H_2 amount (%)	Transmitance at point A (%)	Transmitance at point B (%)	Transmitance at point C (%)
0	5.64564	8.71782	11.91731
12.5	0.23008	0.23331	0.21106
22.5	0.55752	0.61174	0.49967

Reflectance spectra

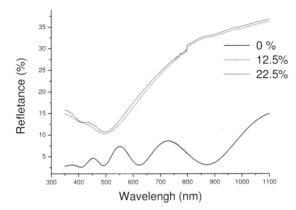

Figure 6. Spectra of light reflectance for films deposited in a plasma atmosphere in the presence of various $H_2/(Ar + N_2 + H_2)$ ratios.

The reflectance spectrum for films deposited in a plasma atmosphere without H_2 differed from those of films obtained in plasma atmosphere with H_2 (Figure 6). This result suggests that the phase of the film grown in plasma without hydrogen is less stoichiometer than others. Moreover, the reflectance spectra for films deposited in presence of 12.5 and 22.5% of H_2 in the plasma atmosphere were similar to those obtained by magnetron sputtering deposition and reported in the literature [16]. The displacement in wavelength of the maximal transmittance and minimal reflectance resulted from the difference in lattice parameter.

Ellipsometry results

The ellipsomtry results (Table IV) provide evidence that the film deposited in plasma atmosphere containing 12.5 or 22.5% of H_2 had similar thickness of approximately 21 nm. It was not possible to find a mathematical function to calculate optical constants and thickness for film obtained without hydrogen, because there was an increase in the N/Ti ratio in film, which does not allow to make an adjust for the stoichiometry TiN.

Table IV. Thickness (φ) and composition values of films deposited in the presence of H_2 gas in the plasma atmosphere.

H_2 amount (%)	φ (nm)	TiN (%)	Titanium (%)	Porosity (%)
12.5	21.4 ± 5	40.7	16.9	42.4
22.5	21.1 ± 5	3.9	15.9	44.1

The refraction indices and extinction coefficients (Figure 7) were similar. These small variations in refraction indices result in transmittance and reflectance spectra modifications. It occurred because of variation in the content nitrogen in lattice of titanium nitride films.

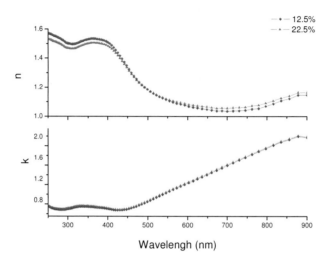

Figure 7. Dispersion plots, real part n (λ) and imaginary part k (λ), obtained using spectral ellipsometry analysis for film deposited in plasma atmosphere with various $H_2/(Ar + N_2 + H_2)$ ratios.

Conclusions

The cathodic cage plasma technique provides deposition of titanium nitride thin films on glass substrates. These TiN films have nano-size particles. The microstructure and optical properties of the deposited films depend on the plasma species density.

The luminous intensities of the Hα and N_2^+ species are not directly dependent on the flow rate of H_2 gas. The maximum value of the luminous intensity was obtained when the H_2 gas containing in plasma atmosphere was 12.5%; the luminous intensity decreased as the H_2 gas containing in the plasma increased to 22.5 %.

Several aspects of the deposited film correlated to the luminous intensity of Hα species. Higher film deposition rate and higher crystallinity were obtained when the $I_{H\alpha}/I_{Ar}$ ratio was higher. The deposited film roughness increased with increasing luminous intensity of Hα. Furthermore, the transmittance decreased when the H_2 gas content in plasma increased while reflectance decreased.

References

1. C. Alves Jr. et al. "Use of cathodic cage in plasma nitriding," *Surface and coatings Technology*, 201 (2006), 2450-2454.
2. R. R. M. Sousa et al. "Uniformity of temperature in cathodic cage technique in nitriding of austenitic stainless steel AISI 316," *Surface Engineering*, 24 (2008), 313-318.

3. R. R. M. Sousa et al. "Cathodic cage nitriding of samples with different dimensions," *Materials Science and Engineering A*, 465 (2007), 223-227.
4. S. C. Gallo, H. Dong, "On the fundamental mechanisms of active screen plasma nitriding," *Vacuum*, 84 (2010), 321-325.
5. Z. Peng, H. Miao, L. Qi, S. Yang, C. Liu, "Hard and wear-resistant titanium nitride coatings for cemented carbide cutting tools by pulsed high energy density plasma," *Acta Materialia*, 51 (2003), 3085-3094.
6. Ph. Roquiny, F. Bodart, G. Terwagne, "Colour control of titanium nitride coatings produced by reactive magnetron sputtering at temperature less 100°C," *Surface and Coatings Technology*, 116-119 (1999), 278-283.
7. G. B. Smith, A. Ben-David, P. D. Swift. "A new type of TiN coating combining broad band visible transparency and solar control," *Renewable Energy*, 22 (2001), 79-84.
8. Y. Cheng, Y. F. Zheng, "Formation of TiN on biomedical TiNi shape memory alloy by PIIID," *Materials Science and Engineering A*, 434 (2006), 99-104.
9. M. Wittmer, J. R. Noser, H. Melchior, "Characteristics of TiN gate metal-oxide-semiconductor field effect transistors," *Journal of Applied Physical*, 54 (1983), 1423-1428.
10. M. B. Sankar et al. "Optimization of plasma parameters for high hate deposition of titanium nitride films as protective coating on bell-metal by reactive sputtering in cylindrical magnetron device," *Applied Surface Science*, 254 (2008), 5760-5765.
11. M. Tamaki, Y. Tomii, M. Yamoto, "The role of hydrogen in plasma nitriding: Hydrogen behavior in the titanium nitride layer," *Plasma and Ions,* 3 (2000), 33-39.
12. Y. Hirohata, N. Tsuchiya, T. Hino, "Effect of mixing of hydrogen into nitrogen plasma," *Applied Surface Science*, 169-170 (2001), 612-616.
13. C. A. Figueroa, F. Alvarez, "On the hydrogen etching mechanism in plasma nitriding of metals," *Applied Surface Science*, 253 (2006), 1806-1809.
14. J. M. Priest, M. J. Baldwin, M. P. Fewel, "The action of hydrogen in low pressure r.f. –plasma nitriding," *Surface and Coatings Technology*, 145 (2001), 152-163.
15. I. Safi, "Recent aspects concerning DC reactive magnetron sputtering on thin films: a review," *Surface and Coating Technology*, 127 (2000), 203-217.
16. A. J. Perry, M. Georgson, W. D. Sproul, "Variation in the reflectance of TiN, ZrN and HfN," *Thin Solid Films*, 157 (1988), 255-265.

CATHODE RAY TUBE GLASSES IN GLASS CERAMICS

M. Reben* and J. Wasylak

AGH University of Science and Technology, Faculty of Materials Science and Ceramics,
al. Mickiewicza 30, 30-059 Cracow, Poland
*e-mail;manuelar@agh.edu.pl

ABSTRACT

A series of sintered glass-ceramic and glass-ceramics formed by bulk crystallization of Cathode Ray Tube (CRT) glass cullet mixed with calsiglass and calumite was investigated. Glass-ceramics materials prepared by the controlled crystallization of glass have a variety of established uses depending on their uniform reproducible fine-grain microstructure, absence of porosity and wide-ranging properties that can be tailored *via* composition and heat treatment. The laboratory results demonstrated that it is possible to produce a glazed tile, where the glaze deployed CRT glass cullet instead of commercial ceramic frits. Glass-ceramics were obtained at low temperature and short reaction times with the lead-free panel glass as one component.

The crystallization experiments were carried out on the basis of differential thermal analysis (DTA), X-ray diffraction (XRD), and scanning electron microscopy (SEM). The resulting glass-ceramic had sufficiently large compressive strength for practical use.

1. INTRODUCTION

The aim of this study was to find possible applications, in which glass materials from Cathode Ray Tubes (CRT) can be utilized in ceramic and glass industry. Color display faceplates (panels) of CRT are typically made from a lead-free barium–strontium glass [1-3]. In ceramic industry the CRT glasses, however, could be applicable in many processes as secondary raw materials The comparison of ceramic raw materials to the composition of CRT glass materials gives one possibility to introduce secondary raw material to the manufacturing process [4-6].. The sintering method has recently found to bring additional advantages in the development of unusual crystal phases or highly dense glass-ceramic matrix composites [7,8]. The research on sintered glass-ceramics, developed by "sinter-crystallization", is focused on a relatively new processing, exhibiting advantages in both vitrification and crystallization when compared to the traditional methods. When fine glass powders (typically <80 μm) are used, the crystallization associated to the viscous flow sintering is generally fast (30-60 min at 900-1000°C), owing to a surface mechanism of nucleation, active even without nucleating agents (TiO_2, Cr_2O_3, etc.) in the glass formulation, and it may be achieved even for unusual crystal phases (such as alkali feldspars and feldspathoids. A key benefit of sintered glass-ceramics relies on rapid crystallization [9,10].

The powder sinter-crystallization technique is considered an alternative for the production of glass-ceramics, which allows the production of specimens with complicated shape and different sizes [11-13]. Moreover, no nucleation step is required and parent glasses with low purity and degree of homogenization can be used. During sinter-crystallization, the densification and the crystallization take place in the same temperature interval. As a result, when the crystallization trend is too high, the

sintering rate may be considerably reduced, leading to residual porosity and decreasing in the mechanical properties. For this reason glasses with lower crystallization
rate are usually used [14].

2. METHODS AND PROCEDURES

2.1. Glass-ceramics formed by bulk crystallization

Cleaned panel glass from dismantling plants of TV kinescopes was used. The as-received panel glass was dry milled for 3h in a laboratory ball mill and sieved under 80 μm. The glass composition were prepared by dry mixing of CRT panel glass with different amounts of calumite and calsiglass. All the chemicals were mixed properly to ensure the homogeneity. The glasses have been obtained by melting 50g batches in platinum crucibles in an electric furnace at the temperature 1450°C in air atmosphere. The melts were poured out onto a steel plate forming a layer about 2 to 5mm thick. The chemical analysis of raw materials was performed by inductively coupled plasma. The composition of the investigated glasses is listed in Table I. The ability of the obtained glasses to crystallization was determined by DTA/DSC measurements conducted on the Perkin-Elmer DTA-7. The ability of glasses for crystallization was measured by the values of the thermal stability parameter of glasses ($\Delta T = T_{cryst.} - T_g$). Glasses revealing the crystallizations events were selected for further thermal treatment. To obtain glass-ceramics they were subjected to heating for 5h at the temperature of the maximum crystallization events, respectively. The kind and the size of the formed crystallites were examined by XRD and SEM methods, respectively.

Table I. Glass compositions (wt%)

Glass ID	Mixture (wt%)		
	CRT panel glass	Calsiglass	Calumite
MC1	80	20	-
MC2	70	30	-
MC3	60	40	-
MC1C	80	-	20
MC2C	70	-	30
MC3C	60	-	40

2.2. Sintered glass-ceramic

One composition MC2C was considered to sintering process (see Table II). This composition corresponds to mixtures of wastes such as: CRT glass cullet and calumite. Sintering experiments were performed on square compacts (5x5 cm obtained by uniaxial pressing (at 75 MPa) of fine powders in a square steel die at the room temperature, without any binder. The compacts were subjected sintering treatments, consisting of a heating stage at 10°C/min, a holding stage at about 910°C for different times (in furnace). The samples were carefully polished to a 6 μm finish, by using abrasive papers and diamond paste. Two point bending were performed by using an automatic pressure test machine YAW-3000 C TYPE, TIME GROUP INC. Powdered glass-ceramics were investigated by X-ray diffraction.

3. RESULTS AND DISCUSSION

The chemical compositions of the CRT panel glass, calumite and calsiglass are presented in Table II. Calumite and calsiglass are granular material that are 99% glassy in nature. It has traditionally been viewed as an alumina source, but it is a valuable source of all the major glassmaking oxides.

Table II. The chemical composition of used raw materials (wt%).

Oxides	CRT panel glass	calsiglass	calumite
SiO_2	61.82	39.60	37.8
CaO	-	**43.20**	37.2
MgO	0.01	**7.30**	12.7
Al_2O_3	2.03	**6.73**	9.0
Fe_2O_3	0.07	0.15	0.23
TiO_2	0.06	0. 20	0.87
Na_2O	**7.95**	0.60	0.30
K_2O	**7.35**	0.37	0.68
BaO	**7.27**	-	-
SrO	**10.80**	-	-
ZrO_2	1.70	-	-
SO_3	-	-	0.08
Others	0.94	1.85	1.14

By DTA curves performed on the quenched glasses two kind of thermal events were carried out: an infection point (T_g) corresponding to glass transition temperature, an exothermic peaks ($T_{cryst.}$) indicating the crystallization (Fig.1). In case of CRT glass panel, the thermal behavior can be explained by the relevant quantities of alkaline oxides (~15 wt%) (Table III). The DTA curve of CRT glass panel did not present exothermal events, indicating that the glass does not crystallize in the temperature range studied, as confirmed by the low percentages of CaO, MgO and Fe_2O_3, typical modifier oxides that induce crystallization (Fig.1.). The increase of CaO and MgO content in the examined glasses causes an increase of the transformation temperature T_g and increase of the specific heat (ΔCp) accompanying the glass transition region (Table III). Moreover due to the increasing of CaO and MgO content the temperature of the maximum events of crystallization is shifted towards a higher temperatures. This is evidence of increasing ability of the glass for crystallization, manifested by decreasing value of the index of thermal stability of the glass ΔT (Table III). Temperature of the reversible effect of glassy state transition depends on the cation type and increased with presence of cations having high ionic strength (Ca-O, Mg-O) (Table II.). Effect of glassy state transformation is influenced by a ratio of its non-ordered structure, expressed by a ratio of ordering of the bond forming elements. It has been proved that the influence is expressed by the different value of the specific heat ΔC_p during the transition. Introduction of modifiers Ca^{2+} and Mg^{2+} instead of decreasing amount of the Si^{4+} ions into glasses results in gradual increase of the parameters, which characterize glassy state transformation (T_g, ΔCp,) (Table III). Modifier in form of Mg^{2+} ions, which is present in the calumite and calsiglass influences the increase of the T_g to greater extend than Ca^{2+} ions. The presented changes of parameters, which characterize the glassy state transition confirm the influence onto the transition of the chemical bonding type present in the glass structure. Changes of the parameters, which characterize the glassy state transition caused by introduction of the modifiers are related to breaking of the chemical bonding, or to displacements of atom groups in order to transit from non-ordered amorphous state to crystalline state, or in order to relax the internal stresses caused by non-ordered structure of the glass.

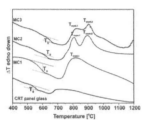

a) CRT+ calsiglass

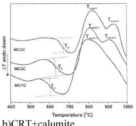

b)CRT+calumite

Figure1. DTA curves of glasses.

Table III. Thermal characteristic of glasses.

ID	T_g [^{0}C]	ΔCp [J·g$^{-1\cdot0}$C^{-1}]	$T_{cryst.1}$ [^{0}C]	ΔT_1 [^{0}C]	$T_{cryst.2}$ [^{0}C]	ΔT_2 [^{0}C]
MC1	600	0,223	816	216	-	-
MC2	619	0,318	814	195	909	290
MC3	655	0.429	818	163	913	258
MC1C	603	0,231	798	195	-	-
MC2C	628	0,458	800	172	920	292
MC3C	668	0,513	803	135	925	257

The kind of crystallizing phases of investigated glasses after the heat treatment were selectively analyzed by using a powder X-ray diffraction (XRD) (Fig.2a). XRD measurements of glass-ceramic MC1C-MC3C (CRT+calumite) obtained by heat treating the glass at its first crystallization temperature showed several sharp diffraction peaks overlapped on the amorphous hump. It is easy to identify that the crystal phase is wollastonite ($CaSiO_3$) and diopsyde strontian ($Ca_{0.75}Sr_{0.15}Mg_{1.1}(Si_2O_6)$). In case of glass MC1C after thermal treatment at the 798^{0}C, $CaSiO_3$ was the only crystal phase separated from the vitreous matrix which appeared as a surface crystallization. With the calumite content increased (MC2C and MC3C glasses) two kind of crystalline phases appeared after the heat treatment. Wollastonite ($CaSiO_3$) as a surface crystallization and diopsyde strontian ($Ca_{0.75}Sr_{0.15}Mg_{1.1}(Si_2O_6)$ as a bulk crystallization. Due to the increasing of CaO and MgO content the increasing ability of the glass for crystallization was observed what was manifested by decreasing value of the index of thermal stability of the glass ΔT (Table III). Increasing CaO and MgO content improves crystallinity (i.e. increasing intensity of peaks). In case of glass-ceramic MC3 (CRT+calsiglass) obtained by heat treating the glass at its first and second crystallization effect temperature crystal phase is wollastonite (CaSiO₃). In the glass–ceramics with 3h crystallization times at the 820^0C and 903^0C, CaSiO₃ was the only crystal phase separated from the vitreous matrix (Fig.2b).

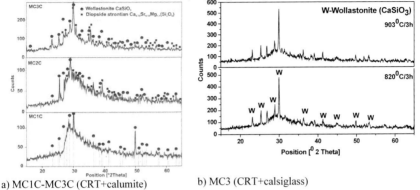

a) MC1C-MC3C (CRT+calumite)
b) MC3 (CRT+calsiglass)
Figure 2. XRD patterns of glass ceramic.

The morphology of the crystals formed after the heat treatment of glass MC3C and MC3 was studied by SEM microscopy. The influence of CaO on the microstructure of GCs is shown in Fig. 3. The microstructures of glass MC3C after the heat treatment at 803^0C/5h and MC3 heat-treated at 820°C/3h. showed well formed crystals (Fig.3).

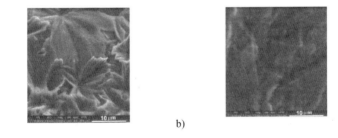

a)   b)

Figure.3. SEM images of glasses MC3 after the thermal treatment at 820°C/3h and MC3C after thermal treatment at the 803^0C/5h (b).

As reported by Table III, one of investigated glass (MC2C) exhibited a second crystallization temperature close to 920°C. The temperature 910° C was chosen as reference temperature for sintering treatments of fine glass powders. Fig.4 reports XRD results of conventional sintering (10°C/min heating rate) of MC2C fine powder at the temperature 910° C for 0, 5, 10 h. One compact was sintered at 1000 ° C for 5h. Prolonged treatments caused minor variation of peaks in the X-ray diffraction. Glass-ceramics from MC2C fine powder are based on three main crystal phases, as illustrated in Fig.4. These phases correspond to diopside (calcium-magnesium silicate, $CaMgSi_2O_6$), calcium silicate

(CaSiO$_3$) and akermanite (Ca$_2$MgSi$_2$O$_7$). Fig. 4 reports the results of conventional sintering (10°C/min heating rate): after 5 h the crystallization was practically completed. It is interesting to note that for 0 h sintering (after reaching sintering temperature), a very slight crystal precipitation is visible. Prolonged heat treatment of compacts suggest consecutive transformation of the parent glassy material into crystalline phases. The degree of crystallinity of formed phases change with time evolution. The XRD pattern of the sample obtained at to 910°C/5h is composed of more visible akermanite phase. A further heat treatment time increase up to 10h enhances the intensity of diopside (calcium-magnesium silicate, CaMgSi$_2$O$_6$) and calcium silicate (CaSiO$_3$) phases are observed. Structural rearrangements might take place in the earlier stage of crystallization process. An increase of sintering temperature up to 1000°C caused the appearance of only one phase corresponds to a calcium silicate CaSiO$_3$, the amount of residual glass is also observed.

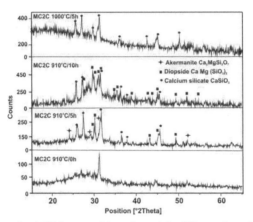

Fig.4. XRD patterns for MC2C composition sintered for different times (10°C/min heating rate);

Physical and mechanical properties of glass-ceramics obtained by sintering CRT glass cullet mixed with are listed in Table IV. Densities of the glass-ceramics were in the range of 2.54–2.61 g/cm^3, and increased with prolonged treatments time. Porosity decreased with an increase of sintering time. It may be noted that the bending strength was significantly lowered by the porosity. For mechanical properties, four-point bending strength of glass–ceramics is a function of different heat treatment time. The achieved strength corresponds to the limit for other glass-ceramics from wastes (the strength of Slagsitalls is in the range of 90-130 MPa) [15,16], although the reported process is much simpler. The residual porosity of MC2C composition sintered at 910°C for 10 h specifically intended to favour the crystallization, did not exceed 3% . The presence of crystalline phases particularly diopside (calcium-magnesium silicate, CaMgSi$_2$O$_6$) and calcium silicate (CaSiO$_3$) in compacts MC2C after 5 and 10 h of sintering increases bending strength. The heat treatment of the MC2C at the sintering temperature 1000°C, causes increasing the amount of amorphous phase in the materials, and tendency to foaming what affects the mechanical properties.

Table. IV. Mechanical and physical properties of the MC2C sintered glass-ceramics

Composition	MC2C	MC2C	MC2C	MC2C
Sintering temperature [°C] and holding time [h]	910/5	910/10	910/0	1000/5
Elastic modulus [GPa]	73.1	81.2	56.8	48
Bending strength [MPa]	98	104	78	63
Porosity (vol%)	6	3	12	15

4. CONCLUSIONS.

Based on the results obtained from this study, the following conclusions can be reached:
1. The CRT panel glass with calumite and calsiglass during heating demonstrates two crystallization events, while in case of CRT panel glass crystallization effect does not appear.
2. Obtained glasses are characterized by greater crystallization ability, due to the highest CaO and MgO content.
3. The mechanical properties of sintered glass-ceramics exceed those of porcelain stoneware
tiles, obtained by conventional sintering treatments, the achieved products could find profitable applications in the field of floor and wall tiles. Moreover, low cost and availability of the raw materials make them very attractive from the economic point of view.

ACKNOWLEDGEMENTS

The work was supported by Grant No. N R08 0025 10, of the Ministry of Science and Informatisation of Poland.

5. REFERENCES

1. C.S. Poon, "Management of CRT glass from discarded computer monitors and TV sets," Waste Management, 28 (9) (2008), 1499- 1499.

2.. F. Andreola, "Cathode ray tube glass recycling: an example of clean technology", Waste Management & Research, 23 (4) (2005), 314-321.

3. F. Andreola, et al., "CRT glass state of the art: A case study: Recycling in ceramic glazes," Journal of the European Ceramic Society, 27, (2-3) (2007), 1623-1629.

4. A. Karamanov, G. Taglieri, and M. Pelino, "Iron-rich sintered glass-ceramics from industrial wastes", J. Am. Ceram. Soc, (82) (1999), 3012–3016.

5. F. Andreola, L. Barbieri, A. Corradi, A. M. Ferrari, I. Lancellotti and P.Neri, "Recycling of EOL CRT glass into ceramic glaze formulations and its environmental impact by LCA approach", (12) (6) (2007), The International Journal of Life Cycle Assessment.

6.E. Bernardo, R. Castellan, S. Hreglich, "Sintered glass-ceramics from mixtures of wastes", Ceram. Int. (33) (2007) 27-33.

7. F. Andreola, L. Barbieri, F. Bondioli, I.Lancellotti, P Miselli and A. M. Ferrari, "Recycling of Screen Glass Into New Traditional Ceramic Materials", International Journal of Applied Ceramic Technology, 7 (6) (2010) 909-917.

8. E. Bernardo, F. Andreola, L. Barbieri, I.Lancellotti, "Sintered glass-ceramics and glass-ceramic matrix composites from CRT panel glass", J. Am. Ceram. Soc. 88 (2005), 1886- 1891.

9.E. Bernardo, M. Varrasso, F. Cadamuro, S. Hreglich, "Vitrification of wastes and preparation of chemically stable sintered glass-ceramic products", J. Non-Cryst. Sol. 352 (2006), 4017-4023.

10.G. Brusatin, E. Bernardo, G. Scarinci, "Sintered glass-ceramics from waste inert glass", Proc. IV International Workshop VARIREI (Valorization and recycling of industrial residues), L'Aquila, Italy, June 2003.

11. E. Bernardo, E. Bonomo, A. Dattoli, "Optimisation of sintered glass–ceramics from an industrial waste glass", Ceram. Int. 36 (2010) 1675–1680.

12. E. Bernardo, L. Esposito, E. Rambaldi, A. Tucci, Y. Pontikes, G.N. Angelopoulos, "Sintered esseneite–wollastonite–plagioclase glass–ceramics from vitrified waste",J. Eur. Ceram. Soc. 29 (2009) 2921–2927.

13. E. Bernardo, "Micro- and Macro-cellular Sintered Glass-ceramics from Wastes," J. Eur. Ceram. Soc., 27 (2007) 2415-2422.

14. L. Maccarini Schabbach, F. Andreola, E. Karamanova, I. Lancellotti, A. Karamanov, L. Barbieri , "Integrated approach to establish the sinter-crystallization ability of glasses from secondary raw material", Journal of Non-Crystalline Solids 357 (2011) 10–17.

15. P. Colombo, G. Brusatin, E. Bernardo, G. Scarinci, "Inertization and reuse of waste materials by vitrification and fabrication of glass-based products", Current Opinion in Solid State and Materials Science 7 (2003), 225-239.

16. R.D. Rawlings, J.P. Wu, A.R. Beccaccini, "Glass-ceramics: Their production from wastes - A Review", J. Mat. Sci. 41 (2006), 733-761.

APPLICATION OF ALUM FROM KANKARA KAOLINITE IN CATALYSIS: A PRELIMINARY REPORT

L.C. Edomwonyi-Otu[1,2], B.O. Aderemi[2], A.S. Ahmed[2], N.J. Coville[4], M. Maaza[3]

[1] Chemical Engineering Department, University College London, Torrington Place, London. WC1E 7JE, UK.
[2] Chemical Engineering Department, Ahmadu Bello University, Zaria, Nigeria 870001
[3] NanoSciences Laboratory, Materials Research Department, iThemba LABS – National Research Foundation, Somerset West 7129, South Africa
[4] School of Chemistry, University of the Witwatersrand, Private Bag X3, P.O. Wits, 2050, South Africa

ABSTRACT

This paper presents a preliminary report on the preparation and testing of a catalyst formulated from kaolin matrix for the oxidation of benzyl alcohol to benzaldehyde. The vast deposit of kaolinite in Nigeria has remained untapped with insignificant research effort in this direction. The bimetallic catalyst (Pt-Ag) on partially decomposed alum having a BET surface area of 32.345 m^2/g, was found to be 95 % selective within the reaction time of 30 hours. The TEM result shows a good metal dispersion for this system with an average particle size of 8nm, which confer on it the character of a nanocatalyst, the result of a careful preparation procedure. The main attraction of this catalyst system is the ease of preparation and the resulting activity and selectivity.

INTRODUCTION

Several types of catalysts have been known and are still been developed to meet emerging industrial needs, from the simple enzymes in wine/food fermentation (in use for several centuries) to the complex enzymes/catalysts used in complex synthesis such as in the pharmaceutical industries, food processing and manufacturing and even in environmental pollution control. Examples of catalysts are finely divided iron in the Haber process for ammonia production, nitrogen dioxide in the production of sulfur trioxide from sulfur dioxide and oxygen, platinum in automobile catalytic converters for the reduction of poisonous carbon monoxide and nitrogen oxides to carbon dioxide, nitrogen and water (an important process in air pollution prevention and control), zeolites in oil refining and petrochemicals, enzymes in bio-catalysis (Henry et al, 1997) for drug production in the pharmaceutical industries, amongst several others.

The effectiveness of most of these catalysts are enhanced by materials called promoters or supports some of which includes alum, alumina, silica, activated carbon, zeolites, clay etc. They possess good characteristic thermal stability, mechanical strength, large support/surface area which they contribute to enhance the catalysts. Some of these supports (alumina, zeolites) can themselves act as catalysts depending on their method and purpose of preparation (Bartholomew, 2005).

One of the most important reactions in organic chemistry and by extension in chemical and or petrochemical processes is the oxidation of alcohols to carbonyl compounds because of their huge commercial significance (Rhodium.org). This article therefore presents a preliminary report on the potentials of using alum (Edomwonyi-Otu and Aderemi, 2009; 2010), a product of

Kankara kaolinite (one of Nigeria's kaolin deposits), as a support in catalyzing the conversion of benzyl alcohol to benzaldehyde. Alum has also been reported as a catalytic support (Aderemi and Hameed, 2009). The main aim of the synthesis of the Pt-Ag/alum catalysts, now used in this reaction, is in the reduction of NOx in automobile exhausts. The authors are not aware of any previous preparation, characterization and application of alum from Kankara kaolinite in catalysis.

EXPERIMENTAL

Preparation
        Ammonium alum was produced from Kankara kaolinite as reported in our previous investigation (Aderemi et al, 2009; Edomwonyi-Otu and Aderemi, 2009; 2010). In a typical preparation, a calculated amount of platinum Pt and silver Ag precursor salts (Tetraammineplatinum(II) nitrate 99 %, Strem Chemicals, USA ; and Silver nitrate 99.9 %, BDH London. These chemicals were used as received without further purification) that will give the desired amount of Pt and Ag on the dried catalyst was added to10g of alum in 50ml solution in a beaker (Pyrex). The mixture was gently stirred while been heated at 80°C in a magnetic heat stirrer (Adams 180, UK) for 2 hours, and later transferred to a furnace (Nobertherm 3000, Germany) where it was program-heated at 250°C for 4 hours and then at 500°C for 3 hours and later cooled to room temperature before it was grinded to powder and kept air-free.

Characterization
        The catalyst sample so prepared was subjected to some characterization techniques using BET surface area analyzer (Micromeritis, Tristar 3000), Transmission electron microscopy (Technai T20, FETEM), X-ray diffractometry (XRD) in a θ-2θ mode with CuKa 1 (AXS Bruker, λ=1.54056Å), FTIR-ATR spectroscopy via Perkin Elmer Spectrum 1000 FT-IR Spectrometer, Temperature programmed reduction (TPR) and Particle induced X-ray emission spectroscopy (GEO-PIXE) to determine its surface area, micrograph, crystallography, bond type/vibrations, reduction nature and elemental composition respectively.

Catalyst testing
        0.1g of catalyst sample was activated for 1 hour at 140oC in a round bottom flask (Pyrex) reactor placed in an oven. 1ml Benzyl alcohol (Aldrich, Germany) plus 10ml Dioxan 1,4 (Aldrich, Germany) were added to the activated catalyst in the reactor and placed in an oil bath maintained at 80°C. The benzyl alcohol and Dioxan 1,4 were used as received without further purification. The reaction mixture was stirred continuously at 200rpm while high purity oxygen gas (Air-Liquid, SA) was bubbled into it via an oxygen balloon with a controlled flow. The reaction was left for about 30 hours during which time aliquots were continuously taken to be able to determine whether the reaction had progressed and/or completed. The aliquots were analyzed using a gas chromatograph (Varian Autograph, Varian 3000 GC) connected to a PC.

RESULTS AND DISCUSSION
        Table 1 shows the major elemental composition of the catalyst sample. The measurement inaccuracy of the PIXE is about 1% which implies that the values obtained are a true reflection of the catalyst composition which was intended to be 5%Ag and 3%Pt by weight. The difference

can be attributed to the purity level of the precursor salts and the alum, as well as decomposition conditions. It can be observed that the material contains a high percentage of sulfur (about 12.65 wt%), which was intended to provide a resistance to sulfur poisoning in the NOx reduction reaction for which this catalyst was originally designed (Bartholomew, 2005). The presence of phosphorus is an impurity which is highly associated with products of sulfuric acids. It should be noted that the heating/decomposition temperature for the preparation of this sample is not sufficient to completely decompose the structural sulfate as well as the phosphates present in the alum (Maczura et al, 1978).

Table 1 Elemental composition of synthesized catalyst sample by Geo-PIXE

Elements	Al	Si	P	S	K	Fe	Zn	Ag	Pt
Composition ppm)	166334	8750	15546	126495	1556	1122	5	48491	26467

Table 2 shows the result of the BET analysis. It can observed that the catalyst has a well-developed pore structure owing to the large pore size (larger by a high order of magnitude than the molecules of the reactants; benzyl alcohol, oxygen and the product benzaldehyde) and surface area which are key ingredients in the activity of a catalyst. The surface though good enough is low when compared to the traditional alumina based catalysts (Mastiller, 1978; Richardson, 1992; Coelho et al, 2007; Nampi et al, 2010). This low value could be attributed to the incomplete decomposition of the alum matrix to obtain alumina, resulting in the presence of sulfates and phosphates (table 1 referred) which may have taken up some of the spaces within the structure thereby reducing the total available space within the final product. This can be corroborated by the BET result obtained when a similar sample was calcined at 800°C for 4 hours and resulted in a surface area of about 190 m^2/g reported in one our previous reports (Edomwonyi-Otu et al, 2011a). From the result of the oxidation reaction, the pore structure and surface can be said to be good enough in this regard.

Table 2 Result of BET analysis

Analysis	BET surface area (m2/g)	Pore volume (cm3/g)	Pore size (nm)
	32.019	0.353123	44.11423

Figure 1 shows the crystallography of the catalyst sample. The peaks of Pt and Ag metals can be observed as given by the PDF files of the XRD analysis. Their peaks are observed to be close to each other because they belong to the same FCC crystal structure, for example Pt (2θ = 46.244 and 67.456) and Ag (2θ = 44.278 and 64.427). The presence of these metals in their free state is required for desired catalytic activities of the catalyst which was observed in the result of the oxidation reaction in which the catalyst was tested (Richardson, 1992). It can also be observed that there is the presence of Pt-Ag alloy peaks (2θ = 39 and 81.3) which may be the result of the high decomposition temperature (about 500°C). The incomplete decomposition may also explain why the individual metal peak did not disappear completely as reported in a similar investigation (Zhang et al, 2000). The presence of a weak silver oxide peak was observed near the sulfate peak with highest intensity. The silver oxide may be due to the incomplete decomposition of the silver sulfate formed as a product of a double decomposition of silver nitrate and sulfuric acid.

The presence of sulfates and phosphates are also detected by the analysis confirming the result of the PIXE analysis. For comparison, a sample of alum dried at the same condition was also analyzed just to help distinguish and confirm the presence of the active ingredient in the catalyst sample. It should be noted that the noise present at the background of the diffractogram is the result of the incomplete decomposition of the alum as already mentioned before, which could have resulted in the observed surface area. Some chemically bonded water is still been observed as OH, which as mention, has not been completely decomposed.

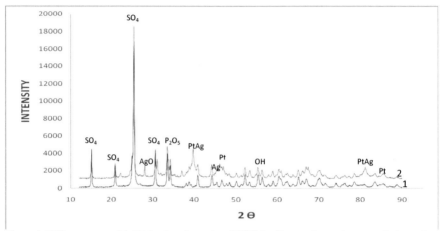

Figure 1 Diffractogram of **1**: Plain alum heated to 500°C for 3hours; **2**: catalyst sample heated to 500°C for 3hours

Further work need to be carried out to quantify the effect (positive or negative) of these sulfate, phosphate and hydroxyl species in this catalyst sample with regards to this reaction. Their presence was also further confirmed by the result of the FTIR-ATR that showed vibrational frequencies of OH and SO4 bands at 600 – 700/cm and 1050 – 1300/cm respectively, as observed in figure 2 which also agree with some other investigators (Andler and Kerr. 1965; Brown and Hope, 1996; Safaa and Mohamad, 2007) The protrusion at the side of the sulfate band can be attributed to a Ag-O stretching/vibrational band and it is corroborated by the weak peak of AgO observed near the strong sulfate peak in the diffractogram already discussed. The absence of any platinum band in the spectra also shows that the platinum compounds were completely decomposed to give the metal at the treatment condition. When the catalysts was calcined at 650°C for 4 hours, sintering of the platinum and silver metal was observed. The thermal stability of the catalysts is with regards to its reactivity and selectivity is been investigated.

Figure 3 shows different TEM micrograph of the catalyst sample taken at different spots. The fairly uniform spread/distribution of the active ingredients (Pt and Ag metals) is noteworthy, while the average particle size was measured to be about 8nm for both metal species, which confer the quality/character of a nanocatalyst on this sample (Bandyopadhyay, 2008; Nampi et al, 2010). The uniform distribution of active metals on their supports is highly desirable for catalytic activity and can account for the high activity and selectivity of this catalyst as observed in the oxidation of benzyl alcohol to benzaldehyde. This distribution may be the result of the careful preparation procedure undertaken in the course of this investigation and can be improved upon to obtain a better dispersion with highly reduced clustering of particles which is observed mostly in figure 3c. The clustering can also be attributed to the high concentration of the active ingredients present which also suggests the possibility of using far lower concentrations for this particular reaction. These results show the great potentials in the application of Kankara kaolinite products in different catalytic processes.

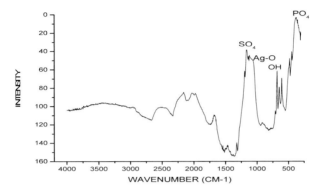

Figure 2 FTIR-ATR spectra of catalyst sample showing the vibrational bands of non-decomposed species present

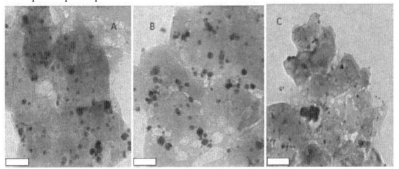

Figure 3 TEM micrographs (A,B & C) of catalyst sample showing the distribution of active Pt and Ag species on the dried alum support (dimension of embedded white rectangle is 50 nm).

The chromatograph from the GC showed the amount of the products present in the analyzed mixture. It revealed about 95% conversion of the starting material and also represented about 95% selectivity to benzaldehyde, while there was still the presence of insignificant amount of benzyl alcohol and other noises due to impurities. It should be noted that the time taken to achieve this level of activity is less than that taken by the conventional chromium (VI) reagents, dimethyl sulfoxide, permanganates, periodates or N-chlorosuccinimmide (NCS) catalyst used in this process (Hudlicky, 1990). These solid supported chromium reagents are also known to require difficult and laborious preparation, activation by subjecting the supported reagents to high temperature for prolonged periods of time, have a longer reaction time, produce low yields, and require elevated temperature for the reaction to work (Rhodium.org). This reaction(s) is of huge commercial importance due to the large number of derivable products such as agro chemical industries, dyestuffs, perfumes and other household products (Buonomenna and Drioli, 2008)

CONCLUSION
From the results of this investigation, the following conclusions can be made;

1. The treatment temperature is not sufficient to completely decompose the functional groups present in the catalyst sample as revealed by most of the characterization techniques

2. The surface area and pore size for this catalyst sample was adequate for the oxidation of benzyl alcohol to benzaldehyde, though can be improved upon.

3. The average particle size of 8nm confers the character of nanocatalysts on this catalyst sample and thus shows possibility for varied applications.

4. Careful preparation procedure resulted in catalyst with very good dispersion of active ingredients which most likely resulted in a selectivity of 95%.

5. Kankara kaolinite and its products hold great promise in catalytic applications with the ultimate result of economic contribution.

ACKNOWLEDGEMENT
The authors wish to acknowledge financial support from Nano Sciences African Network (NANOAFNET) and Petroleum Technology Development Fund (PTDF) for part of this work

REFERENCES

1. B.O. Aderemi L. C. Edomwonyi-Otu and S.S. Adefila, A new approach to metakaolin dealumination. *Australian Journal of Basic and Applied Sciences* (AJBAS). INSI net Publication. 3(3), 2243-2248 (2009)

2. B.O. Aderemi and B.H. Hameed, Alum as a heterogeneous catalyst for the trans-esterification of palm oil, *App Cat. A: General* 370 54-58 (2009).

3. H.H. Andler and P.F. Kerr, Variations in infrared spectra, molecular symmetry and site symmetry of sulfate minerals. The American Mineralogist, Vol, 50, January-February (1965).

4. A.K, Bandyopadhyay, Nanomaterials. New Age international Publishers 136-166 (2008).

5. C. Bartholomew, Catalyst Deactivation And Regeneration; In Kirk Othmer Encyclopedia of Chemical Technology Wiley Interscience New York, vol. 5, 256-262 (2005).

6. G.M. Brown and G.A. Hope, A SERS study of SO2-/Cl - ion adsorption at a copper electrode in-situ *J. of Electroanalytical Chem* 405 211-216 (1996).

7. M.G. Buonomenna and E. Drioli, Selective oxidation of benzyl alcohols to benzaldehydes using a membrane contactor. *Chem. Eng. Trans* 13 303-310 (2008).

8. A.C.V. Coelho, H.S. Santos, P.K. Kiyohara, K.N.P. Marcos and P.S. Santos, Surface area, crystal morphology and characterization of transition alumina powders from a new gibbsite precursor. *Mat. Res.* 10 (2) 183-189 (2007).

9. L.C. Edomwonyi-Otu and B.O. Aderemi, Alums from Kankara kaolin. *J. of Res. in Eng* (IRDI). 6 (1), 105-111 (2009).

10. L.C. Edomwonyi-Otu and B.O. Aderemi, Comparative study of alum production from some Nigerian kaolin. In book of abstracts (No. 14). International conference on Materials Science and Technology, (MS&T'10), Houston Tx, USA (2010)

11. R.B. Henry, E.H. Arthur and T.T. George, Biochemical engineering; In Perry handbook of chemical engineering. McGraw-Hill Book Co section 24, 3- 19 (1997).

12. M. Hudlicky, Oxidations in Organic Chemistry; ACS:Washington, DC, (1990).

13. C. Ohn, Interpretation of Infrared Spectra, A Practical Approach; In R.A. Meyers (Ed.) Encyclopedia of Analytical Chemistry John Wiley & Sons Ltd, Chichester 10815–10837 (2000).

14. Kirk Othmer, Encyclopedia of Chemical technology Wiley Interscience New York, vol 3 240-279 (1997).

15. G. MacZura, K.P. Goodboy, J.J. Koenig, Aluminum Sulfate and Alums In: Kirk-Othmer (Ed.), Encyclopedia of Chemical Technology, Wiley Interscience New York, vol. 2, 245-250 (1978).

16. C.M. Marstiller, Aluminum Oxide (Alumina)" In: Kirk-Othmer (Ed.), Encyclopedia of Chemical Technology, Wiley Interscience New York, vol. 2, 219-240 (1978).

17. P. P. Nampi, P. Moothetty, W. Wunderlich, F.J. Berry, M. Mortimer, N.J. Creamer and K.G. Warrier, High-surface-area alumina-silica nanocatalysts prepared by a hybrid sol-gel route using a boehmite precursor. *J. of Am. Ceram. Soc*, 93(12) 4047–4052 (2010).

18. J.T. Richardson, Principles of Catalyst Development. Plenum Press, New York. 25-37, 50-54, 60-73 (1992)

19. Rhodium.org. A simple and efficient reagent for oxidation of benzyl alcohols to benzaldehydes. Synthetic Communications 31(9), 1389-1397 (2001).

20. K.H.K. Safaa and A. A. Mohamad, Application of Vibrational Spectroscopy in Identification of the Composition of the Urinary Stones. *J of App Sci. Res*, 3(5) 387-391 (2007).

21. J.L. Wood, Isomer." Microsoft® Encarta® [DVD]. Redmond, WA: Microsoft Corporation, (2009a)

22. Q. Zhang, J. Li, X. Liu and Q. Zhu. Synergistic effect of Pd and Ag dispersed on Al_2O_3 in the selective hydrogenation of acetylene. *App Cat. A: General* 197 221-228 (2000)

GRAIN BOUNDARY RESISTIVITY IN YTTRIA-STABILIZED ZIRCONIA

Jun Wang and Hans Conrad
Materials Science and Engineering Dept.
North Carolina State University
Raleigh, NC 27695-7907 USA

ABSTRACT

Three topics pertaining to the effect of grain size d on the grain boundary resistivity ρ_b in yttria-stabilized zirconia are considered: (a) the pertinent literature, (b) our recent isothermal annealing experiments and (c) development of a new model based on the effect of an electric field on grain growth. The grain size dependence of the bulk resistivity determined from the annealing tests was in reasonable accord with those in the literature from impedance spectroscopy. Two regimes occurred in the grain size dependence of the bulk resistivity, one for $d > \sim 0.4 \mu m$, the other for smaller grain sizes. Employing the brick layer model, the grain boundary resistivity in the latter regime was approximately an order of magnitude larger than that within the grain interior. Our new model gave values for the space charge potential and grain boundary energy in accord with expectations and measurements. It is concluded that the space charge is the major factor responsible for the grain boundary resistivity.
Key Words: zirconia, grain boundary, electric field, resistivity

INTRODUCTION

It is well-known that the processing and properties of crystalline ceramics are dependent on the grain size[1]. The grain size (GS) is usually controlled by thermo-mechanical treatments and the addition of solutes. In recent years it has been found that an applied dc electric field can also affect the grain growth rate[2-14]. In the case of yttria-stabilized tetragonal zirconia polycrystals (3Y-TZP), a modest DC electric field retards grain growth (Fig.1) and in turn influences the rates of annealing[2,5,6], sintering[6-9] and plastic deformation [2,10-12]. An applied electric field thus provides an additional means for controlling the grain size in the processing and properties of ceramics.

Figure 1. Effect of an applied dc electric field on grain size in 3Y-TZP: (a) and (b) SEM micrographs of fully-sintered (1500 °C) specimens sintered without and with field $E_0 = 14$ V/cm, respectively, (c) effect of electric field strength on the grain size ratio d_E/d processed with field to that without for various processing conditions. From H. Conrad and Di Yang, Mater. Sci. Eng A 528(2011)8523[9].

Because of its relatively high ionic conductivity, good mechanical properties and reasonable cost, polycrystalline yttria-stabilized zirconia (Y-ZP) is an attractive material for solid oxide fuel

Corresponding Author: Hans Conrad. Tel:(919)515-7443, Fax: (919)515-7724. Email: hans_conrad@ncsu.edu.

cells (SOFC) and oxygen sensors [15, 16]. The conductivity however is known to decrease with decrease in GS[15-17]. Our understanding of the mechanism by which this occurs is still incomplete. The objective of our paper is to provide additional information and understanding of this subject. Three topics are addressed: (a) the pertinent literature, (b) our recent experiments pertaining to the subject and (c) a new model for the governing mechanism.

PERTINENT LITERATURE
1. Character of the Grain Boundaries
1.1Stereology: The 3-D stereology of the grain microstructure is important in any consideration of the effect of the grain boundaries on processing and properties. Serial sectioning (Fig.2a)[18] (and chemical dissolution) of polycrystalline metals have revealed that the 3-D geometric shape of the grains is generally a tetrakaidecahedron[19]. The shape of the grains in the fracture surface of ceramics (see for example Fig.2b[20]) indicates that this 3-D geometric form also applies to the grains in these materials.

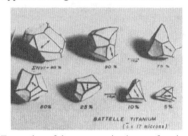 

Figure 2. Examples of the geometric shape of grains in polycrystals: (a) cumulative frequency of the 3-D grain shape in titanium determined by serial sectioning. From K. Okazaki and H. Conrad, Metall. Trans. 3(1972)2411[18] and (b) SEM micrograph of the fracture surface of 3Y-TZP showing geometry of the grains. From J. Obare, D. Griffin and H. Conrad, J. Mater. Sci. 47(2012) 5141[20].

The grain size in ceramics is often given as the mean linear intercept grain size $\bar{d}$ determined on a 2-D planar surface, the individual intercept lengths usually having an approximately log-normal size distribution [19,20]. Han and Kim [21] have pointed out that the 3-D intercept length corresponding to the 2-D measurements is given by $\bar{d}_{3-D} = K\bar{d}$, where K =1.5(or 1.62), 1.78 and 2.25 for sphere, tetrakaidecahedron and cube geometrics, respectively. Smith and Guttman [22] have shown that for a uniform grain size the grain boundary area A_b per unit volume $(A_b/V) = 2/\bar{d}$.

1.2 Space Charge and Solute Segregation: As a result of the difference in the Gibbs free energy for the formation of cation and anion vacancies a space charge occurs at the grain boundaries in ionic ceramics [23-27]; see Fig.3. The width of the space charge zone (δ_b) is of the order of 1-10 nanometers. Segregation of solutes to the boundary will affect the magnitude (and can even reverse the polarity) of the space charge electrostatic potential, depending on the valence of the solute compared to the host[28-30].

The degree of segregation of a solute to the grain boundary depends on temperature and the magnitude of the space charge potential. Early considerations by Mc Lean [31] gave the following for the temperature dependence of the segregation.

$$\frac{c_b}{c} = \frac{\exp(-\Delta G_a/kT)}{1+c\exp(-\Delta G_a/kT)} \tag{1}$$

where C_b is the solute concentration at the grain boundaries, C is the bulk concentration and ΔG_a is the Gibbs free energy of adsorption. Guo and Maier [32] proposed the following relation for the effect of the space charge potential on segregation

$$\frac{C_b(x)}{C} = \exp\left(\frac{-Ze\Delta\emptyset_i(x)}{kT}\right) \qquad (2)$$

where $C_b(x)$ is the solute concentration in the space charge zone, x is the distance from the GB core, Z is the valence of the solute and $\Delta\emptyset_i$ is the electrostatic potential at the GB referenced to that in the grain interior. Additional calculations of the degree of solute segregation to the grain boundary in ionic ceramics have been made by Yan et al.[28] and by Grönhagen and Argen[33].

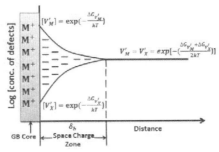

Figure 3. Space charge at the grain boundary in an MX compound and the corresponding charge distribution when the Gibbs free energy for formation of anion vacancies $\Delta G_{V_X'}$ is greater than that $(\Delta G_{V_M'})$ for cation vacancies.

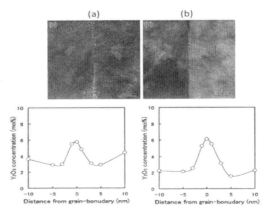

Figure 4. Segregation of yttria at the grain boundaries in 3Y-TZP sintered at : (a) 1300 °C (d =0.2 μm) and (b) 1500 °C (d=0.5 μm). Grain size d measured by the planimeter method. From K. Matsui et al, Acta Mater. 56(2008) 1315[34].

Regarding yttria-stabilized zirconia polycrystals (Y-ZP), it is well-established that yttria segregates to the GBs; see for example Fig.4 from Matsui et al [34]. To be noted in Fig.4 is that the degree of yttria segregation is ∼ 3 mol% above the concentration in the grain interior and that the total width of the segregation zone (both sides of the grain boundary core) is ∼10nm. Hwang and Chen [30] concluded that the segregation of yttria to the GBs in Y-ZP gave a positive space charge potential and further that the space charge was the major factor contributing to the GB energy γ_b rather than the ion size misfit. Calculations by Conrad and Yang [9] based on the effect of an applied DC electric field on the grain growth rate gave that the space charge accounted for 94% of the grain boundary energy. Finally, in any analysis involving the segregation of solutes to the GBs in ceramics the so-called "grain boundary complexions"[35-37] should be considered.

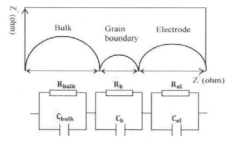

Figure 5. Schematic of a typical impedance spectroscopy plot for an ion-conducting polycrystalline ceramic and the equivalent circuits.

2 Analyses

2.1 Conductivity: The conductivity of yttria-stabilized zirconias has been determined employing both the conventional 4-point probe method (AC and DC) and impedance spectroscopy[15-17,32,38-56], most measurements being by impedance spectroscopy. By the impedance method both the GB and grain interior conductivities can be determined; see Fig.5. The impedance spectroscopy measurements are however limited for the most part to temperatures below 600 °C. An example of the temperature dependence of the *bulk* conductivity of 8Y-SZ determined by the 4-point probe method is shown in Fig.6 [57]. Impedance spectroscopy measurements by these authors on the same materials were in accord with the 4-point probe values.

The temperature dependence of the conductivity σ in Y-SZ determined by both the 4-point probe and impedance spectroscopy methods is in accord with the Nernst-Einstein equation

$$\sigma = nq\mu = (A/T)\exp(\Delta H / kT) \tag{3}$$

where n is the ion carrier concentration, q is the charge (in coulombs) and μ is the mobility of the charge carriers (in $cm^2s^{-1}V^{-1}$). The magnitude of ΔH (the enthalpy pertaining to the charge carrier mobility) given by the temperature dependence of the conductivity in Fig.6 (and in other studies on Y-SP) is of the order of 1.0 eV, which is similar to that (1.0-1.2 eV) for oxygen ion migration [58]. The mobility of this ion is therefore generally assumed to be the mechanism governing conductivity in yttria-stabilized zirconia. Support for this assumption is provided by electrotransport studies [59].

Conductivity measurements on ion conducting ceramics give that the bulk resistivity $\rho(= \sigma^{-1})$ consists of the sum of two components, namely

$$\rho = \rho_g + \rho_b^* \tag{4}$$

where ρ_g corresponds to the grain interior and ρ_b^* is the *grain boundary contribution to the bulk resistivity*. To determine the magnitude of the actual GB resistivity ρ_b (the so-called "specific grain boundary resistivity") from ρ_b^*, investigators have employed the so-called brick layer model [60-65] for the geometry of the grain boundary microstructure(Fig.7). Employing the brick layer model and assuming that the conductivity of both the GBs parallel and perpendicular

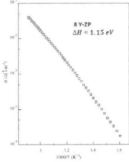

Figure 6. Arrhenius plot of the bulk conductivity of polycrystalline 8 mol % yttria-stabilized zirconia sintered in air for 24h at 1600 °C measured by 4- point probe dc. From M. Kurumada, H. Hara and E. Iguchi, Acta Mater. 53(2005) 4839 [61].

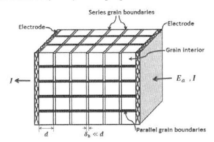

Figure 7. Schematic of the brick layer model for calculation of the grain boundary conductivity.

to the applied electric field vector (as well as the grain interior) obey the Nernst-Einstein equation, Näfe [62] developed the following equation for the temperature dependence of the bulk ionic conductivity

$$\sigma T = \frac{(2\delta_b/d)A_b\exp(-\Delta H_b/kT)+A_g\exp(-\Delta H_g/kT)}{1+(\delta_b/d)(A_g/A_b)\exp[(\Delta H_b-\Delta H_g)/kT]} \tag{5}$$

where δ_b is the GB width including the space charge, d is the grain size, A_g and A_b are the respective grain interior and grain boundary pre-exponentials in the Arrhenius equation and ΔH_g and ΔH_b the respective activation enthalpies. For the case where $(\delta_b/d)(\sigma_g/\sigma_b) \ll 1$, Eq.5 gives for the bulk resistivity at a constant temperature

$$\rho = \rho_g + (\rho_b\delta_b/d) \tag{6}$$

i.e. the bulk resistivity increases with decrease in grain size. The form of Eq.6 has also been considered by other investigators [60,61,63-66]. Impedance spectroscopy measurements on Y-SZ in terms of Eq.6 [15-17,32,38-56] have reported $\rho_b/\rho_g \approx 10^1 - 10^3$ (the ratio tending to increase with decrease in grain size), $\Delta H_b \approx 1.0\sim1.2\ eV$ and $\Delta H_g \approx 0.8\sim1.1\ eV$, all only slightly, if any, dependence on the yttria concentration.

2.2 *Grain Growth*: Conrad [6] has considered the effect of an applied electric field E_a on grain growth in terms of the thermodynamic absolute reaction rate concept; see Fig.8. In this concept the reverse ionic jumps are considered as well as the forward jumps. This then gives for the grain growth rate $\dot{d}$ based on the solute drag grain growth rate theory [67,68]

$$\dot{d} = \frac{A_g^*}{kT d^m} \exp\left(-\frac{\Delta G_s}{kT}\right)\left[1 - \exp\left(-\frac{\Delta G_d}{kT}\right)\right] \tag{7}$$

where A_g^* is a constant, ΔG_s is the Gibbs free energy for diffusion of the pertinent solute ion, ΔG_d is the Gibbs free energy corresponding to the driving force for grain growth and m=1-2 is the GS exponent. Integration of Eq.7 and with $d^n \gg d_0^n$ and A^* and $\Delta G_s \neq f(E)$[6,8] one obtains for the effect of electric field on the grain size

$$\left(\frac{d_E}{d}\right)^n = \frac{1-\exp(-\Delta G_{d,E}/kT)}{1-\exp(-\Delta G_d/kT)} \tag{8}$$

where the subscript E refers to the value with field and n=m+1. $\Delta G_{d,E} = \left(\Delta G_d - \delta\Delta G_{d,E}\right)$ with $\delta\Delta G_{d,E}$ being the decrease in ΔG_d by the field. Taking the experimental value $d_E/d = 0.40$ at $E = E_c$ (26 V/cm) (Fig.1) and $\Delta G_{d,E} = 0.064\Delta G_d$ at $E = E_c$[9], solution of Eq.8 gave $\Delta G_d (= e\emptyset_i) = 0.36 - 0.54\ eV$ for n =2 and 3, respectively; where $e\emptyset_i$ is the space charge potential. Further, taking the GB energy $\gamma_b = \Delta G_d/\Omega^{2/3}$, where Ω is the atomic volume, gave $\gamma_b = 0.54 - 0.81\ J/m^2$, which is in reasonable accord with measurements [69]. Also, the calculated value of $e\emptyset_i(= \Delta G_d)$ is in accord with expectations [25,32].

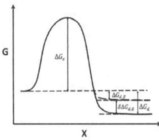

Figure 8. Schematic of the thermodynamic absolute reaction rate concept applied to the effect of an electric field on grain growth rate in 3 Y-TZP. From H. Conrad, Emerging Mater. Res., ICE(2012)[6].

OUR RECENT WORK

It was noted in the Introduction that a modest applied electric field ($E_a = 2 - 30$ V/cm) retards grain growth in 3Y-TZP; see Fig.1. Hence, the variation of the bulk resistivity ρ with GS can be obtained by performing isothermal annealing tests without and with an applied electric field and measuring the grain growth and corresponding electric current as a function of annealing time. Such tests were performed on fully-sintered (relative density $\rho_r \geq 99.5\%$) 3Y-TZP at 1400 °C without and with applied DC electric field E= 18.1 V/cm[70]; the results are presented in Fig.9. The material (powder from Tosoh), powder compaction, sintering procedure and the nature of the electrode connections were the same as those given in [6-9]. The grain sizes shown in Fig.9 are mean linear intercept values ($\bar{d}$) measured on SEM micrographs of thermally-etched cross sections at each of three locations along the specimen gage section: (a) ~5 mm below the upper positive (+) electrode, (b) ~5 mm above the lower negative (−) electrode and (c) midway between. Approximately 200 intercept counts with a resolution of 5 nm were made on each of the three micrographs for a given annealing time, making a total of ~600 intercepts for each reported grain size. The error bars in Fig.9 represent the scatter which occurred between the three locations. There was no consistent trend in $\bar{d}$ from the positive to the negative electrode.

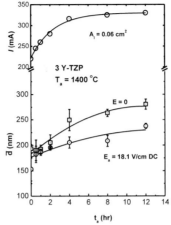

Figure 9. Grain growth and the corresponding electric current during isothermal annealing at 1400 °C without and with an applied electric field $E_a = 18.1$ V/cm. Data from J. Wang et al (2012) [70].

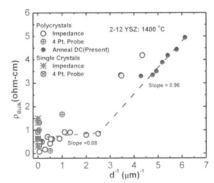

Figure 10. Bulk resistivity determined by various methods of yttria-stabilized zirconia single crystals and polycrystals vs the reciprocal of the grain size. The data points for single crystals are from [50-55], those for the polycrystals from [39-49].

Taking the current density $j = I/A$, where A($=0.06$ cm^2) is the cross-section area of the gage section, a plot of the bulk resistivity $\rho(= E_a/j)$ vs the reciprocal of the grain size $\bar{d}^{-1}$ determined from the data in Fig.9 is presented in Fig.10. Included are data points taken from the literature [15,16,32,38-55] for single and polycrystalline specimens and employing 4-point probe(DC and AC) and impedance spectroscopy methods. The plotted impedance values are linear extrapolations to 1400 °C of the lower temperature Arrhenius plots. To be noted in Fig.10 are the following: (a) there exist two regimes, one for d $> 4\mu m$ and another for smaller grain size, the latter having a slope about an order of magnitude greater than the former, (b) the intercept for the polycrystalline specimens is in reasonable accord with resistivity of single crystals, (c) the values obtained from impedance spectroscopy measurements are in general accord with those obtained by the conventional 4-point probe, and (d) the values obtained from the present isothermal annealing tests are in reasonable accord with those obtained by impedance spectroscopy.

Assuming that the bulk resistivity ρ in each grain size regime in Fig.10 is in keeping with Eq.6, the least-squares value of the slope for d > $4\mu m$ is 8×10^{-8}(ohm-cm)m and that for the smaller grain sizes is 96×10^{-8}(ohm-cm)m. Taking the slope equal to $\rho_b\delta_b$ with δ_b=10 nm(Fig.4) one obtains ρ_b= 8 ohm-cm and 96 ohm-cm respectively for the two regimes. Further, taking the intercept (0.6 ohm-cm) in Fig.10 for ρ_g, one obtains for the ratio ρ_b/ρ_g = 13.3 and 160, respectively for the two regimes. These values of ρ_b/ρ_g are in accord with those obtained by impedance spectroscopy employing the brick layer model. The results in Fig.10 thus indicate that measurements of the effect of an electric field on grain growth provides an additional means for determining grain boundary resistivity. The occurrence of the two GS regimes in Fig.10 is however not clear. They indicate that the product $\rho_b\delta_b$ increases with decrease in grain size, but do not define which of the two parameters is mainly responsible. The review by Miyayama and Yanagide[71] of the influence of the grain size on the *grain boundary resistivity* of yttria-stabilized zirconias also exhibited two regimes. This suggests that the variation of ρ_b with d is the major factor leading to the two regimes in Fig.10.

From a knowledge of the electric current density j during the annealing tests and the magnitude of the GB resistivity ρ_b determined employing the brick layer model one can obtain the value for the electric field across the GB (E_b) through the relation

$$E_b = j\rho_b \qquad (9)$$

The magnitude of E_b so obtained is plotted vs $(\bar{d})^{-1}$ in Fig.11. It is seen that E_b increases with increase in GS.

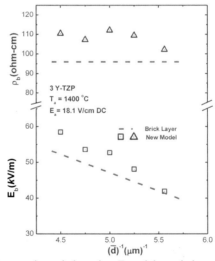

Figure 11. Electric field across the grain boundary E_b and the grain boundary resistivity ρ_b determined from the isothermal annealing tests employing the brick layer model. Included are the values employing our model based on the effect of electric field on grain growth.

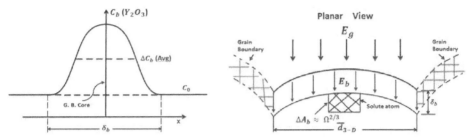

Fig.12 Model proposed for grain boundary resistivity in yttria-stabilized zirconia based on the effect of electric field on grain growth.

In our prior work [6,9] the retardation of grain growth in 3Y-TZP by an electric field was attributed to a reduction in the driving force (i.e., the grain boundary energy γ_b) by the interaction of the field with the space charge. We will now develop a model for the nature of the interaction between the space charge and the applied field. The model is shown schematically in Fig.12. In this model we assume that the grain growth rate $\dot{d}$ without field is given by the conventional solute drag grain growth rate equation[67,68]

$$\dot{d} = \frac{AD_s\gamma_b}{kTd^m} \tag{10}$$

where D_s is the solute diffusion coefficient, γ_b the GB energy and m=1 the grain size exponent. With application of an electric field we take for the GB energy

$$\gamma_b = \gamma_b^0 - Ne\,\emptyset/\Omega^{2/3} \tag{11}$$

where γ_b^0 is the GB energy without field, $Ne = \delta_b d_E \Delta C_b(avg)\Delta Z$ is the total electric charge within the grain boundary zone, $\emptyset$ is the space charge potential, Ω is the atomic volume. δ_b is the space charge width, $d_E = 1.78\bar{d}_E$ is the mean 3-D grain size with field, $\Delta C_b(avg)$ is the average solute segregation at the GB and ΔZ is the difference in valence between the host and solute. Dividing the grain growth rate with an applied field by that without then gives

$$\frac{\dot{d}_E = \frac{AD_s}{kTd_E}(\gamma_b^0 - \frac{Ne\emptyset_b}{\Omega^{2/3}})}{\dot{d} = \frac{AD_s\gamma_b}{kTd}} \tag{12}$$

Since A and D_s are not a function of E_a[8,9], Eq.12 gives

$$Ne\emptyset_b = [1-(\frac{d_E d_E}{dd})]\gamma_b^0\Omega^{2/3} \tag{13}$$

Further, taking $Ne\emptyset = NeE_b\delta_b$, where $\emptyset = E_b\delta_b$ with E_b being the electric field across the grain boundary, one obtains

$$E_b = \frac{[1-(\frac{d_E d_E}{dd})]\gamma_b^0\Omega^{4/3}}{(\delta_b^2 d_E\Delta C_b(avg)\Delta Z)} \tag{14}$$

and in turn the GB resistivity

$$\rho_b = E_b/j \tag{15}$$

where j is current density.

Thus, knowing the values of γ_b, δ_b and the average concentration of solute in the space charge $\Delta C_b(avg)$, one can obtain the magnitudes of E_b and ρ_b from the annealing tests with and without electric field. Taking γ_b=0.62 J/m² at 1400 °C[69], δ_b= 10 nm(Fig.4), $\Delta C_b(avg)$ =1.5 mol. % (Fig.4) and $d_E(3-D) = 1.78\bar{d}_E$ [21], one obtains the values of E_b and ρ_b vs $(\bar{d})^{-1}$ included in Fig.11. It is seen that these values of E_b and ρ_b and their GS dependence are in reasonable accord with those obtained applying the brick layer model to the annealing data. Thus, the model proposed in Fig.12 provides a reasonable description of the governing conditions relating to grain boundary resistivity in 3Y-TZP.

Accepting that the retardation of grain growth in 3Y-TZP by an electric field results from a reduction in γ_b by the field, and that γ_b becomes essentially nil when $E = E_c$ (Fig.1), one can obtain the magnitude of the Gibbs free energy ΔG_d for the driving force when $E_a = 0$ and in turn the value of γ_b^0 from the results determined from the annealing tests in Fig.11. To obtain ΔG_d we take

$$\Delta G_d = (E_c/E_a)E_b\delta_b Ne \qquad (16)$$

where the ratio (E_c/E_a) normalizes the data to the condition that the applied electric field $E_a = E_c$. To obtain γ_b^0 we take

$$\gamma_b^0 = \Delta G_d/\Omega^{2/3} \qquad (17)$$

The calculated values of ΔG_d and γ_b^0 based on the magnitude of E_b at the grain size $\bar{d}=0.20\ \mu m$ obtained employing the brick layer and grain growth models are listed in Table 1. Similar values are obtained for the other values of $\bar{d}$. Included are the values of $\Delta G_d (= e\emptyset_i)$ obtained from the thermodynamic absolute reaction rate analysis of the effect of an electric field on the grain growth rate [6]. It is seen that there exists reasonable agreement in the values obtained employing the three models. Moreover, the values of both the space charge potential energy $e\emptyset_i$ and the GB energy γ_b^0 without field are in accord with expectations and measurements [25-27,69].

Table 1 Magnitude of the parameters pertaining to the grain boundary resistivity in 3Y-TZP with $\bar{d} = 0.20\ \mu m$.

Process: Model	$\rho_b(ohm-m)$	$E_b(kV/m)$	$\Delta G_d(eV)$	$\gamma_b^0(J/m^2)$
Anneal(1400 °C): Brick Layer	96	45.1	0.32	0.48
Anneal(1400 °C): Grain Growth	109	51.3	0.37	0.55
Sinter(1400 °C): Absolute React. Rate	–	–	0.36	0.54

Note: $\delta_b = 10$ nm [6], $\Delta G_b(avg) = 1.5\ mol\%$[6], $\Delta Z =1.0$[30], grain size exponent m =1, $\gamma_b^0 = 1.215 - 0.358 \times 10^{-3}\ T(J/m^2)$[69].

SUMMARY AND CONCLUSIONS

It is shown that the retarding effect of an applied electric field ($E_a < {\sim}30$ V/cm) on grain growth in yttria-stabilized zirconia (and the corresponding electric current) provides a means for determining the grain boundary resistivity ρ_b, the grain boundary energy γ_b and the grain boundary space charge potential energy $e\emptyset_i$. The magnitude of γ_b^0(without field) determined in this manner is in accord with direct measurements; the magnitude of the grain boundary resistivity ρ_b is in accord with impedance spectroscopy measurements employing the brick layer model, and the magnitude of $e\emptyset_i$ with expectations and measurements.

Two regimes appear to exist for the effect of grain size on the bulk resistivity ρ_{Bulk}, one for the mean linear intercept grain size $\bar{d} > {\sim}0.4\ \mu m$ and another for $\bar{d} < {\sim}0.4\ \mu m$, the later exhibiting an order of magnitude greater resistivity than the former. Further work is need to confirm the existence of the two regimes and the reason for them.

It is concluded that the space charge potential ($e\emptyset_i \approx 0.35eV$ for 3Y-TZP) is the primary factor responsible for γ_b^0 and ρ_b in this material.

ACKNOWLEDGEMENTS
This research was funded by NSF Grant No. DMR-1002751, Dr. Lynnette Madsen, Manager Ceramics Program, Materials Science Division. The authors wish to acknowledge stimulating correspondence with Professor Helfried Näfe, University of Stuttgart, Germany.

REFERENCES
[1]W. D. Kingery, H. K. Bowen, D. R. Uhlmann, **Introduction to Ceramics**, Wiley, New York (1976).
[2]Di Yang, H. Conrad, Influence of an Electric Field on the Superplastic Deformation of 3Y-TZP, Scripta Mater., 36,1431-1435(1997).
[3]Di Yang, H. Conrad, Influence of an Electric Field on Grain Growth in Extruded NaCl, Scripta Mater., 38, 1443-1448(1998).
[4]Di Yang, H. Conrad, Retardation of Grain Growth and Cavitation by an Electric Field During Superplastic Deformation of Ultrafine-grained 3Y-TZP at 1450-1600 °C, J. Mater. Sci., 43, 4475-4482(2008).
[5]S. Ghosh, A. H. Chokshi, P. Lee, R. Raj, A Huge Effect of Weak dc Electric Field on Grain Growth in Zirconia, J. Am. Ceram. Soc. 92,1856-1859(2009).
[6]Hans Conrad, "Retardation of Grain Growth in Nanocrystalline Zirconia by an Electric Field", in H. Wang Ed. Acta Mater., Gold Metal Symp., 2011, Thomas Telford Ltd., London, in print.
[7]Di Yang, R. Raj, H. Conrad, "Enhanced Sintering Rate of Zirconia (3Y-TZP) Through the Effect of a Weak dc Electric Field on Grain Growth", J. Am. Ceram. Soc., 432935-2937(2010).
[8]Di Yang, H. Conrad, "Enhanced Sintering Rate and Finer Grain Size in Yttria-stabilized Zirconia (3Y-TZP) with Combined dc Electric Field and Increased Heating Rate", Mater. Sci. Eng. A 528,1221-1225(2011).
[9]Hans Conrad, Di Yang, Dependence of the Sintering Rate and Related Grain Size of Yttria-stabilized Polycrystalline Zirconia (3Y-TZP) on the Strength of An Applied dc Electric Field, Mater. Sci. Eng. A 528,8523-8529(2011).
[10]H. Conrad, Di Yang, Effect of dc Electric Field on the Tensile Deformation of Ultrafine-Grained 3Y-TZP at 1450-1600 °C, Acta Mater. 556,789-6797(2007).
[11]H. Conrad, Di Yang, P. Becher, Effect of an Applied Electric Field on the Flow Stress of Ultrafine-grained 2.5Y-TZP at High Temperature, Mater. Sci. Eng. A 477,358-365(2008).
[12]H. Conrad, Di Yang, Influence of an Applied dc Electric Field on the Plastic Deformation Kinetics of Oxide Ceramics, Phil. Mag., 90,1141-1157(2010).
[13]J.- W. Jeong, J.-H. Han, D.-Y. Kim, Effect of Electric Field on the Migration of Grain Boundaries in Alumina, J. Am. Ceram. Soc., 839,15-918(2000).
[14]J.-I. Choi, J.-H. Han, D.-Y. Kim, Effect of Titania and Lithia Doping on the Boundary Migration of Alumina Under an Electric Field, J. Am. Ceram. Soc., 86,640-643(2003).
[15]M. C. Martin, M. L. Mecartney, Grain Boundary Ionic Conductivity of Yttrium Stabilized Zirconic as a Function of Silica Content and Grain Size, Solid State Ionics, 161,67-79(2003).
[16]S. Hui, J. Roller, S. Yick, X. Zhang, C. Decés-Petit, Y. Xie, R. Marc, D. Ghosh, A Brief Review of the Ionic Conductivity Enhancement for Selected Oxide Electrolytes, J. Power Sources, 172,493-502(2007).
[17]S. H. Chu, M. A. Seitz, The ac Electric Behavior of Polycrystalline ZrO_2-CaO, J. Solid State Chem., 23,297-314(1978).
[18]K. Okazaki, H. Conrad, Recrystallization and Grain Growth in Titanium: I Characterization of the Structure, Metall. Trans. 3,2411-2421(1972).

[19]F. Schückher, "Grain Size", **Quantitative Microscopy**, R. Dehoff, F. Rhines, eds., McGraw-Hill, New York, 201-265(1968).

[20]J. Obare, W. Griffin, H. Conrad, Effects of Heating Rate and dc Electric Field During Sintering on the Grain Size Distribution in Fully Sintered Tetragonal Zirconia Polycrystals Stabilized with 3 % Molar Yttria (3Y-TZP), J. Mater. Sci. 475,141-5147(2012).

[21]J. H. Han, D. Y. Kim, Analysis of the Proportionality Constant Correlating the Mean Intercept Length to the Average Grain Size, Acta Metall. Mater. 43,3185-3188(1995).

[22]C. Smith, L. Guttman, Measuremtn of Internal Boundaries in Three-dimensional Structures by Random Sectioning, Trans. AIME, 197,81-87(1953).

[23]K. L. Kliewer, J. S. Koehler, Space Charge in Ionic Crystals. I General Approach with Application to NaCl, Phys. Rev., 140, A1226-A1240(1965).

[24]K. L. Kliewer, Space Charge in Ionic Crystals. II The Electron Affinity and Impurity Accumulation, Phys. Rev. 140,A1241-1246(1965).

[25]W. D. Kingery, Plausible Concepts Necessary and Sufficient for Interpretation of Ceramic Grain-boundary Phenomena: I Grain Boundary Characterize Structure and Electrostatic Potential, J .Am. Ceram. Soc., 57,1-8(1979).

[26]J.-H. Han, D.-Y. Kim, Interaction and Chemistry of Defects at the Grain Boundaries of Ceramics, J. Am. Ceram. Soc., 84,539-550(2001).

[27]S.-J. I. Kang, **Sintering**, Elsevier, New York, pp181-196(2005).

[28]M. F. Yan, R. M. Cannon, H. K. Bowen, Space Charge, Elastic Field and Dipole Contributions to Equilibrium Solute Segregation at Interface, J. Appl. Phys. 54,764-778(1983).

[29]W. C. Johnson, Grain Boundary Segregation in Ceramics, Metall. Trans A, 8A,1413-1422(1977).

[30]S.- L. Hwang, I.-W. Chen, "Grain Size Control of Tetragonal Zirconia Polycrystals Using the Space Charge Concept, J. Am. Ceram. Soc. 73,3269-3277(1990).

[31]D. Mc Lean, **Grain Boundaries in Metals**, Clarendon Press, Oxford (1957).

[32]X. Guo, J. Maier, Grain Boundary Blocking Effect in Zirconia: a Schottky Barrier Analysis, J. Electrochem. Soc., 148E,121(2001).

[33]K. Grönhagen, J. Ágren, Grain Boundary Segregation and Dynamic Solute Drag Theory-Aphase-field Approach, Acta Mater. 55,955-960(2007).

[34]K. Matsui, H. Yoshida, Y. Ikuhara, Grain Boundary Structure and Microstructure Development Mechanism in 2~8 mol% Yttria-stabilized Zirconia Polycrystals, Acta Mater., 56,1315-1325(2008).

35 S. J. Dillon, M. P. Harmer, "Multiple grain boundary transitions in ceramics: A case study of alumina", Acta Mater., 55(2007) 5247-5254.

[36]S. J. Dillon, M. Tang, W. C. Carter, M. P. Harmer, Complexion: A New Concept for Kinetic Engineering in Materials Science, Acta Mater. 55,6208-6218(2007).

[37]S. J. Dillon, M. P. Harmer, Demystsfying the Role of Sintering Additives with "Complexion",J. Euro. Ceram. Soc. 28,1485-1493(2008).

[38]S. P. S. Badwal, J. Drennan, Yttria-zirconia: Effect of Microstructure on Conductivity, J. Mater. Sci., 22,3231-3239(1987).

[39]M. Weller, H. Schubert,"Internal Friction, Dielectric Loss, and Ionic Conductivity of Tetragonal ZrO2-3% Y2O3 (Y-TZP), J. Am. Ceram. Soc., 22,573-577(1987).

[40]S. P. S. Badwal, J. Drennan, Yttria-zirconia: Effect of Microstructure on Conductivity, J. Mater. Sci., 22,3231-3239(1987)

[41]S. P. S. Badwal, F. T. Ciacchi, M. V. Swain, V. Zelizko, Creep Deformation and the Grain-Boundary Resistivity of Tetragonal Zirconia Polycrystalline Materials, J. Am. Ceram. Soc., 73,2505-2507(1990).

[42]T. Tsubakino, H. Ikeda, H. Maeda, B. Zang, Change of Conductivity Accompanied by Phase Transformation in Partially Stabilized Zirconia, Mass and Charge Transport in Cramics, Am. Ceram. Soc. pp.213-224(1996).

[43]P. S. Badwal, F. T. Ciacchi, V. Zelizko, The Effect of Alumina Addition on the Conductivity, Microstructure and Mechanical Strength of Zirconia-yttria Electrolytes, Ionics, 4,25-32(1998).

[44]A. Pimenov, J. Ullrich, P. Lunkenheimer, A. Loidl, C. H. Rüscher, Ionic Conductivity and Relaxations in $ZrO_2–Y_2O_3$ Solid Solutions, Solid State Ionics, 108,111-118(1998).

[45]T. Uchikoshi, Y. Sakka, K. Hiraga, Effect of Silica Doping on the Electrical Conductivity of 3 mol % Yttria-Stabilized Tetragonal Zirconia Prepared by Colloidal Processing, Jnl. Electroceram, 4:S1,113-120(1999).

[46]I. Kosacki, V. potrovsky, H. Anderson, Electrical Conductivity in ZrO_2-Y, mrs Symp. Proc. Vol.548,505-510(1999).

[47]R. Ramamorthy, D. Sundararaman, S. Ramasamy, Ionic Conductivity Studies of Ultrafine-grained Yttria Stabilized Zirconia Polymorphs, Solid State Ionics, 123,271-278(1999).

[48]M. Kurumada, H. Hara, E. Iguchi, Oxygen Vacancies Contributing to Intragranular Electrical Conduction of Yttria-stabilized Zirconia (YSZ) Ceramics, Acta Mater., 53,4839-4846(2005).

[49]Y. Shiratori, F. Tietz, H. Penkalle, J. He, D. Stöver, Influence of Impurities on the Conductivity of Composites in the System $(3YSZ)_{1-X}$-$(MgO)_X$, J. Power Sources 14,832-42(2005).

[50]S. Ikeda, O. Sakurai, K. Uematsu, N. Mizutani, M. Kato, Electrical Conductivity of Yttria-stabilized Zirconia Single Crystals, J .Mater. Sci. 20,4593-4600(1985).

[51]J. D. Solier, I. Cachadiña, A. Dominguez-Rodriguez, Ionic Conductivity of ZrO_2-12 mol% Y_2O_3 Single Crystals, Phys. Rev. B 48,3704-3712(1993-II).

[52]A. Drago, C. Chang, A. Franklin, J, Bethin, A Grain Boundary in Yttria-stabilized Zirconia, Sci. Technol. ZrO_2, 184-195(1993).

[53]A. Pimehov, J. Ullrich, P. Lunkenheimer, A. Loidl, C. Rüscher, Ionic Conductivity and Relaxation in ZrO_2-Y_2O Solid Solutions, Solid State Ionics 109,111-118(1998).

[54]I. Kosaki, C. Rouleau, P. Becher, J. Bentley, D. Lowndes, Nanoscale Effects on the Ionic Conductivity of Highly Textured YSZ Thin Films, Solid State Inoics 176,1319-1326(2005).

[55]S. Azad, O. Marina, C. Wang, L. Saraf, V. Shutthanandan, D. Mccready, A. EI-Azab, J. Jaffe, M. Engelhard, C. Peden, S. Thevuthasan, Nanoscale Effects on Ion Conductance of Layer-by-layer Structures of Gadolinia-doped Ceria and Zirconia", Appl. Phys. Lett. 86,131906-1-131906-3(2005).

[56]X. Guo, E. Vasco, S. Mi, K. Szot, E. Wachsman, R. Waser, Ionic Conduction in Zirconia Films of Nanometer Thickness, Acta Mater. 53,5161-5166(2005).

[57]M. Kurumada, H. Hara, E. Iguchi, Oxygen Vacancies Contributing to Intragranular Electric Conduction of Yttria-stabilized Zirconia (YSZ) Ceramics, ACta Mater. 53,4839-4846(2005).

[58]E. Watson, D. Cherniak, Oxygen diffusion in zircon, Earth and Planetary Sci. Lett. 148,527-544(1997).

[59]D. Bray, U. Merten, Transport Numbers in Stabilized Zirconia, J. Electrochem. Soc. 111,447-452(1964).

[60]J. E. Baurle, Study of Solid Electrolyte Polarization by a Complex Admittance Method, J. Phys. Chem. Solids 30,2657-2670(1969).

[61]D. Wang, A. Nowick, The Grain Boundary Effect in Doped Ceria Solid Electrolytes, J. Solid State Chem. 35,325-333(1980).

[62]H. Näfe, Ionic Conductivity of ThO_2- and ZrO_2-based Electrolytes Between 300 and 2000K, Solid State Ionics, 13,255-263(1984).

[63]R. Gerhardt, A. Nowick, Grain-boundary Effect in Ceria Doped with Trivalent Cations: , Electrical Measurements, J.Am. Ceram. Soc. 69,641-646(1986).

[64]S. Hui, J. Roller, S. Yick, X. Zhang, C. Decés-Petit, Y. Xie, R. Marc, D. Ghosh, A Brief Review of the Ionic Conductivity Enhancement for Selected Oxide Electrolytes, J. Power Sources, 172,493-502(2007).

[65]J .MacDonald, **Impedance Spectroscopy: Emphasizing Materials and Systems**, Wiley-Intersciences, New York (1987).

[66]S. Haile, D. West, J. Campbell, The Role of Microstructure and Processing on the Proton Conducting Properties of Gadolinium-doped Barium Cerate, J. Mater. Res. 13,1576-1595(1998).

[67]K. Lücke, H.-P. Stüwe, On the Theory of Grain Boundary Motion", in **Recovery and Recrystallization of Metals**, L. Himmel, Ed. Gordon and Breach, New York,pp.171-210(1963).

[68]J. Cahn, "The Impurity-drag Effect in Grain Boundary Motion, Acta Metall. 10,789-798(1962).

[69]A. Tsoga, P. Nikolopoulos, Surface and Grain Boundary Energies in Yttria-stabilized Zirconia (YSZ-8 mol%), J. Mater. Sci. 31,5409-5413(1996).

[70]J. Wang, A. Du, Di, Yang,R. Raj, H. Conrad, Effect of a dc Electric Field on Grain Growth and Corresponding Electric Current in 3 mol% Yttria-stabilized Zirconia During Isothermal Annealing at 1400 °C", unpublished research (2012).

[71]M. Miyayama, H. Yamacida, Dependence of Grain-boundary Resistivity on Grain-boundary Density in Yttria-stabilized Zirconia, Comm. Am. Ceram. Soc, C-194-C-195(Oct. 1994).

MULTISCALE THERMAL PROCESSES IN HIGH VOLTAGE CONSOLIDATION OF POWDERS

Evgeny G. Grigoryev, Eugene A. Olevsky
NRNU MEPhI
Moscow, Russia

ABSTRACT
Compressed metal powders have a very large electrical resistance due to the oxide layers on grains. A transition from an insulating to a conducting state of powder sample is observed as the applied current is increased. Experiments show that this transition comes from an electro-thermal coupling in the vicinity of the microcontacts between particles where microwelding and electric microexplosion occurs. This paper presents multiscale simulation results of the thermal processes occurring under high voltage consolidation of powder materials. The thermal processes in the contact zones of powder particles are characterized by significant spatial inhomogeneity and time dependence during this sintering. Their analysis and finding of the main regularities in the behavior of the powder material in the contact zones makes it possible to establish the optimal pulse electrical current parameters. The simulation results demonstrate the correlation between the modeling and experimental data for compacted powders of heat-resistant alloy.

INTRODUCTION

Unique opportunities high voltage consolidation (HVC) methods of powder materials are reflected in the ever-growing number of scientific publications. A wide range of electrical parameters of the impact on the powder creates large number of these methods[1]. Advantages of these methods can be implemented with optimum electric consolidation parameters as intensively electro-thermal effect on powder material can lead to instability of the compaction process, formation of heterogeneous material structure and even to the destruction of the sintered specimens and technological equipment. The time dependence of thermal processes at the inter-particle contacts plays a key role in the electro-pulse powder consolidation.

Nature of thermal processes in the inter-particle contacts has a significant influence on the temperature distribution throughout the volume of the consolidated material. The experimental results show that if insufficient local Joule heat is generated in contacts and this leads to weak inter-particle bonding and lower final density and mechanical strength of consolidated specimen. But on the other hand there is an upper level for the local Joule heating of inter-particle contacts beyond which the powder material disintegrates like an exploding wire[2]. Therefore an optimum electric pulse current amplitude and pulse time length are necessary to generate sufficient heat for producing of strong inter-particle joining. The goal of the present work is to investigate and evaluate how pulse electric current parameters influence the temperature distribution in the conductive powder particles during HVC. It has been established from experimental research that there is a qualitatively different thermal regimes of inter-particle contacts during electric pulse consolidation. The low-voltage electric pulse sintering is characterized by two different stages of consolidation process. The first stage has a relatively low pressure on a powder sample and uses a sequence of direct current pulses with a certain amplitude and duration with pauses between pulses. The second stage has the simultaneous impact of more powerful current pulses (in some cases together with a constant electric current) and applies of high mechanical pressure to the sample. It is very important the right choice for the applied pressure and pulse current parameters (amplitude, duration and the interval between pulses). Because under optimum process parameters the surface oxides films are cleared from the powder particles by electrical explosion of the primary conductivity contact spots. In this case, it is implemented fairly uniform heating of the total specimen. At lower pressures it is formed the local conductive channels in the powder material having a higher temperature than the rest of the sample volume. This leads to the formation of inhomogeneous structures consolidated material in the finishing stage of the process. At pressures above the optimal values of the first stage, the temperature of inter-particle contacts is insufficient high to destroy surface films of particles, which also affects the properties of the consolidated sample.

Thermal processes in inter-particle contacts at HVC are associated with the action of a single much more powerful pulse electric current (in contrast to the low-voltage sintering). High voltage applied to the sample at the beginning of the process, usually provides the breakdown of surface oxide films in the entire volume of the sample homogeneously. The magnitude of the applied pressure, defines the initial electric resistance of the inter-particle contacts, and consequently, the power thermal sources in the contacts between powder particles.

EXPERIMENTAL RESULTS

The nonlinear contact electrical resistance of powder compact depends on the properties of surface films on powder particles, applied pressure to them and the pulse current parameters. The experimental conductivity results of commercial powders Fe, Ti, Cu versus applied pressure are displayed on Figure 1. It can be seen that the conductivity of the investigated powders depends on the pressure by a power law.

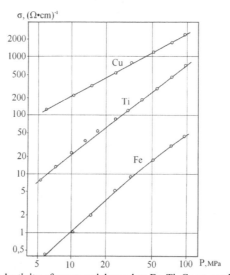

Figure 1. Conductivity of commercial powders Fe, Ti, Cu vs. applied pressure.

The experimental results[3] obtained for the manner in which the instantaneous resistance of Cu powder column oxidized for different periods of time is demonstrated on Figure 2. The resistance of powder column shows an initial sharp reduction which is consistent with breakdown of insulating oxide layers on particles. In the subsequent stage of consolidation after breakdown of oxide layer the powder resistance slightly increases, this may be due to increasing temperature. The instantaneous resistance of a column depends on the thickness of an oxide film as shown on Figure 2.

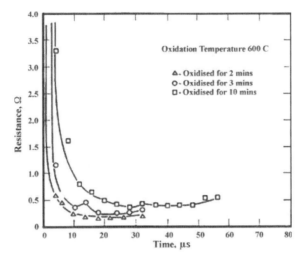

Figure 2. Instantaneous resistance of oxidized Cu powders during high-voltage consolidation[3].

Mechanical pressure in the inter-particle contacts also affects the intensity of thermal sources. Figures 3 and 4 show the contact areas of spherical Mo powder particles (~ 150 μm), formed by the action of various intensity sources[2].

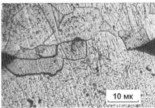

Figure 3. Plastic deformation of contact between Mo particles.

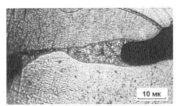

Figure 4. Melting of contact between Mo particles.

The action of the source of lower power locally enhances the plasticity of the material and causes the subsequent intense deformation of the contact region (Figure 3). Action more powerful source of leads to local melting of inter-particle contact and high speed crystallization (Figure 4). If the power of contact source exceeds the upper limit, it causes an electric thermal explosion of the inter-particle contact.

In the general case the distribution of the density and pressure may be inhomogeneous in the green compact before the impact of the electric pulse current. This is the cause of the different temperature gradients in the scale of powder particles as well as scale of the whole sintered specimen. Research and analysis of such multi-scale thermal processes is especially important for high-voltage pulse electric current sintering of conductive powder materials.

SIMULATION RESULTS AND DISCUSSION

The high energy density in the particle contact zones causes a change in the aggregate state of the material (from a solid to a liquid and, partially, a dense low-temperature plasma). The physics processes in the contact zones are characterized by spatial inhomogeneity and time dependence. The analysis and finding the main regularities in the behaviour of the material in the contact zones of particles makes it possible to establish the optimal high voltage electric pulse electric current parameters. A mathematical model of the physical processes occurring under high voltage consolidation in a powder both during compacting as a whole and taking into account the processes in the particle contact zones was proposed in[4]. The system of equations describing the processes under high voltage consolidation is based on the mass, momentum, and energy conservation laws and the electrodynamics equations for consolidated powder materials[4]. The general scheme of high voltage consolidation processes is as follows. A current pulse passing through a powder and punches - electrodes strongly heats only the powder material without significant punches - electrodes heating, because the powder resistivity greatly exceeds that of the electrode material. The intense heating of the powder significantly decreases its resistance to plastic deformation, and, under the action of an external mechanical pressure, it is consolidated with a characteristic rate, dependent on the mechanical pressure system. Simultaneously, heat sink from the powder to the punches and die occurs due to the thermal conduction. The time of energy injection to the powder is determined by the current pulse width: $t_0 < 10^{-3}$ s. The time of formation of a consolidated material from the powder t_1, depends on the loading system and lies in the range $2 \times 10^{-3} < t_1 < 2 \times 10^{-2}$ s. The cooling time of the consolidated material, t_2, is determined by the thermal conductivity of the materials and the characteristic size of the compacted sample: $t_2 \sim 2.5$ s. In this case, the time scales of the processes obey the following relation:

$$t_0 < t_1 \ll t_2 \tag{1}$$

Thus, Joule heat is released in the inter-particle contacts at the green density distribution of the powder sample. Thermal processes simulation in the inter-particle contacts identifies parameters of the pulse current, at which there is localization of heat (blow-up process) in the contact areas. The analysis blow-up temperature process (localization of heat) is similar to that described in[5].

Time dependence of current density in the inter-particle contact has the form:

$$j(t) = j_0 \cdot \exp(-\beta t) \sin(\omega t) \tag{2}$$

(j_0 – amplitude, ω – frequency and $1/\beta$ – duration of the pulse current, t – time).

Numerical calculations were performed in dimensionless variables. The designations are given in Table I.

Table I. The dimensionless parameters of thermal processes

Parameter	Parameter description
γ, c, χ, T_m	density, heat capacity, thermal diffusivity, melting temperature
ρ_0	specific contact resistance
R	characteristic size of the powder particles
$t_R = R^2/\chi$	characteristic time of heat diffusion in the particle
$\theta = T/T_m$	dimensionless temperature
$\tau = t/t_R, x = r/R$	dimensionless time, dimensionless coordinate
$\varepsilon = (\rho_0 j_0^2 t_R)/(\gamma c T_m)$	dimensionless power of heat source
$\Omega = \omega t_R$	dimensionless frequency of pulse current density
$\delta = 2\beta t_R$	dimensionless duration of pulse heat source
$f = \varepsilon e^{-\delta \tau} \sin^2(\Omega \tau)$	dimensionless instantaneous power of the heat source

Figure 5 shows the time dependence of the dimensionless power of heat source in the inter-particle contact.

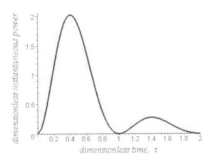

Figure 5. The dimensionless instantaneous power of heat source.

Figure 6 illustrates specific features of the behavior of the dimensionless temperature θ in the inter-particle contact versus dimensionless time τ.

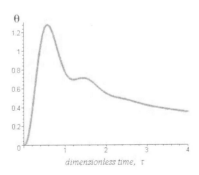

Figure 6. The inter-particle contact surface temperature vs. dimensionless time.

Figure 7 illustrates the localization of the spatial temperature distribution in the contact region for the two different time points (x - dimensionless distance from the inter-particle contact surface) for dimensionless parameters: $\varepsilon = 5$; $\Omega = 3$; $\delta = 2$. These values of the dimensionless parameters correspond to high-voltage pulse electric current sintering conditions in which there is partial melting of inter-particle contacts.

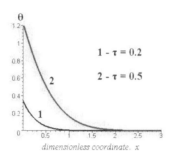

Figure 7. Instantaneous spatial temperature distributions.

Comparison of simulation results with experimental measurements of the sample temperature was carried out on Fe and Cu powders[6]. This comparison showed that the calculated surface temperature of the specimen during the cooling process coincides with the experimental measured temperature.

Simulation of thermal processes in the inter-particle contacts has established the critical amplitude of the pulse current density, at which there is an electric thermal explosion of contact:

$$j_0 < \sqrt{\frac{2\xi\sigma}{\rho h}} T_b^2 \tag{3}$$

Here: σ – the Stefan-Boltzmann constant, T_b – boiling point material, ρ – the electrical resistivity of contact spot, h – thickness of contact area, $\xi \leq 1$.

The Figure 8 shows the experimental points and calculated line from (3) for the current amplitude of the explosion contact as a function of particle size.

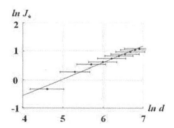

Figure 8. Experimental points and theoretical line for the pulse current amplitude of the contact explosion as a function of particle size.

CONCLUSIONS

Simulation of high-voltage pulse electric current sintering revealed two different time scales of thermal processes: at the first stage of energy input into powder and the finishing stage of cooling consolidated material. The numerical results have established the possibility of localization of heat in the inter-particle contacts at certain parameters of the high-voltage pulse electric current. It was found an upper level for the high voltage pulse current amplitude beyond which the inter-particle contacts in powder material disintegrate by electric thermal exploding.

REFERENCES

[1] R. Orru, R. Licheri, A. M. Locci, A. Cincotti, and Cao G, Consolidation/synthesis of materials by electric current activated / assisted sintering, *Materials Science and Engineering*, **R 63**, 127-287 (2009).

[2] S. A. Balankin, S. S. Bashlykov, L. P. Gorbachev, E. G Grigoryev, D. M. Skorov, and V. A. Yartsev, Thermal processes at electro-pulsed compaction of powders, *Physics and Chemistry of Materials Processing*, **2**, 124-129 (1984).

[3] M. Alitavoli, and A. Darvizeh, High rate electrical discharge compaction of powders under controlled oxidation, *Journal of Materials Processing Technology*, **209**, 3542–3549 (2009).

[4] E. G. Grigoryev, Modeling of physical processes of electro-discharge compaction of powder materials, *Proceedings of Scientific Session of MEPhI-2007*, **9**, 121-122 (2007).

[5] A.A. Samarskii A.A., V.A. Galaktionov, S.P. Kurdyumov, and A.P. Mikhailov, *Blow-up in Quasilinear Parabolic Equations, Berlin: Walter de Gruyter* (1995).

[6] E. G. Grigoryev, Kinetics of densification processes of powder materials under electropulse sintering, *The Arabian Journal for Science and Engineering*, **34**, 29-33 (2009).

Nanotechnology

NANOTECHNOLOGY DEVELOPMENT IN ARAB STATES

Bassam Alfeeli[1], Ma'moun Al-Rawashdeh[2]

[1]Kuwait Institute for Scientific Research, Safat, Kuwait
[2]Eindhoven University of Technology, Eindhoven, Netherlands

ABSTRACT
A growing number of initiatives on nanotechnology research, education and industry have been recently launched by several Arab States to quickly build scientific capacity and track the worldwide developments in nanotechnology. Some countries, namely the oil rich States, have allocated large funds to support these initiatives. This comprehensive commitment is intended to serve national interests in energy, water and food supply, medicine, and local industry. Other Arab States are also pursuing nanotechnology, however with fewer funds but with more human resources. This study assesses current status of nanotechnology in the Arab Republic of Egypt, Hashemite Kingdom of Jordan, Kingdom of Saudi Arabia, State of Kuwait, State of Qatar, and United Arab Emirates. The study is aimed at having an overview of the status of existing, underdevelopment, and planed educational and research programs relevant to nanotechnology. The overview also includes nanotechnology research focus areas, challenges, and opportunities.

INTRODUCTION

Developing countries such as the Arab States have recently realized their lag in taking part of the international effort to develop nanotechnology. In fact, Arab States are entering a new phase of political reform and economic development. With emphasis on science and technology (S&T) for promoting sustainable development, a growing number of nanotechnology research, education and industry initiatives have been recently launched by several Arab States to quickly build scientific capacity and track the worldwide developments in nanotechnology. This study assesses current status of nanotechnology in Arab States. The countries examined are Arab Republic of Egypt, Hashemite Kingdom of Jordan, Kingdom of Saudi Arabia, State of Kuwait, State of Qatar, and United Arab Emirates. The study documents the approach of the different countries along three parameters: national commitment, funding and infrastructure. The descriptive overview covers the status of existing, underdevelopment, and planed educational and research programs relevant to nanotechnology. This study drew information from publically available official documents and reports as well as material collected from official websites of institutions discussed. All reproduced text in this article has been released into the public domain by its authors.

BACKGROUND

The Arab world consists of 22 countries covering about 10% of the world's land and is home to about 300 million representing 4.5% of the world's population. The UNESCO Science Report[1] groups the Arab States into three groups in terms of per capita income. The first group is characterized by an almost total economic dependence on oil (Bahrain, Emirates, Kuwait, Oman, Qatar, and Saudi), with gross domestic product (GDP) per capita income being the highest in Qatar and the lowest in Oman. Around 11% of the Arab population belongs to this group of States. The higher education system and science, technology and innovation (STI) in these States are new but developing rapidly. The second group consists of Algeria, Egypt, Iraq, Jordan, Lebanon, the Libya, Morocco, Syria and Tunisia. Here, GDP per capita is the highest in the Libyan and the lowest in Egypt. Although the States in this category have modest oil reserves, with the notable exception of Iraq and Libya, they possess relatively mature higher education

199

infrastructure which includes some of the oldest universities in the Arab world. The population of this group amounts to around 70% of the population in the Arab world. The third group is characterized by limited or underdeveloped natural resources and an equally small supply of trained human resources. States in this group also possess some of the lowest GDP per capita in the world, which classifies them as least developed countries. They are Comoros, Djibouti, Mauritania, Sudan and Yemen. This group of countries represents around 19% of the total population of the Arab world.

According to Sawahel[2], on average, Arab States allocate less than 0.2% of their GDP on research compared to 1.6% in East Asia countries and 2.6% in developed countries. Only 6.2% of the combined Arab States budget is spent on fundamental research. The budget allocated for purchasing armament in these States surpasses health, education, and research budgets combined. However, in recent years the Arab society and their leaders recognized the importance of education and STI. For example, the share of engineering and science students in the total number of higher education students increased by 51% in Bahrain, 29% in Kuwait, 21% in Emirates, and 19% in Oman[3]. Moreover, at the annual Arab summit and for the past 5 consecutive years (Sudan 2006, Saudi 2007, Syria 2008, Qatar 2009, and Libya 2010 summits) the Arab leaders adopted several decisions to establish science-based economy and knowledge society. One of the decisions was to increase the expenditure on research and development (R&D) to about 2.5% of GDP[3]. They also agreed to make education and scientific research permanent items at all future Arab summits. The Arab States don't only share common language and culture but also R&D priorities which include water and energy. The traditional sector of agriculture and the relatively new fields of information and communication technologies (ICTs), nanotechnology and biotechnology are also viewed as priority research areas as stated in the 20th Arab League summit held in Syria 2008.

Nevertheless, there have been a number of challenges that faces the development of S&T in the Arab States: (1) lack of supportive governmental policies; (2) limited and inconsistent R&D funding; (3) a small scientific research community; and (4) limited venture capital. In particular, there has been no national commitment to establish organizations dedicated to devising strategies that address S&T. There is a need for a policy-making body that: formulate national policies, link R&D with national development priorities, and execute technological development programs that fit the national economic and social objectives. Improving innovation requires a political decision and must be supported by a clear vision. Moreover, in the past the Arab States lacked the foundation for advanced science and modern technology base. Constrains on R&D in the Arab States are not only limited to the weak institutional structure and infrastructure, but also inefficient administrative arrangement. This stresses the need to develop the human capacity. Additionally, there is a small number of Arab scientists. Their contribution accounts for only 1% of the world's scientific production[2]. For the case of nanotechnology development, the Arab States lack the critical mass of researchers and scientist specializing in nanotechnology to effectively enhance their innovation capability[4]. Although this is gradually changing, the desire for scientific enquiry and activity in the Arab society is yet to be strengthened.

CURRENT STATUS

Egypt

Egypt does not have formally enacted national science policy[5]. However, the amount of funds allocated to R&D and the number of well-established scientific institutions indicate the existence of an implicit national policy[6]. In 2007, a presidential decree was issued to establish the Science and Technology Development Fund (STDF) to support the scientific research and technological development. In the following year, another presidential decree was issued to establish the Higher Council for Science and Technology (HCST). The council is chaired by the Prime Minister and includes the minster of higher education and scientific research and a group of prominent Egyptian scientists. HCST was established to promote R&D in the country and identify priority research areas which include: health, water resources, renewable energy, food and agriculture, and space technology. It didn't explicitly state nanotechnology as a priority. In 2009, the so called Specialized Scientific Councils (SCCs) were established to provide scientific advices, studies and strategic plans to support the policy makers and the community[7].

As R&D continues to expand, efforts are being made to promot nanotechnology development in Egypt. Efforts to establish micro/nano research and fabrication facility started as early as 2003. This is marked by the establishment of Yousef Jameel Science and Technology Research Center (YJ-STRC) at the American University in Cairo (AUC). YJ-STRC was the fruit of the generous support ($8 million over 5 years) of a Saudi businessman and AUC alumnus Yousef Jameel. His vision was to create nanotechnology center of excellence at AUC. The center houses class-100 clean room and state-of-the-art fabrication and characterization equipment. It should be noted that AUC offers Masters of Science in Nanotechnology and Doctor of Philosophy in Applied Sciences with specializations in Nanotechnology. To date, YJ-STRC has secured $13 million in funding and recruited 17 high profile faculty members with diverse backgrounds. Furthermore, with the support of 8 postdoctoral fellows, 13 doctoral students, 20 master's students, and 4 technical staff, YJ-STRC conducts its work through 6 research groups: micro- and nano-systems, nanostructured materials, surface chemistry, biotechnology, environmental science and engineering, and novel diagnostics and therapeutics. The research groups are serviced by well-equipped research facilities that include: micro- and nano-systems fabrication, materials synthesis, biotechnology, surface chemistry.

In 2006, the Nile University (NU) was established, by support from international and national companies represented by the Egyptian Foundation for Technological Education Development (EFTED), as a not-for-profit, privately-owned, and autonomously-managed university. NU is the first academic institution in Egypt to be founded by a partnership between the private sector (EFTED), government, business, and industry. NU was allocated around 0.5 km^2 of land and two buildings by the government. NU has 6 faculty members working on nanotechnology related research and offers a Master of Science in Nanoscience and Technology degree. NU also houses two nanotechnology centers. The Center for Nanotechnology (CNT) which was establish based on collaboration efforts with Northwestern University in U.S. CNT researchers work on printed electronics, membrane technology, and renewable energy. The other center is Nanoelectronics Integrated Systems Center (NISC) which was funded by: Intel, Mentor Graphics, British Petroleum, European Union, Cypress Semiconductor Corp., Egyptian Information Technology Industry Development Agency (ITIDA), STDF, and National Telecom Regulatory Authority. NISC is pursuing research in areas that include: high performance integrated circuits (ICs), computer aided design ICs, low power circuit design, hardware for wireless sensor network, MEMs, and sensor and actuator design.

In an attempt to capture the currently underutilized human potentials in Egypt, IBM teamed up with ITIDA and STDF to create Egypt's first national research laboratory in 2008, the Egypt-IBM Nanotechnology Research Center (EGNC). The idea was to have IBM experts work with local scientists and engineers on advanced nanoscience and nanotechnology projects. With $30 million in seed money for the first three years, EGNC is focusing on solar energy technology and water desalination. The startup work force of EGNC was about 100 employees but expected to grow up to 1,000 within the next few years. Current research areas include: thin-film silicon photovoltaics, spin-on carbon–based electrodes for thin film photovoltaics, energy recovery from concentrator photovoltaics for desalination, and computational modeling and simulation.

Emirates

Although Emirates does not have formally enacted national science policy, recently, it took several measures to promote R&D within the country. In 2010, the government released "UAE Vision 2021", a document which outlined the vision for UAE in all fields until 2021 when the country will celebrate its 50th anniversary[8]. The document lists the promotion of innovation and R&D as one of the country's priorities. There is no specific mention of nanotechnology as a national priority. Nevertheless, it states that the Emiratis will continue to attract the best talent from around the world and offer fulfilling employment and an attractive place to live to retain the finest and most productive workers and entrepreneurs. The plan also calls for knowledge-based, highly productive and competitive economy that will rival the best in the world by investing in science, technology, and R&D. According to the document, this goal will be achieved by supporting practical programs such as start-up incubators and cultivating a culture of risk-taking where hard work, boldness and innovation are rewarded.

In 2000, the ruler of the Emirate of Sharjah established the Arab Science and Technology Foundation (ASTF) with a $6 million donation. ASTF mission is to identify and support scientific research activities in the Arab world. A foundation dedicated solely to Emirates, the Emirates Foundation (EF), was established by the Emirate of Abu Dhabi in 2005. EF supports five core programs: education, S&T, environment, arts & culture, and social development. In 2007, another foundation was established by the ruler of Dubai, the Mohammed bin Rashid Al Maktoum Foundation (MBRF), with a $10 billion endowment toward the development of a knowledge-based society. A foundation specifically dedicated to R&D, the National Research Foundation (NRF), was established in 2008 with a $30 million budget. The vision of NRF is to support research activities, and create competitive research environment and innovation system in the Emirates. NRF is tasked with the introduction of research excellence centers, strategic initiatives, and competitive awards and grants.

The United Arab Emirates University (UAEU) is the first national university in Emirates. The number of colleges grew to 9 from the original 4 when it was established in 1976. As of 2011, UAEU have over 650 faculty members and offers both undergraduate and graduate degrees to over 12,000 students. It was announced in 2009 that the Emirates Centre for Nanosciences and Nanoengineering will be established within the College of Engineering at UAEU by a $10 million grant from NRF. At the time of the announcement, there were 18 scientists working on nanotechnology research projects. The center's director stated that another 10 researchers will be recruited for the new center. The research at this center will be aimed toward cancer treatment, solar energy, and building materials.

The Higher Colleges of Technology (HCT), the largest higher educational institution in Emirates, was founded in 1988. With 13 colleges and 17 campuses throughout the country, HCT offers more than 90 programs at different levels in applied communications, business, engineering, information technology, health sciences and education. English is the official language of instruction and faculty members are recruited from around the world. In 2006, the Center of Excellence for Applied Research and Training (CERT) began as the commercial arm of HCT. The $35 million investment made CERT grow to be the largest investor in the discovery and commercialization of technology in the Middle East. According to its website, CERT is the only supercomputing center in the Middle East region. CERT's Blue Gene supercomputer offers 5.7 teraflops calculating speed for use in biotechnology, nanotechnology, and genetics research as well as oil and gas simulation.

Masdar Institute of Science and Technology (MIST) is the world's first graduate-level-only university. MIST was established in 2007 with a goal to become a world-class research-driven university, focusing on advanced energy and sustainable technologies. MIST has strong ties with the Massachusetts Institute of Technology (MIT) which has supported its development and aim to become a world-class institution. MIST is the first part of the wider Masdar City Master plan to be realized a prototypical and sustainable city, one in which residents and commuters can enjoy the highest quality of life with the lowest environmental footprint. It should be noted that the $16 billion Masdar enterprise is a wholly-owned subsidiary of the Mubadala Development Company.

MIST commenced teaching in 2009 with 92 students from 22 countries and is planning to reach student population of about 800. Accepted students are offered a full tuition scholarship, monthly stipend, travel reimbursement, personal laptop, textbooks, and accommodation. Currently, MIST has 9 faculty members working in nanotechnology areas with diverse expertise. Supported by 20 students and facilities that include clean room (under construction), these faculty members work on projects related to: photovoltaic devices, biodegradable nanocomposite materials, nanoparticles, nanofluids, nanoscale transport in thermoelectric materials, nanostructured materials and their applications in emerging technologies, microfabrication and nanofabrication, low power high-performance nanoelectronics, nanophotonics, and nanomemory technologies.

Ras Al Khaimah Center for Advanced Materials (RAK-CAM) was established in 2007 by the Ruler of the Emirate of Ras Al Khaimah. RAK-CAM aims to position Ras Al Khaimah as a key contributor to the long-term technological development of the Emirates as a leader in advanced materials research. RAK-CAM research areas include: nanomaterials for diverse applications, inorganic and hybrid materials, materials for environmental remediation and hydrocarbon processing (catalysis and separations), materials for water purification and conservation, materials for solar energy applications and energy storage systems, advanced structural materials, ceramics and composites, polymeric materials and polymer nanocomposites, and biomaterials and biofuel technologies. Each research area is supported by 4-5 permanent scientists along with technical and administrative staff as well as short-term post-doctoral researchers. RAK-CAM's research facilities include integrated state-of-the-art systems for materials synthesis and preparation, analysis, testing and characterization, together with an advanced research computing capability. The materials characterization facility serves both the local and regional industry and will seek industrial collaborations in joint research projects.

Another Abu Dhabi Government initiative was the establishment of the Khalifa University of Science, Technology and Research (KUSTAR) in 2007. KUSTR recently announced it will set up a nanotechnology research center. The aim of the center is to play a leading role in the establishment of nanotechnology research, development, and industry in Abu Dhabi and the UAE. The center will be dedicated to research on theoretical and experimental nanotechnology with strong emphasis on the performance attributes and functional demonstration of nanomaterials/systems, by assembling/integrating sophisticated materials, composite materials, and structures that translate to devices and systems at the macroscale. The research center is also intended to develop materials/solutions for applications in power/optoelectronic, aerospace, and diagnostic monitoring.

Jordan

In the early sixties, Jordan realized the importance of S&T to socio-economic development of the country and that led to the establishment of the Scientific Research Council in 1961. The council was responsible for planning, promoting, and financing research; identifying national research priorities; promotion of scientific research culture; and enhancing S&T cooperation with other countries. In 1970 the Late King Hussein Bin Talal and his brother Prince El Hassan (Crown Prince at that time) established the Royal Scientific Society (RSS) which later became the largest applied research institution, consultancy, and technical support service provider in Jordan. Seven years later, the Directorate of Science & Technology as a part of the National Planning Council was established to prepare the science & technology policy, plans, and programs. In 1978 and under the patronage of King and Crown Prince, Jordan's Science & Technology Policy Conference was held. This event was a turning point in reviewing the major issues facing the country in organizing and orienting its scientific and technological efforts.. The conference recommended the institutionalization of science & technology activities under a national umbrella. This event led to the birth of the Higher Council for Science and Technology (HCST) in 1987. HCST is chaired by Prince El Hassan Bin Talal and was established to build a national S&T base to contribute to the achievement of national development objectives. HCST has developed Jordan's first national S&T policy which was adopted in 1995 and is responsible for its implementation. Regarding Nanotechnology, HCST has listed Nanotechnology as a priority in its scientific research priorities for the Years 2011-2020[9].

From the start of his ruling in the year 2000, King Abdullah II has paid a special attention to the higher education and in specific to science and technology. In 2001 The King Abdullah II Fund for Development (KAFD) was created to encourage innovation and growth in Jordan's public and private sectors. The KAFD with the support of the king organizes Petra Conference of Nobel Laureates which is a regular event trying to shed more attention related to science, technology, economy and peace towered Jordan and the nearby region. In 2008, KAFD supported a nanotechnology workshop in Jordan which attracted more than 70 participates from Jordan and the Arabic world whom did research related to nanotechnology. The workshop was organized by the University of Illinois at Urbana-Champaign, the University of Jordan, and the Saudi King Abdullah Institute for Nanotechnology. Additionally, in 2006 his majesty the king has established KADDB industrial park which is a specialized park providing high-level support, service and high-security to host advanced defense and military industries. Applying nanotechnology applications in this industrial park is set as a priority research.

There are a few and relatively limited capital available for science funding in Jordan. For example, the budget of HCST in 2006 was just US$2.1 million[10]. In 1994, the Industrial Scientific Research and Development Fund (IRDF) were established with the objective to increasing the competitiveness of Jordanian industries through the utilization of science and technology. Around 80% of IRDF budget comes from the government and the rest comes from the private sector. Later on, the National Fund for Enterprises Support (NAFES) was established in 2001 as one of the components of Jordan-Japan industrial development program, under the umbrella of HCST. In the 2005, the HCST has established the Scientific Research Support Fund (SRF) as the arm which distributes the government allocated budget for research according to its priority research list. In 2011, the Applied Scientific Research Fund (ASRF) was established as a non-government, non-profit organization which was created by Mr. Samih Darwazah, founder of Hikma Pharmaceuticals, to promote the development of applied science and engineering ideas. ASRF issues 3-6 new grants annually of $15,000 - $150,000 to fund colleges and universities in fields related to medicine, natural sciences, technology and others. In November 2012, ASRF will host the Micro Flow Chemistry and Biology (MIFCAB) workshop. MIFCAB has more than 25 renowned international speakers; many of them are top experts in the field of Nanotechnology. This initiative aim to educate academic and industrial participants from the MENA region on topics related to Microreactor, Microfluidics, and Nanotechnology.

Recently, Jordan has created several centers that capitalize on advanced technologies including nanotechnology. In 1999, the Hamdi Mango Center for Scientific Research (HMCSR) was established in Jordan University (established in 1962) as a result of gracious grant by a Jordanian businessman, the late Mr. Ali Mango. HMCSR conduct original research in the fields of material science, nanotechnology, biotechnology, pharmaceutical solution and drug discovery

HMCSR houses the Material Science and Nanotechnology Laboratories which conduct research in three areas: superconductor material, colloid and surface chemistry, and natural geomaterials for construction and industrial applications. These research areas are supported by more than ten researchers. Additionally HCST has established National Nanotechnology Centre of Jordan (NANCEJ) by the end of 2009 as a response of IBM plan to build a Nanotechnology center in Jordan similar to that in Egypt and Saudi Arabia. In 2010, HCST has approved the establishment of a nanotechnology center at Jordan University of Science & Technology (JUST). That center has more than 30 affiliated faculty members from JUST University with varied backgrounds, chemical, biomedical, mechanical, and electrical engineering; and as well applied science such as chemistry, biology and biotechnology, and physics. It was structured into three divisions: Integrated nano systems, modeling and simulation, and nano structured material and characterization. The research in that center was planned to cover wide range of applications such as industrial microbiology, defense/security, microelectronics fabrication, nanobio, and computational modeling. However, no recent information was founded on the status of allocated budget, or progress of the IBM plan, or over the NANCEJ.

Kuwait

Interest in S&T started in Kuwait soon after it gained its independence from the British Empire in 1961 as demonstrated by the establishment of Kuwait University (KU) in 1966, Kuwait Institute for Economic and Social Planning in the Middle East (now Arab Planning Institute) also in 1966, Kuwait Institute for Scientific Research (KISR) in 1967, Kuwait Science Club in 1974, Kuwait Foundation for the Advancement of Sciences (KFAS) in 1976 (chaired by head of state, the Emir of Kuwait), Ministry of Higher Education 1988, Center for Research and Studies on Kuwait in 1992, the Scientific Center and the Kuwait Inventors Bureau in 2000, Dasman Centre for Research and Treatment of Diabetes in 2006, annual Kuwait Science Fair since 2008, and Sabah Al-Ahmad's Center for Giftedness & Creativity in 2009. By late 70s Kuwait had a renewable energy (solar and wind) research program and even initiated a nuclear energy program for seawater desalination and electricity production. Although lacking a formally enacted national science policy, such development made Kuwait a leading hub for S&T in the region in the 60s, 70s and early 80s. However, during mid-80s, the country suffered from terrorism (bombing of public areas, hijack of Kuwait Airways airplane, and attempt to assassinate head of state. The act of terrorism reached its climax in 1990 when Kuwait was invaded by its northern neighbor, Republic of Iraq. At the end of the seven month-long Iraqi occupation, around 773 Kuwaiti oil wells were set ablaze by the Iraqi army resulting in a major environmental and economic catastrophe. Kuwait's infrastructure was severely damaged during the Gulf War. Moreover, Kuwait has been recently rocked by a series of political crises. In the last six years, eight governments have resigned and the parliament has been dissolved four times. This took a toll on the country's development and delayed many vital projects.

KISR took the initiative of preparing a draft for national science policy, including studying the national base, and the needs of different sectors in the field of S&T in Kuwait. The document entitled "National Policy for Science, Technology and Innovation of the State of Kuwait" was reviewed by experts from Kuwait, Arab and foreign countries including the United Nations Economic and Social Commission for Western Asia (UN-ESCWA) and finalized for submission to the competent authorities in Kuwait in 2007[11]. This made Kuwait one step ahead of most Arab States toward national science policy framework. The national policy draft acknowledged the importance of nanotechnology for Kuwait's S&T development.

In 2010, a $250 million initiative was launched by the Emir of Kuwait to support R&D projects in renewable energy, peaceful uses of nuclear energy, food science, water resources, etc. KISR was able to secure about $50 million of the initiative budget to establish nanotechnology center. KISR's interest in nanotechnology dates back to 2006 when it held Kuwait's first nanotechnology conference. The focus areas of KISR Nanotechnology Research Center (KNRC) research focus areas include renewable energy systems (photovoltaic, fuel cell, and hydrogen storage), construction materials (high performance concrete), surface protection coating materials (corrosion and erosion resistant, self-cleaning, and antibacterial), catalyst materials (oil production and refining), water purification and desalination, and chemical and physical sensing technologies. KNRC will house state-of-the-art 360 m^2 clean room facility equipped fabrication and characterization tools, materials synthesis laboratory, modeling and simulation laboratory, and chemical and physical properties characterization facilities.

Back in 1976 KU College of Science established an Electron Microscopy Unit (EMU) which evolved over the years to be known now as Nanoscopy Science Center (NSC). The center offers many services for biological and material science research. NSC is managed by five faculty members and seven fulltime staff members. The college of engineering and petroleum at KU established in 2007 the Kuwait University Nanotechnology Research Facility (KUNRF) to service nanotechnology R&D in the college. KUNRF is managed by four faculty members, employs four technicians, and three fulltime professional research assistants. The facility houses a clean room with several fabrication and characterization tools. Most of KU nanotechnology research fall in the fields of photovoltaic, nano-electronics, biotechnology, and advanced materials.

Qatar

According to its national development strategy 2011-2016 document[12], Qatar has recently invested considerable resources in R&D. An outstanding infrastructure is in place for scientific research, with programs to draw potential researchers and build partnerships with universities and businesses. However, similar to most Arab States, Qatar does not have formally enacted national science policy. The country's R&D took a sharp turn when Qatar's current head of state (the Emir of Qatar) assumed power in 1995. In the same year, he established Qatar Foundation for Education, Science and Community Development (QF). The aim of this foundation is to unlock the human potentials through its three pillars of education, science & research, and community development. Since its inception, Qatar's First Lady acted as QF's chairperson and has been its driving force. In 2006, the Emir of Qatar pledged to allocate 2.8% of Qatar's GDP to science and research which translates to about $1.5 billion per year. QF is tasked to manage this budget. Moreover, in the same year, he issued a decree to establish the general secretariat for development planning to coordinate plans, strategies and policies in support of Qatar's National Vision 2030 (QNV2030). Approved in 2008, Qatar's long-term development strategy defines broad future trends, sets goals, and provides the framework for Qatar's National Development Strategy. QNV 2030 rests on four pillars: human development, social development, economic development, and environmental development. However, there is no explicit mention of nanotechnology in QNV 2030 or in the Qatar National Development Strategy 2011-2016.

Qatar has invested heavily in developing its capabilities for scientific innovation and research. Particularly, QF has established a broad range of research centers within Qatar University and Education City, as well as research opportunities in scientific and technical areas; policy, social and business areas; innovative design and culture and heritage. QF also paid attention to education programs on scientific research at the university as well as K–12 level. In addition to graduate level research programs, undergraduate research experience program and specialized math and science tracks in secondary schools were established.

The Qatar National Research Fund was established in 2006 to accelerate quality R&D by providing research grants to a wide range of beneficiaries. Qatar Science and Technology Park (QSTP) was established in 2009 to attract investments from several international businesses for frontier R&D. European Aeronautic Defense and Space Company (EADS), ExxonMobil, General Electric (GE), Microsoft, Shell, Total and others have already committed $225 million of R&D investment at QSTP.

Education City is QF's flagship project. Located on the outskirts of Doha, the capital of Qatar, the city covers 14 km^2 and houses educational and research facilities from primary school to graduate level. Education City aims to be the center of educational excellence in the region. It is conceived of as a forum where universities share research and forge relationships with businesses and institutions in public and private sectors. Education City is home to branch campuses of six international universities which include Weill Cornell Medical College, Virginia Commonwealth University School of the Arts, Carnegie Mellon University offering programs in business administration, biological sciences, computational biology, computer science and information systems, Texas A&M's School of Engineering, Georgetown University School of Foreign Service, and Northwestern University School of Communication and Medill School of Journalism.

QF also planning to establish several applied research centers of excellence in Qatar including: Center for Genomic and Proteomics Medicine, Center for Stem Cells Research, Center for Molecular Imaging Research, Center for Infectious Diseases, Center for Bioinformatics and Data Mining, Center for Applied Nanotechnology, and Center for Environmental Research.

Current nanotechnology R&D activities in Qatar are faculty members driven with some research focused on: catalyst for natural gas liquefaction, nanoparticles for cancer treatment, and nanomaterials synthesis including functional nanofibers for protective textile applications, water filtration, and biomedical applications.

Saudi

Saudi leaders have recognized the importance of harnessing S&T for their developmental needs as early as 1985 when they established the King Abdul Aziz City for Science and Technology (KACST) as Saudi's principal agency for promoting scientific and technological R&D. KACST was directed by its charter to propose a national policy for the development of S&T. However, preparations for the national policy for S&T did not start until mid-1997. KACST in cooperation with the Ministry of Economy and Planning (MOEP) developed a long-term national policy for S&T. In July 2002 the Council of Ministers approved the National Policy for Science and Technology (NPST) under the name of "The Comprehensive Long-Term National Policy for Science and Technology" making Saudi one of the few Arab States to have such policy. Saudi's NPST included a timetable for gradually increasing sources of funding R&D, which is to reach 1.6% of the GDP in 2020. KACST was also put in charge of supervising the implementation of the policy.

KACST is also responsible for five years strategic and implementation plans for 11 technologies: water, oil & gas, petrochemicals, nanotechnology, biotechnology, information technology, electronics, communication & photonics, space and aeronautics, energy, environment, and advanced materials. Each plan establishes a mission and vision, identifies stakeholders and users, and determines the highest priority technical areas for Saudi.

The mission of the National Nanotechnology Program (NNP) is to ensure that Saudi is a major player within the international community in R&D of nanotechnologies. NNP will foster academic excellence and ensure that world-class R&D facilities are available to all parts of the economy, from academic institutions to industry. NNP is envisioned to create a multidisciplinary program leveraging all branches of science in order to build competence and capability in nanotechnologies that will help to ensure the future competitiveness of Saudi[13].

Currently, much of the expertise and many of the facilities for conducting nanotechnology research are located at KACST and the following universities: King Fahd University of Petroleum and Minerals (KFUPM) established in 1963, King Abdul Aziz University (KAU) established in 1967, Riyadh University established in 1957 but was renamed to King Saud University (KSU) in 1982, King Abdullah University of Science and Technology (KAUST) established in 2009 with $10 billion endowment (6th largest in the world), King Khalid University (KKU) established in 1998, King Faisal University (KFU) established in 1975, and Taibah University (TaibahU) established in 2003. It is estimated that approximately 30 research projects in the field of nanotechnology have been launched at the above universities and research institutes. Industry is also take advantage of nanotechnology research. Local companies such as Saudi national oil company (established as Arabian American Oil Company, known now as Saudi ARAMCO) and Saudi Basic Industries Corporation have devoted resources to conducting nanotechnology research. The two companies alone have launched more than 20 research projects in the field of nanotechnology. To support this research, they have employed more than 20 PhDs on staff with expertise applicable to nanotechnology research. Much of this research has been in materials and synthesis. While the application of this research has often been aligned with the industrial and economic needs, for instance, looking at improving fossil fuel extraction with nanomaterials, some research has looked at other nanotechnology applications such as: structural materials and coatings, biotechnology, catalysis and membranes, sensors and measurement, electronics and magnetics, energy and environment.

Saudi has established several nanotechnology research centers even before the full realization of the NNP. The Center of Excellence in Nanotechnology (CENT) at KFUPM was established in 2005. CENT currently employs five faculty members, one research scientist, three post-doctoral fellows, one lecturer, two engineers, two scientists, and two administrative staff. CENT also has twenty two affiliated faculty members from physics, chemistry, and engineering departments within KFUPM. CENT's main research focus is on catalysis and photo-catalysis, nanostructured chemical sensors, and carbon nanotubes production and applications. It also conducts activities in the field of anticorrosion processes, biotechnology, environment, and solar cells. CENT is equipped with various state-of-the-art instruments ranging from compositional analyses and physical properties measurements to sintering and synthesis of wide ranges of materials.

The Center of Nanotechnology (CNT) at KAU was established in 2006. The multidisciplinary center work covers several science and engineering areas such as: engineering, pharmacology, medical sciences, genetic engineering (Artificial DNA), basic sciences e.g. physics, chemistry and biology, material science, MEMS devices, computational nanotechnology, fabrication & assembly of different nanomaterials, advanced materials including polymers and semiconductor nanomaterials, and safety and health effect of nanoparticles. CNT carry out its research activities through ten research groups. Each group consists of affiliated researchers from different departments and collages at KAU. The groups name and their size are: nanomaterial (36 researchers), nanofabrication (30 researchers), nanocomposites (23 researchers), nanobiotechnology (55 researchers), nano drug delivery (49 researchers), nanomedicine (61 researchers), nanotechnology-based renewable energy (28 researchers), nanodevices & nanosystems (26 researchers), nanotechnology for desalination & water treatment (30 researchers), nano-computation and simulation (17 researchers). CNT supports its research with several facilities which include: nanomeasurements laboratory, nanomaterials synthesis laboratory, nanofabrication laboratory, and microscopy laboratory.

In 2009, the Center of Excellence of Nano-manufacuturing Applications (CENA) was established at KACST. CENA is a research consortium between KACST, Intel, and selected universities in the Middle East and North Africa (MENA) region. The objectives of the consortium are to build up regional human capacity and reverse the brain drain in the MENA region. Research in CENA is focused on three main themes: (1) fabrication of MEMS, sensors, and integration of CMOS-MEMS, (2) fabrication of autonomous RF sensors and applications, and (3) nano-materials synthesis, processing, and characterization. CENA is planning to construct 3500 m^2 center with state-of-the-art equipment to enable research in nanoprocessing, III-V materials and devices, and optoelectronics.

One of KAUST's core facilities is the Advanced Nanofabrication, Imaging and Characterization (ANIC) facility. ANIC is dedicated to providing the instrumentation, technical expertise, and team-teaching environment to stimulate collaborative research in nanoscale technology. ANIC is a complex of multidisciplinary laboratories that supports research across many different departments within KAUST. ANIC supports not only materials and device research in physics, electrical engineering, mechanical engineering and chemistry, but it also facilitates research interaction and collaboration between the physical, chemical, biological and medical disciplines. ANIC laboratories are grouped into two sub-facilities, advanced nanofabrication sub-facility and imaging and characterization sub-facility. The nanofabrication sub-facility occupies 2000 m^2 of Class-1000 clean room space with multiple bays at Class-100. It includes capabilities for device fabrication and characterizations with a wide range of equipment. The sub-facility is divided into 5 modules, namely: lithography and mask making, deposition and thermal diffusion, wet and dry etching, metrology, and package. This sub-facility is managed by 4 research scientists, 3 research engineers, and 7 technicians. The imaging and characterization sub-facility has comprehensive capabilities for scanning, transmission, confocal, and Raman microscopy, magnetic and thermal measurements, and other instrumentation for materials characterization. It also houses a NMR laboratory which comprises of a suite of 10 NMR spectrometers for solution-based and solid-state samples, together with comprehensive sample preparation facilities for the study of macromolecular structures and spatial distributions, dynamics in solution, and chemical composition of small features in solid-state samples. The sub-facility is divided into 9 modules, namely TEM, SEM, optical microscopy, surface science, XRD and X-ray fluorescence, thermal analysis, low temperature physics laboratory, NMR spectroscopy, and microwave testing laboratory. This sub-facility is managed by 10 research scientists, 2 research engineers, and 7 technicians.

Most recently, King Abdullah Institute for Nanotechnology (KAIN) was established in 2010 at KSU. KAIN's conducts a wide range of activities such as R&D and applied activities in energy, water treatment and desalination, telecommunications, medicine and pharmaceutical, food and environment, and manufacturing and nanomaterials as well as modeling and simulation of nanomaterials, education and training, and economic and industrial studies. To support these research activities, KAIN is planning to employ 15 group leaders, 30 researchers, 30 assistant researchers, 15 technicians, and 60 graduate students. A 13,000 m^2 has been designated to KAIN in which 8,000 m^2 will be used for building laboratories, administrative and researcher's offices, warehouses, workstations, and service areas. A budget of $20 million has been allocated to establish a clean room equipped with state-of-the-art equipment for nanotechnology research activities.

CONCLUSION

Arab States, though share a common language, similar cultures, and even R&D priorities, they vary in terms of wealth, political structure, and S&T development stage. It's clear that governance plays an important role in the advancement of S&T. In our six countries sample, three have parliamentary system (Egypt, Jordan, and Kuwait) and the other three have absolute monarchy. Generally speaking, S&T infrastructure development started earlier in Egypt, Jordan, and Kuwait. Saudi, Emirates, and Qatar had to wait for the executive decision but experienced exceptionally rapid growth and surpassed the other three in many areas especially nanotechnology infrastructure development. With NPST and NNP in place, Saudi quickly achieved advanced nanotechnology R&D stage relative to others. Egypt and Jordan rich abundance of human resources allowed them to lead the scientific production in nanotechnology even with its limited infrastructure and financial resources. Emirates and Qatar are rising stars in nanotechnology and anticipated to produce significant R&D in nanotechnology once they build their human capacity. There are tremendous opportunities in nanotechnology R&D for students, researchers, technology providers such as materials suppliers and equipment manufacturers, as well as startup companies in the six countries investigated.

REFERENCES
[1]UNESCO Science Report 2010: The Current Status of Science around the World, United Nations Educational, Scientific and Cultural Organization, (2010).
[2]Wagdy Sawahel, Building scientific and technological talent in the Broader Middle East and North Africa region (BMENA) Arab capability status & present development and proposed action plan, IDB Science Development Network, (2010).
[3]Wagdy A. Sawahel, Higher education and science & technology in IDB member countries Present development and future prospects, IDB science development network, (2008).
[4]Muhammad Yahaya, Muhamad Mat Salleh, Imran Ho-Abdullah, and Yap Chi Chin, Roadmap for Achieving Excellence in Higher Education in Nanotechnology, Islamic Development Bank, (2009).
[5]Peter Hahn and Gerd Meier zu Köcker, The Egyptian Innovation System: An Exploratory Study with Specific Focus on Egyptian Technology Transfer and Innovation Centres, Institute for Innovation and Technology, (2008).
[6]Science and technology policy, research management and planning in the Arab Republic of Egypt, U.S. National Academy of Sciences, (1976).
[7]Science, Technology and Innovation (STI) System in Egypt, Academy of Scientific Research & Technology, (2010).
[8]UAE Vision 2021, UAE Federal Cabinet, (2010).
[9]Scientific Research Priorities in Jordan for the Years 2011-2020, Higher Council for Science and Technology, (2011).
[10]P. Larzillière, Research in Context: Scientific Production and Researchers Experience in Jordan, *Science, Technology and Society*, **15** (2), 309 (2010).
[11]Nader Al-Awadhi and Yousuf Al-Sultan, National Policy for Science, Technology and Innovation of the State of Kuwait, Kuwait Institute for Scientific Research (2007).
[12]Qatar National Development Strategy 2011~2016, Qatar General Secretariat for Development Planning, (2011).
[13]Strategic Priorities for Nanotechnology Program, King Abdulaziz City for Science and Technology and Kingdom of Saudi Arabia Ministry of Economy and Planning, (2007).

VISCOSITY OF ETHYLENE GLYCOL+ WATER BASED Al₂O₃ NANOFLUIDS WITH ADDITION OF SDBS DISPERSANT

Babak LotfizadehDehkordi[*a], Salim. N. Kazi[a], Mohd Hamdi[b]

[a] Department of Mechanical Engineering, Faculty of Engineering, University of Malaya, 50603 Kuala Lumpur, Malaysia.
[b] Department of Engineering Design and Manufacture, University of Malaya, 50603 Kuala Lumpur, Malaysia.
[*] Corresponding author email: babakld@siswa.um.edu.my

ABSTRACT

During the last decade many studies have been conducted on the thermophysical properties of nanofluids as a new medium in heating and cooling systems. However, few studies are available on the viscosity of nanofluids especially in the presence of a dispersant. In heat transfer systems, the viscosity of fluid is an essential factor for pumping power estimation. This study focused on the viscosity of Al_2O_3 nanoparticles, at concentrations of 0.01-1.0% dispersed in the mixture of ethylene glycol+water (mass ratio of 60:40) with addition of sodium dodeobcylbenzenesulfonate (SDBS). Results indicate that nanofluids with higher Al_2O_3 nanoparticle concentrations have higher viscosities. It is also interesting that, nanofluids with low volume concentrations showing almost same viscosity values, which might be due to well dispersed nanoparticles as a result of the SDBS dispersant. However, at high concentrations the SDBS dispersant negatively influences the viscosity of nanofluids. In addition, the viscosity of the nanofluid significantly decreases with rising temperature.

INTRODUCTION

Enhancement of energy efficiency is the major concern of present science and technology. Therefore many investigations conducted to improve the thermal properties of the heat transfer fluids by adding nano-sized particles to the base fluid, which is called a nanofluid. Recently, many studies have been conducted to investigate the thermophysical properties of nanofluids where the thermal conductivity of the nanofluids attracts many attentions [1-4]. Although the viscosity of the nanofluid is an essential factor for nanofluids applications, there is limited literature on the viscosity of nanofluids [5].

For the first time, Choi et al. [6] stated the concept of nanofluid as the new class of nanotechnology by suspending tiny nanoparticles in conventional heat transfer liquids to enhance the thermal conductivity of them. In addition, nanoparticles have overcome the challenges of using micrometer and millimeter sized particles with respect to stability, clogging and erosion in the channels [1, 6].Various experimental and analytical studies were performed on the thermophysical characteristic of nanofluids [7-10].

General trend reveals enhancement of viscosity of nanofluids with further addition of nanoparticle and its reduction with rise of temperature. There are discrepancies between experimental results, and theoretical models are almost futile for viscosity predication [11-13]. Lee et al. [14] studied the effective viscosity of Al$_2$O$_3$-water nanofluids for low volume concentration without any dispersant. They observed reduction of effective viscosity with a rise of temperature and decrease of Al$_2$O$_3$ concentration.

Limited investigations were performed on the effective thermal conductivity and viscosity of metal oxides nanofluids at low volume concentrations [14]. The present investigation was conducted to determine the viscosity of alumina nanofluids from 0.01 up to 1 vol.% concentrations with temperature influence. The mixture of ethylene glycol and water (EG-W) by mass ratio of 60:40 was used as the base liquid, which is a common base liquid in the cold regions. Anionic dispersant, SDBS, was added to the mixture of EG-W to investigate the advantages of higher stability and its influence on the effective viscosity of Al$_2$O$_3$ nanofluids. SDBS is also considered as an organic dispersant, which has considerable advantages over inorganic dispersants [15].

2. EXPERIMENTAL METHOD

Preparation of nanofluids is fundamental for their investigation, because of the worth of the stability of nanofluids for running experiments and applications. The two-step method preparation which has shown good results for oxide nanosuspensions [1], was used in this study. Base liquid, was obtained by mixing proportionately water with ethylene glycol (manufactured by Fluka Company). An anionic dispersant SDBS (provided by Sigma Aldrich company) with a mass ratio of SDBS to alumina nanoparticle of 1:1 [16], was added to the EG-W base liquid. Then, alumina nanoparticle with an average size of 13 nm (provided by Sigma Aldrich Company) was added to the mixture of EG-W and SDBS to prepare the nanofluids.

Ultrasonication of the nanofluid samples was implemented for breaking down of the agglomerated particles. An ultrasonic cleaner (manufactured by E-Chrom Tech.) with amplitude of 40 KHz and 350 W, was applied in the present work. In addition, Zeta potential of four dilute nanofluids, with different sonication times (30, 60, 90, 180 min) was measured by Malvern ZS NanoSanalyzer (Malvern Instrument Inc, London, UK) to obtain the optimized sonication time.

The KD2Pro which works on the basis of a transient hot wire was employed for thermal conductivity measurements.

The Vibro viscometer (SV-10 Sin wave, A&D Company) with accuracy of 3% was used for viscosity measurements. It works on the basis of a pair of spring plates with a sensitive plate at the end, which is subjected to a resonance vibration of driving current in reverse phase (like tuning-fork).

3. RESULTS AND DISCUSSION

3.1. Zeta potential

Fig. 1 shows the zeta potential of the alumina nanofluids, which is scattered at the negatively charged state. Initially, by increasing the sonication time the absolute value was increased up to 44 mV, which is considered a good stability [17]. Further, by increasing the sonication time the absolute value was decreased, which represents the stability reduction of the nanofluid. It can be declared that at the first stage ultrasonic vibration had broken down the agglomerated particles and dispersed them well in the base liquid, which ultimately increases the stability of the nanofluid. However, at the second stage more sonication time retards the repulsive forces between nanoparticles and reduces the stability of the nanofluid. The one and a half hour is identified as an optimized sonication time for the preparation of nanofluids.

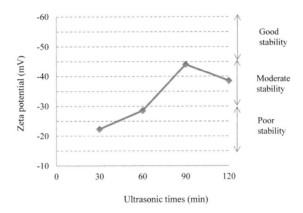

Fig. 1. Zeta potential of alumina nanofluids versus ultrasonic times

3.2. Benchmark study

Before conducting experiments, the viscosity of the EG-W based liquid was measured for comparison with available data [18] and also for estimation of the accuracy of viscometer apparatuses. Fig. 2 shows the comparison between measured viscosity data with the 5% specific error bar and trend of ASHRAE data, which illustrates a fine agreement between them. A maximum deviation of 2.9% is observed at 25 °C in comparison to ASHRAE standard [18].

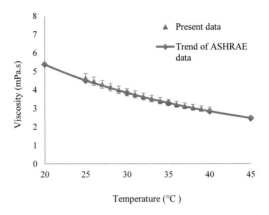

Fig. 2. Comparison of present viscosity data with ASHRAE data [18] for EG-W base liquid.

3.3. Influence of SDBS on viscosity of EG-W

The mixture of SDBS and EG-W was investigated to indentify the influence of SDBS on the viscosity of the base liquid. Fig. 3 shows the effect of SDBS concentration on the relative viscosity of EG-W. It shows that the viscosity of the EG-W mixture is strongly dependent on the loading of SDBS dispersant. Minor enhancement of relative viscosity was observed at low concentrations (<1 wt%) of SDBS. A maximum viscosity enhancement of 16% is observed at 4 wt% concentration of SDBS dispersant, which might be due to formation of foams and contamination of the fluid medium. Therefore, it can be inferred that high concentrations of SDBS dispersant increase the viscosity of the nanofluid.

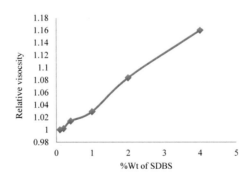

Fig. 3. SDBS influence on viscosity of EG-W base liquid.

3.4. Effective viscosity

Effective viscosity of alumina nanofluids was measured at 0.01-1 vol.% concentrations and temperature range of 25-40 °C (Fig. 4). Results indicate that nanofluids with higher alumina nanoparticle concentrations have higher viscosities. It is also interesting that nanofluids with low volume concentrations (<0.1 vol.%) showing almost same viscosity values with the rise of temperature. In addition, viscosity of nanofluids significantly decreases with rising of temperature at various concentrations. This shows that loading of the alumina nanoparticles increases the friction and flowing resistance of the fluids which ultimately causes an increase of viscosity. Rising the temperature creates a higher space, due to particles motion. Therefore, the density of the nanofluids decreases, which causes the reduction of shear stress as reported by others [19].

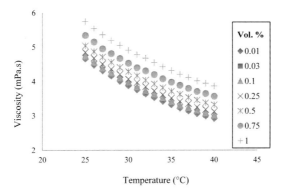

Fig. 4. Viscosity of alumina nanoparticles in EG-W mixture as a function of temperature.

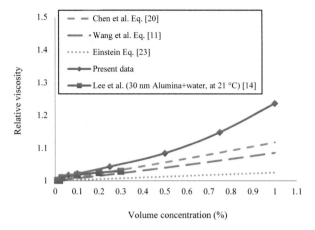

Fig. 5. Relative viscosity of alumina in EG-W mixture nanofluid at 25 °C.

Fig. 5 presents comparison between relative viscosities of present experimental data and theoretical correlations [11, 20, 21] as a function of volume concentration for alumina nanoparticles dispersed in the EG-W at 25 °C. The results show growth of relative viscosity with increasing alumina volume concentration from 0.01 to 1 vol.% in EG-W base fluid. A maximum enhancement of about 23% is achieved at 1 vol.% concentration. Present experimental data show higher relative viscosity over predicted values from theoretical models. In addition, a comparison between present experimental data and works by Lee et al. [14] with alumina water nanofluid at 21 °C (without any dispersant and surfactant) shows higher relative viscosity of present data. However, nonlinear enhancement is observed in both investigations which were against linear prediction of theoretical models. This might be a sign of homogeneous dispersion of nanoparticles in the base fluid which causes a shorter interparticle distance and therefore more interaction between nanoparticles in nanofluid suspension [14]. The higher viscosity values that were obtained in the current experiment in comparison to the data of Lee et al. [14] might be due to the difference in base liquids, particle size, temperature and also presence of SDBS dispersant in present work.

4. CONCLUSION

The viscosity of the alumina nanofluid having EG-W by mass ratio of 60:40 as the base fluid has been studied in the presence of the SDBS dispersant. One and a half hour running of ultrasonic bath was the optimized sonication time.

Low concentrations of SDBS had negligible effect on the viscosity of the EG-W base fluid, while at higher concentrations (>1 wt%) the SDBS leads to enhancement of viscosity which influences application of SDBS for the nanofluids preparation. It is clear from the results that the relative viscosity is highly temperature dependent. Present investigation shows nonlinear enhancement of viscosity of nanofluids with loading of alumina nanoparticles, while estimation of various theoretical correlations represent linear and lower enhancement of viscosity with particles loading. Enhancement of viscosity of present data over modeling prediction is higher which is explained by the presence of SDBS. At low concentrations (<0.1 vol.%) of nanofluids, SDBS influences better dispersion of particles in suspension. However, at higher concentrations (>0.1 vol.%), addition of SDBS produces foam in the nanofluid, that has negative effects and increases the viscosity of the nanofluids. Hence, further studies needed to optimize the amount of SDBS for the best stability and thermal conductivity of the nanosuspension.

REFERENCE

1.   Choi, S.U.S., *Nanofluids: from vision to reality through research.* J. Heat Transfer, 2009. **131**: p. 033106.
2.   Chandratilleke, T., D. Jagannatha, and R. Narayanaswamy, *Heat transfer enhancement in microchannels with cross-flow synthetic jets.* Int. J. Therm. Sci., 2010. **49**(3): p. 504-513.
3.   Huq, M., A. Aziz-ul Huq, and M.M. Rahman, *Experimental measurements of heat transfer in an internally finned tube.* Int. Commun. Heat Mass Transfer, 1998. **25**(5): p. 619-630.
4.   Ho, C.J., L.C. Wei, and Z.W. Li, *An experimental investigation of forced convective cooling performance of a microchannel heat sink with Al2O3/water nanofluid.* Appl. Therm. Eng., 2010. **30**(2-3): p. 96-103.
5.   Mahbubul, I.M., R. Saidur, and M.A. Amalina, *Latest developments on the viscosity of nanofluids.* International Journal of Heat and Mass Transfer, 2012. **55**(4): p. 874-885.
6.   Choi, S.U.S. *Enhancing thermal conductivity of fluids with nanoparticles.* in *Dev. Appl. Non-Newtonian Flows.* 1995. New York: ASME.
7.   Duangthongsuk, W. and S. Wongwises, *Measurement of temperature-dependent thermal conductivity and viscosity of TiO2-water nanofluids.* Exp. Therm. Fluid Sci., 2009. **33**(4): p. 706-714.
8.   Golubovic, M.N., et al., *Nanofluids and critical heat flux, experimental and analytical study.* Appl. Therm. Eng., 2009. **29**(7): p. 1281-1288.
9.   Sonawane, S., et al., *An experimental investigation of thermo-physical properties and heat transfer performance of Al2O3-Aviation Turbine Fuel nanofluids.* Appl. Therm. Eng., 2011. **31**(14-15): p. 2841-2849.
10.   Teng, T.P., et al., *The effect of alumina/water nanofluid particle size on thermal conductivity.* Appl. Therm. Eng., 2010. **30**(14-15): p. 2213-2218.
11.   Wang, X., X. Xu, and S.U.S. Choi, *Thermal conductivity of nanoparticle-fluid mixture.* J. Thermophys. Heat Transfer, 1999. **13**(4): p. 474-480.
12.   Timofeeva, E.V., et al., *Thermal conductivity and particle agglomeration in alumina nanofluids: experiment and theory.* Phys. Rev. E., 2007. **76**(6): p. 061203.
13.   Chandrasekar, M., S. Suresh, and A. Chandra Bose, *Experimental investigations and theoretical determination of thermal conductivity and viscosity of Al2O3/water nanofluid.* Exp. Therm. Fluid Sci., 2010. **34**(2): p. 210-216.
14.   Lee, J.H., et al., *Effective viscosities and thermal conductivities of aqueous nanofluids containing low volume concentrations of Al2O3 nanoparticles.* Int. J. Heat Mass Transfer, 2008. **51**(11-12): p. 2651-2656.

15. Gocmez, H., *The interaction of organic dispersant with alumina: A molecular modelling approach.* Ceram. Int., 2006. **32**(5): p. 521-525.

16. Wu, S., et al., *Thermal energy storage behavior of Al2O3-H2O nanofluids.* Thermochim. Acta, 2009. **483**(1-2): p. 73-77.

17. Ghadimi, A., R. Saidur, and H. Metselaar, *A review of nanofluid stability properties and characterization in stationary conditions.* Int. J. Heat Mass Transfer, 2011. **54**(17-18): p. 4051-4068.

18. *ASHRAE Handbook, Fundamentals.* 2001, Atlanta, GA: American Society of Heating, Refrigerating and Air-Conditioning Engineers, Inc.

19. Godson, L., D.M. Lal, and S. Wongwises, *Measurement of thermo physical properties of metallic nanofluids for high temperature applications.* Nanoscale Microscale Thermophys. Eng., 2010. **14**(3): p. 152-173.

20. Chen, H., et al., *Rheological behaviour of ethylene glycol based titania nanofluids.* Chem. Phys. Lett., 2007. **444**(4-6): p. 333-337.

21. Einstein, A., *Investigations on the Theory of the Brownian Movement.* 1956, New York: Dover Pubns.

CLUSTERING THEORY EVALUATION FOR THERMAL CONDUCTIVITY ENHANCEMENT OF TITANIA NANOFLUID

Azadeh Ghadimi, Hendrik Simon Cornelis Metselaar

Center of Advanced Material, Department of Mechanical Engineering, Faculty of Engineering, University of Malaya, 50603 Kuala Lumpur, Malaysia

ABSTRACT

Nanofluid as a potential cooling fluid has attracted a wide span of application due to its great capability of heat dissipation. Nanofluid is a suspension of nanoparticles in a base fluid which often requires stabilization processing. In this study, 0.1%wt. of TiO_2 nanoparticles were mixed with distilled water. Three hours ultrasonication was applied to prepare samples. Sodium lauryl sulfate (0.01-0.2) %wt. and pH (10-12) were used as stability controlling parameters in order to verify the relative thermal conductivity and the particle size of nanofluid. Central composite design (CCD) and response surface methodology (RSM) were used to develop a model as well as define the optimum condition with design of experiments (DOE). Amongst the variety of theories, clustering and agglomeration can be partially responsible for increasing thermal conductivity with rising particle size among different nano-suspensions. Results revealed that optimum condition is located at high pH and low SDS concentration in which high thermal conductivity will be accompanied with larger particle sizes.

1. INTRODUCTION

Nanofluid as a dispersion of nanoparticle in a base fluid is being investigated for its enhanced thermophysical properties such as thermal conductivity, viscosity, specific heat, and convective heat transfer coefficients for almost two decades. From previous investigations, nanofluids have been found to exhibit enhanced thermal conductivity which amplifies with increasing volumetric fraction of nanoparticles[1-5]. Diverse benefits of the nanofluid including great energy saving, emission reduction, and pollution reduction, in various applications such as nuclear systems cooling, buildings air-conditioning, space and defense were reported by researchers [1, 6].

Moreover, the thermal conductivity enhancement mechanism is not quite clear after about two decades of experimental and analytical works on nanofluids. Keblinski et al.[7] elaborated on four potential mechanisms of heat transfer enhancement in nanofluids including Brownian motion, molecular-liquid layering, and diffusive propagation of heat in both particles and liquid, and clustering of nanoparticles. At that time, they found out that clustering may have a negative effect on heat transfer enhancement. In addition, Xuan et al.[8] revealed that nanoparticle aggregation reduced the efficiency of the energy transport enhancement of the suspended nanoparticles. However, in 2006, Prasher et al. [9] boosted the probability of two of the theories. They showed that (1) micro-convection caused by the BM (Brownian motion) of the nanoparticles and (2) clustering aggregation of the nanoparticles leading to local percolation behavior are the most appropriate theories. They reviewed the experimental and simulation studies from 1993 to 2005 and finally proved that optimized aggregation size in low volume concentration nanofluids (less than 1%) explains the surprising improvements in thermal conductivity of nanofluids (see Fig. 1). Therefore, not reaching the optimized size for the agglomerates, can be the possible reason of the diversity in the the reason of scattered data among the investigators.

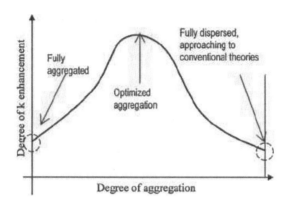

Fig.1   Aggregation effect on the effective thermal conductivity[10]

Prasher (2006) claimed that the probability of aggregation increased with decreasing particle size, at a constant volume concentration. The reason is that a decrease in the average interparticle distance makes the attractive van der Waals force more important. Due to this, we designed some experiments with constant volume concentration in which different dispersion methods show different particle size distributions. Since there is no systematic investigation on the influence of the dispersion method on the particle size distribution and/or the thermal conductivity enhancement, the optimized values of pH and surfactant concentration in the stabilization process of the nanofluid has been monitored in this study. Therefore, the present work is an attempt to optimize the influence of two stabilizing parameters (i.e. pH value and surfactant weight percentage) on the stability response of the particle size along with the relative thermal conductivity response of the nanofluid. The relative thermal conductivity was chosen as a response to show the enhancement of nanofluid thermal conductivity relative to the basefluid (distilled water). The optimization process using RSM with CCD method was performed and the obtained results are presented.

2.   METHODOLOGY
    2.1   Nanofluid material and Preparation
    Our experiments were conducted using 0.1%wt. titania nanoparticles with an average diameter of 25 nm and a specific surface area of 45-55 m^2/g from Sigma Aldrich Company (USA) dispersed in distilled water. An anionic surfactant, SDS in chemical grade, from Sigma

Aldrich (USA), was used to stabilize the suspension. In this research, the value of pH was controlled by adding NaOH and HCl.

A two-step method, which is the most applicable method based on the literature survey (Beck 2008; Wang et al. 2009) was applied to prepare the nano-suspensions. A certain amount of surfactant between 0.01 %wt. and 0.2 %wt. was added to distilled water and mixed by a magnetic stirrer for almost 1.5 h to have a suitable dispersion. In the next step, 0.1 %wt. of titania nanoparticles was added and processed in an ultrasonic bath (Branson 3210) for three hours. The value of pH (10-12) for the suspension was set to the defined value before and after nanoparticle addition.

2.2　Thermal conductivity measurement

Transient hot wire method (THW) is a well-established, accurate, reliable, and robust technique for the thermal conductivity measurement of nanofluids which was applied successfully by Mintsa et al.[11], Wen and Ding[12], Meibodi et al.[13], Moosavi et al.[14], Yeganeh et al.[15], and Abareshi et al.[16]. Therefore, thermal conductivity measurements in this study were done based on the THW method by means of KD2 Pro (Decagon). This thermal analyzer device has 5% accuracy over the temperature range of 5°C to 40°C. The device was calibrated by glycerin and distilled water and the measurements were within the ±5% accuracy. Thermal conductivity of fresh nanofluids was measured right after the preparation.

2.3　Particle size measurement

Dynamic light scattering equipment (Malvern 3000HSA) in nano series was used to measure the particle size distribution (accuracy: 0.1 nm). Since the prepared nanofluid was not clear enough for particle size measurement, the samples were diluted with distilled water to order to be tested by Malvern zeta sizer.

3.　DESIGN OF EXPERIMENT (DOE)

The Design-Expert software (v.8) was used for the statistical design of experiments and data analysis for the nanofluid. In this study, CCD and RSM were applied to optimize the two operating variables, i.e. the value of pH (10-12) and the surfactant concentration (0.01-0.2 %wt.). The range of variables was determined based on the literature as shown in Tables 1 and 2.

3.1　Data analysis

Table 1 shows the coded and real values for the designed experiments based on the CCD methodology. The coded values for SDS %wt. (A) and pH value (B) were set at five levels (-1 (minimum), -0.5, 0 (central), $+0.5$, and $+1$ (maximum)). It is worth mentioning that the thermal conductivity response is the most important parameter to measure the heat transfer of nanofluid.

In the first step of RSM, a proper approximation was implemented to find the relationship between the dependent variable and the set of independent variables. In the next step, the behavior of the system was defined by the following quadratic polynomial equation[17, 18]:

$$Y = \beta_0 + \sum_{i=1}^{k} \beta_i X_i + \sum_{i=1}^{k} \beta_{ii} X_i^2 + \sum_{i<j}^{k} \sum_j^k \beta_{ij} X_i X_j \quad (1)$$

Table 1 Factors and responses used for CCD in model optimization (coded values)

Exp.	Factor A Code	SDS (wt.%)	Factor B Code	pH	Response 1 Particle size nm	Response2 K_{nf}/K_{bf} -
1	0	0.11	-0.5	10.5	271.9	0.994
2	+1	0.20	-1	10.0	232.3	1.006
3	0	0.11	0	11.0	498	1.016
4	+1	0.20	+1	12.0	327	1.006
5	-1	0.01	-1	10.0	295	0.954
6	+0.5	0.15	0	11.0	409.1	1
7	0	0.11	0	11.0	432.4	1.003
8	-0.5	0.06	0	11.0	397	1.013
9	-1	0.01	+1	12.0	350	1.013
10	0	0.11	+0.5	11.5	450	1.008
11	0	0.11	0	11.0	514.2	1.013

Where, ε is the random error, $X_i,...,X_k$ are input factors that influence the predicted response Y, β_0 is the constant and β_i, β_{ii}, β_{ij} are the linear, quadratic, and interaction effects. The fitted equation is expressed as contour plots in order to visualize the relationship between the response and experimental levels of each factor and to infer the optimum conditions. Design Expert (version 8.0.7.1) software package was used to analyze the experimental data. A P-value of less than 0.05 was considered significant.

4.   RESULT AND DISCUSSION
    ANOVA was used to estimate the effects of main variables and their potential interaction on the stability of the considered nanofluid. The experimental results obtained from various homogenization methods of the titania nanofluid were analyzed using multiple regression analysis. The statistical results of the two responses including particle size (nm) and relative thermal conductivity are shown in Table 2.

Table 2   Statistical characteristics of optimum models and ANOVA results

Responses	F	P	LOF F	LOF P	R	$R_{Adj.}$	$R_{Pred.}$	AP	CV
Particle size	15.46	0.0018	1.16	0.5241	0.869	0.812	0.673	10.4	8.44
K	18.78	0.003	0.36	0.7911	0.949	0.8989	0.687	14.971	0.54

    Adequate precision (AP) and coefficient of variance (CV), Table 2, are within the acceptable region which is more than 4 and below 10, respectively. Lack of fit (LOF) measures the error due to deficiency of the model. LOF indicates how well the model fits the data. Strong LOF ($P < 0.05$) is an undesirable property, because it indicates that the model does not fit the data well. It is enviable to have an insignificant LOF ($P > 0.1$). The important outputs of the model are the F-value and associated probability ($P > F$). The higher the F-value, the more likely the model does not adequately fit the data. Similarly, P should be less than 0.05 to demonstrate that the model terms are significant. The F-value and P for all responses are in the acceptable range and therefore, the predicted model is significant [19, 20].

Table 2 also shows that the difference between the adjusted R and predicted R is within the accepted range of 0-0.2. In order to get a quick impression of the overall fit and the prediction power of a constructed model, R, R_{Adj} and R_{Pred} are very convenient. Table 3 represents the statistical results of ANOVA in terms of coded factors for P-values.

Table 3 P-Value of four responses

Model	P-Value	
	Particle size	Thermal conductivity
	Response 1	Response 2
	0.0018	0.0030
A-SDS %wt.	0.2249	0.1443
B-pH	0.0270	0.0025
AB	--------	0.0031
A^2	--------	--------
B^2	0.0005	0.0141
A^2B	--------	--------
AB^2	--------	0.0299
Lack of Fit	0.5241	0.7911

The results obtained by ANOVA imply that the terms B and B^2 have great influence on the stability of TiO_2 nanofluid in terms of particle size as depicted in Table 3. It is concluded that the particle size response completely relies on the %wt. of surfactant rather than the value of pH.

Furthermore, as shown in Table 3, only the coded parameters B, AB, B^2, and AB^2 have significant effect on the thermophysical properties in terms of the thermal conductivity of TiO_2 nanofluid. In other words, any changes in the value of these coded parameters will bring significant change to the thermal conductivity of the nanofluid. It means that K response is more dependent on the value of pH compared to the %wt. of surfactant, however, the interaction between these two factors are a matter of concern in verifying the changes of thermal conductivity even in higher degrees. After eliminating the insignificant parameters, polynomial equations and response surfaces for a particular response in terms of coded factor were produced using RSM.

For the stability characteristics, square root transformation with k set to -150 were required to improve the particle size models. For an experimental design with two factors, the model including linear, quadratic, and cross-terms can be expressed in terms of coded factors as given:

For particle size response (PS):
$$\sqrt{(PS - 150)} = 17.29 - 0.8A + 1.68B - 5.15B^2 \tag{2}$$

For thermal conductivity response:

$$K = 1.01 - 0.013A + 0.014B - 0.015AB - 0.013B^2 + 0.025AB^2 \tag{3}$$

Where A and B are coded factors for %wt. of the surfactant and the value of pH, respectively. Therefore, the gained quadratic models represent the actual process of stabilizing nanosuspension. Fig. 2 represents the linear regression between the predicted and the actual experimental results for all considered responses using RSM.

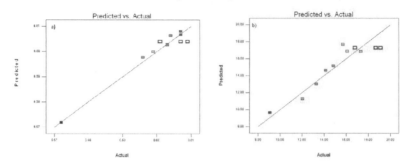

Fig. 2 A parity plot between actual experiment and predicted values using RSM for TiO₂ nanofluid: (a) thermal conductivity, (b) particle size, (The vertical and horizontal axes are predicted output and corresponding targets)

The best linear fit is indicated by a solid line. If there were a perfect fit (outputs exactly equal to targets), the slope would be 1 and the y-intercept would be 0. The correlation coefficient (R-value) between the outputs and targets for all responses are given in Table 2. Fig. 2 indicates that the models were successful in deriving the correlation between the factors and the responses (i.e. particle size, thermal conductivity).

4.1   Influence of the parameters on responses

In general, the ideal situation may be considered as the maximum value of thermal conductivity. In contrast, for particle size response, the best condition may be judged as the minimum value of particle dimension. Fig. 3 illustrates the 3D plots for predicted responses.

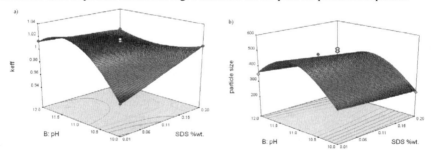

Fig. 3 3D response surface plot for TiO₂ nanofluid: (a) thermal conductivity, (b) particle size

The minimum particle size can be seen at the lowest value of pH (pH=10). In other words, at low surfactant concentration, particle size increases by pH value increment. This phenomenon can be explained by significant quadratic term of B with the negative influence (see Fig. 3b and Eq. (2)).

Since the K response depends on more variables (A, B, AB, B^2, AB^2) than the agglomerate size, determining the enhancement theory is more complex than the particle size. Therefore, the agglomerate size can only be a partial parameter for the K effect. Then, since the terms that are shared between the responses (A, B and B^2) have the same sign we can conclude that they are related. Consequently, it can be inferred that large particle size may construct a chain shape cluster which increases the heat transfer due to the clustering theory by Evans[21].

4.2 Optimization of nanofluids preparation

The main objective of this study is to bring about the optimum TiO$_2$ nanofluid preparation method in terms of pH value and SDS weight concentration. Therefore, maximum thermal conductivity is assumed to reach the most efficient cooling medium for the titania nanofluid. It is a challenging task to define the optimum agglomeration size according to the existing theory of Prashe et al.[9]. Fig. 4 demonstrates the desirability plot for obtained model in 2D and 3D views.

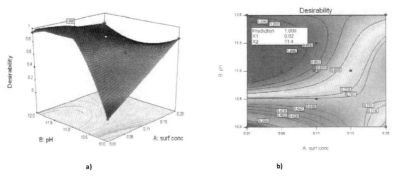

a)   b)

Fig. 4 Desirability plot for TiO$_2$ nanofluid: (a) 3D surface mesh, (b) 2D plot with contours

The desirability is defined as a multiple response method with numerical optimization. In this method a point at which the desirability function is maximized should be found. The goal of optimization is to find a good set of conditions that will meet all the goals, not to get to a desirability value of 1.0.

In this research, the obtained prediction points include 36 points with desirability equal to 1.0. As shown in Fig. 4 the selected point locates at pH value of 11.4 and SDS concentration of 0.02 %wt. The single limiting condition for this point is maximizing thermal conductivity. In this case, we have a preparation method in which enhanced thermal conductivity of nanofluid will be 2.6%. The coagulation size in this condition is 471.36 nm. Therefore, larger agglomerate sizes lead to higher thermal conductivity. Due to the experiment limitations, this set of experiments cannot evaluate the clustering and agglomeration theory by Prasher et al.

5 CONCLUSION

The response parameters which determine the samples' properties were particle size (in nm) and the relative thermal conductivity. Three repeated runs were carried out at the center of experiment in order to measure the reproducibility at different combinations of the process parameters.

The obtained experimental results were exported to Design Expert (v.8) and the effect of the process parameters including surfactant dosage and the value of pH in the stability of TiO_2 nanofluid and the optimum condition for the stability properties were studied using RSM and CCD.

Although, we observed a correlation between aggregate size and thermal conductivity, the behavior of the latter is more complex. The model proposed by Prasher et al.[9] should therefore be improved by adding other factors which are not clear at the present time. The outcomes reveal that bigger radius of gyration results in better enhancement of heat transfer. As the initial average diameter of titanium dioxide was 25 nm and the nanoparticle aggregated size with enhanced thermal conductivity of 2.6% was 463 nm (at 0.1 %wt.), it is concluded that clustering theory is probably one of the roots to describe the increasing thermal conductivity.

ACKNOWLEDGEMENTS

The authors wish to acknowledge the Ministry of Higher Education of Malaysia, Kuala Lumpur, Malaysia, for the financial support under UM.C/HIR/MOHE/ENG/21 (D000021-16001) and express their gratitude to University of Malaya for the financial support grant provided with the reference number of UM.C/241/9.

REFERENCES:

[1]W. Yu, et al., "Review and Comparison of Nanofluid Thermal Conductivity and Heat Transfer Enhancements," *Heat Transfer Eng.*, 29[5] 432-460 (2008).

[2]T.X. Phuoc, M. Massoudi, and R.H. Chen, "Viscosity and thermal conductivity of nanofluids containing multi-walled carbon nanotubes stabilized by chitosan," *INT J THERM SCI*, 50[1] 12-18 (2011).

[3]M.J. Pastoriza-Gallego, et al., "CuO in water nanofluid: Influence of particle size and polydispersity on volumetric behaviour and viscosity," *Fluid Phase Equilibria*, 300[1â€"2] 188-196 (2011).

[4]S.Q. Zhou and R. Ni, "Measurement of the specific heat capacity of water-based Al_2O_3 nanofluid," *APPL PHYS LETT*, 92[9] (2008).

[5]J. Buongiorno, "Convective Transport in Nanofluids," *J HEAT TRANS-T ASME*, 128[3] 240-250 (2006).

[6]X.-Q. Wang and A.S. Mujumdar, "A review on nanofluids - part II: experiments and applications," *Brazilian Journal of Chemical Engineering*, 25 631-648 (2008).

[7]P. Keblinski, et al., "Mechanisms of heat flow in suspensions of nano-sized particles (nanofluids)," *INT J HEAT MASS TRAN*, 45[4] 855-863 (2002).

[8]Y. Xuan, Q. Li, and W. Hu, "Aggregation structure and thermal conductivity of nanofluids,"*AIChE Journal*, 49[4] 1038-1043 (2003).

[9]R. Prasher, P.E. Phelan, and P. Bhattacharya, "Effect of Aggregation Kinetics on the Thermal Conductivity of Nanoscale Colloidal Solutions (Nanofluid)," *Nano Lett.*, 6[7] 1529-1534 (2006).

[10]D. Wen, et al., "Review of nanofluids for heat transfer applications," *Particuology*, 7[2] 141-150 (2009).

[11]H.A. Mintsa, et al., "New temperature dependent thermal conductivity data for water-based nanofluids," *INT J THERM SCI*, 48[2] 363-371 (2009).

[12]D. Wen and Y. Ding, "Experimental investigation into convective heat transfer of nanofluids at the entrance region under laminar flow conditions," *INT J HEAT MASS TRAN*, 47[24] 5181-5188 (2004).

[13]M.E. Meibodi, et al., "The role of different parameters on the stability and thermal conductivity of carbon nanotube/water nanofluids," *INT COMMUN HEAT MASS*, 37[3] 319-323 (2010).

[14]M. Moosavi, E.K. Goharshadi, and A. Youssefi, "Fabrication, characterization, and measurement of some physicochemical properties of ZnO nanofluids," *INT J HEAT FLUID FL*, 31[4] 599-605 (2010).

[15]M. Yeganeh, et al., "Volume fraction and temperature variations of the effective thermal conductivity of nanodiamond fluids in deionized water," *INT J HEAT MASS TRAN*, 53[15-16] 3186-3192 (2010).

[16]M. Abareshi, et al., "Fabrication, characterization and measurement of thermal conductivity of Fe_3O_4 nanofluids," *J MAGN MAGN MATER*, 322[24] 3895-3901 (2010).

[17]R.H. Myers, D.C. Montgomery, and C.M. Anderson-Cook, "Response surface methodology: process and product optimization using designed experiments," Vol. 705, John Wiley & Sons Inc, 2009.

[18]M.Y. Can, Y. Kaya, and O.F. Algur, "Response surface optimization of the removal of nickel from aqueous solution by cone biomass of Pinus sylvestris," Bioresource technology, 97[14] 1761-1765 (2006).

[19]S. Nosrati, N.S. Jayakumar, and M.A. Hashim, "Extraction performance of chromium (VI) with emulsion liquid membrane by Cyanex 923 as carrier using response surface methodology," *DESALINATION*, 266[1-3] 286-290 (2011).

[20]S. Nosrati, N.S. Jayakumar, and M.A. Hashim, "Performance evaluation of supported ionic liquid membrane for removal of phenol," *J HAZARD MATER*, 192[3] 1283-1290 (2011).

[21]W. Evans, et al., "Effect of aggregation and interfacial thermal resistance on thermal conductivity of nanocomposites and colloidal nanofluids," *INT J HEAT MASS TRAN*, 51[5-6] 1431-1438 (2008).

AN ENVIRONMENTALLY-BENIGN ELECTROCHEMICAL METHOD FOR FORMATION OF A CHITOSAN-BASED COATING ON STAINLESS STEEL AS A SUBSTRATE FOR DEPOSITION OF NOBLE METAL NANOPARTICLES

Gary P. Halada[1], Prashant Jha[1], Michael Cuiffo[1], James Ging[1] and Kweku Acquah[2]

[1]Department of Materials Science and Engineering
Stony Brook University, Stony Brook, New York 11794-2275
[2]Department of Chemistry
Worcester State University, Worcester, Massachusetts 01602

ABSTRACT:
   We report on a new method for formation of catalytic nanocomposites surfaces and nanoparticles, with many potential benefits in energy and biomedical applications. This work demonstrates an environmentally-benign process using electrochemistry and post deposition UV irradiation to create a durable, robust layer on a stainless steel (304) electrode surface which serves as a substrate for subsequent deposition of silver nanoparticles. Electron microscopy, vibrational and electron spectroscopies are used to identify the structure and chemistry of the initial chitosan layers, indicating formation of nitro groups through oxidization of a fraction of the amine groups of chitosan which may play a role in enhancing adhesion to the passive film on the stainless steel cathode. By using this surface to synthesize noble metal nanoparticles through a patent pending process which eliminates the use of harsh reducing agents, we propose development of a surface with important functionality for catalytic and biomedical applications.

1. BACKGROUND:

   Chitosan, a linear polysaccharide obtained by deacetylation of chitin from crustaceans, mollusks, insects and fungi, is the second most abundant natural biopolymer (after cellulose). It has found a broad range of applications in the pharmaceutical, food processing and medical industries due to its biocompatibility and non-toxicity, its excellent gel and film-forming ability, and its chemical reactivity as a natural polycation.[1] While the biomedical (e.g. tissue engineering[2]) and pharmaceutical applications (e.g. drug delivery) have been exploited for some time, the potential uses of chitosan-based biomaterials in industry (e.g. as a support for catalysts or substrate in organic electronics) have been hindered by questions of stability, variability in properties, and methods and extent of production.[3] Surfaces for flexible electronics, sensors and device development[4] require both the elasticity of polymeric materials coupled with good toughness and environmental durability as well as the ability to provide a strong interconnect to metallic structures.[5] The potential for chitosan as a hierarchal soft-interconnect for nanoscale components has recently been pointed out.[6] This refers to the ability to use chitosan, processed using a similar electrochemical technique to that which is proposed here, as a means to assemble soft nanoscale components at specific locations.

   In particular, durability and mechanical toughness as well as adhesion to metal substrates have remained challenges to chitosan application, usually requiring coupling agents and complex chemical/thermal processing steps. For example, creation of a chitosan film on metallic substrates used in dental implants[7] required a complex process involving several chemical treatments (including curing at elevated temperature, reaction with a cyano-oxysilane coupling agent and overnight exposure to a glutaraldehyde cross-linking agent) – a process which one would hope to avoid for industrial applications. Chitosan and chitosan loaded with an antibiotic (gentamicin), has been applied to stainless steel bone screws, and found to be biocompatible and to inhibit growth of bacteria.[8] These researchers also used a method for deposition involving silane (to promote adhesion) and glutaraldehyde (a cross-linking agent). We report on a deposition method which does not require adhesion promoters due to the nature of the electrochemical process, which enhances bonding to naturally-formed passive film components (chromates, molybdates) on stainless steel.

Through the use of electrochemical formulation of chitosan-based composites, we reduce the need for enzymes and chemical processing steps involving non-renewable solvents (resulting in a more sustainable process). The method for electrochemically-induced deposition is based upon cathodic polarization of the electrode on which deposition is to occur – a process which has long been shown to create a region of higher pH (due to hydrogen evolution).[9] In the model for electrochemically-driven deposition of chitosan, the mechanism of deposition depends both on generation of a high pH (>6.3) region in the solution near the cathode (which results in deprotonation of amine and deposition) and on electrostatic attraction of positively charged dissolved chitosan as well as other ionic species (i.e. cationic metal ions) in the solution. These methodologies offers distinct advantages over past approaches for chitosan deposition requiring drying or spin casting and the use of potentially hazardous chemicals. By depositing electrochemically, we also gain the added benefit of being able to incorporate reduction of metal ions into the process for design and development of composites and the "green" manufacture of nanoparticles.

1.1 UV surface modification of chitosan

Chemical modification of chitosan utilizes the presence of primary amine and primary hydroxyl as well as secondary hydroxyl groups in the chitosan molecule, which makes it very amenable to further functionalization and modification. Chemical modifications of chitosan have been thoroughly reviewed by Mourya and Inamdar.[10] A recent study by Sionkowska, et al., considered the effects of UV exposure on the surface of properties of chitosan films[11]. While no structural change was noted from UV irradiation, contact angle and surface free energy measurements did show a change in wettability and polarity indicating surface photo-oxidation. The degree of oxidation could not be detected, as surface sensitivity of techniques used to analyze the post-irradiated surface was limited. In this study, we used X-ray photoelectron spectroscopy (XPS), a surface sensitive technique which analyzes C, N and O speciation and chemical environment to a depth of approximately 10 nm, to determine the nature of photo-oxidation. This technique has been found to be especially valuable in our earlier work on UV-induced changes in composite polymer paint films.[12]

2. EXPERIMENTAL

2.1 Electrodeposition of chitosan:

Chitosan solution was made by mixing 1.5g low molecular weight chitosan (sigma Aldrich, 75-85%deacetylated) to 120ml DI water under constant stirring. 1M HCl was then slowly added drop wise until all chitosan was dissolved, which was seen to occur as the solution reached a pH of between 4 and 5. For electrochemical deposition of chitosan, polished type 304 stainless steel (metal composition approx. 19% Cr 9 % Ni, bal. Fe) was chosen as the working electrode, though we have also found that type 316 Mo-bearing stainless steel and even Hastelloy C-22 can also work as cathodes. This process has been detailed by the authors in a previous publication.[13] Pt wire served as the counter electrode. A Gamry Reference 600 potentiostat was used to perform electrochemistry and Gamry Instrument Framework software was used for control and monitoring of voltage and current. Controlled potential coulometry at a voltage of -3.0 V (versus Ag/AgCl reference electrode) was applied for 5 minutes. The resulting hydrogel was rinsed in DI water to remove any chitosan solution entrapped in the resulting hydrogel. The gel was then exposed to UV light (20W at 6 cm) for 10 minutes. In some cases, the film was then peeled from the electrode following liquid nitrogen immersion for one minute so that the thin layer which remains bound to the electrode surface could be further analyzed.

2.2 Scanning Electron Microscopy (SEM)

The SEM system used was a Leo 1550 with a Robinson back scatter Gemini detector with EDAX, BS, AC, EBSP, SE, InLens detectors. Back scattered images were used to investigate morphology of silver particles deposited and an EDS pattern was used to estimate chemical composition of particles formed.

2.3 FTIR Spectroscopy

Fourier Transform Infrared Spectroscopy (FTIR) spectra were obtained using a Nicolet 760 infrared spectrometer modified to collect data in both mid- and far infrared regions. Samples were ground to a fine powder. Spectra were collected using an MCT-A detector with data resolution set to 2 cm^{-1} and summed over 256 scans to improve the signal-to-noise ratio. A Gemini sampling accessory (Spectra-Tech) collected diffuse reflectance data from powder samples. The analysis chamber was purged continuously with doubly dried air to prevent the absorption of water vapor, and a globar-type IR source was used.

2.4 Synchrotron FTIR

Fourier Transformed Infrared microspectroscopy of layers formed on stainless steel was conducted at beamline U10B of the National Synchrotron Light Source at Brookhaven National Laboratory using a Hyperion 3000 IR microscope in 900-4000 cm^{-1} frequency range with a resolution of 4.0 cm^{-1}. Data accumulation and processing were performed using OMNIC for Nicolet Almega version 7.3.

2.5 Raman Spectroscopy

A Nicolet Almega dispersive Raman spectrometer with a 785 nm laser source was used for analysis. Powdered samples or films were placed on quartz slides and Raman microspectroscopy in reflectance mode was used for data acquisition. Data was collected in the 3600-400 cm^{-1} range. An average of 10 scans with 5 sec accumulation time for each exposure was collected. OMNIC for Nicolet Almega software version 7.3 was used to process data. For comparison, chitosan solution at pH3 was dried (without any applied potential) on stainless steel and similarly analyzed.

2.6 X-ray photoelectron spectroscopy

To detect speciation of nitrogen, X-ray photoelectron spectroscopic (XPS) studies were carried out. Sample preparation consisted of peeling off the deposited film from stainless steel substrate after immersion in liquid nitrogen for one minute. The film was removed using a stainless steel razor and mounted on the XPS sample holder using indium foil. XPS measurements were performed using a custom-designed spectrometer that utilized a VG Scientific (Fisons) CLAM2 hemispherical analyzer with lensing, controlled by a VGX900I data acquisition system. An Mg K$\alpha_{1,2}$ (hv=1253.6 eV) X-ray source operating at 20 kV and 10 mA with a 20 eV pass energy was used at a pressure of 10^{-9} Torr. Measurements were taken at a 90° take-off angle with respect to the surface and charge correction was done by referencing to the C 1s line of adventitious carbon (284.6 eV).

3. RESULTS AND DISCUSSION

3.1 Electrodeposition of chitosan

Chitosan dissolves in acidic aqueous solution by addition of a proton to the NH_2 group to form NH_3^+. The process is reversible and on increasing pH higher than 6.3, protonated chitosan deprotonates and precipitates. As reported previously by Fernendes, et al.[14], a localized zone of high pH is created near a cathode held at sufficiently negative potential in water, a method utilized in this process to electrochemically deprotonate protonated chitosan and form deposits on the steel substrate. Stainless steel is chosen as an electrode as it is inexpensive, easy to use, and does not corrode at cathodic potentials. The presence of an ultrathin passive layer surface enriched in chromate on the steel enhances initial chitosan film formation and sorption.

During cathodic polarization, reduction of H^+ ions results in formation of H_2 gas, as indicated by hydrogen bubble formation on the steel substrate. This removal of H^+ ions from solution creates a zone of high pH near the substrate where chitosan is deposited. The film thus formed is shown in the SEM micrograph in Figure 1. Chitosan can be seen to have been deposited in a multilayer structure. The first layer of chitosan is formed on the surface of hydrogen bubbles (indicated in figure 1 as the brightest area due to the greatest penetration of electrons emitted from the substrate below) with diameter around 100 microns. With the continuation of electrolysis and further accumulation of hydrogen, the bubbles coalesce to form larger bubbles of hydrogen 300-400 microns in diameter, which collapse leaving the precipitated chitosan structure behind. An additional layer of chitosan is precipitated on top of the new bubbles formed (the areas with intermediate darkness in Figure 1).

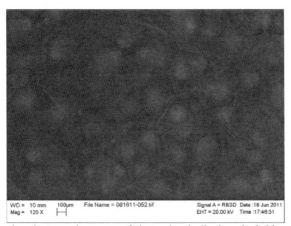

Figure 1 : Scanning electron microscopy of electrochemically deposited chitosan on stainless steel at -3.0V vs Ag/AgCl reference electrode.

The process continues forming still larger bubbles with diameter 300 μm to 1mm forming the top layer of the precipitated chitosan film. The darkest areas (dark due to the blocking of electrons emitted from the substrate below by chitosan) in the figure are those with the maximum amount of chitosan which could have formed by transport of material along the bubble surface (hence making the top of the bubble progressively thinner until it breaks; the resulting fractures in this layer are clearly visible in the picture).

3.2 Chemical characterization

FTIR spectroscopy from crushed powders of the deposited material reveals a number of features consistent with a degree of oxidation occurring during the deposition process. For example, peaks at 1663 cm^{-1}, 1203 cm^{-1} and 970 cm^{-1} indicate oxidation of some of the nitrogen in chitosan as a result of the electrochemical process. Raman spectra of air-dried chitosan film as well as electrochemically modified chitosan are shown in Figure 2. The peak at 2920 cm^{-1} from the air-dried sample most likely results from a combination of asymmetric C-H vibration from CH$_2$OH and C-O stretching vibration from the hydroxyl group. The broad peak at 1634 cm^{-1} is expected to result from a combination of amine N-H deformation and aromatic C=C stretching vibrations. The peak at 1465 cm^{-1} is expected to be from the C-H scissor vibration while the peak at 1384 cm^{-1} is expected to result from CH$_2$ wagging. Peaks at 1325 cm^{-1} and 1188 cm^{-1} can be attributed to C-N stretching from amine group. The peak at 1256 cm^{-1} can be attributed to OH deformation vibration of CH$_2$OH group. The peaks at 1188 cm^{-1} and 820 cm^{-1} are expected to be from symmetrical and asymmetrical C-O-C stretching in cyclic ether while the peaks at 935 cm^{-1} and 468 cm^{-1} are expected to result from C-O-C bonds between monomer units. The peaks at 628 cm^{-1} and 560 cm^{-1} can be attributed to ring vibrations.

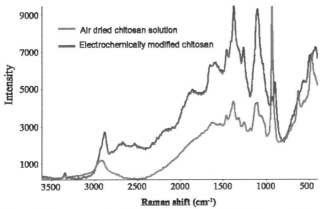

Figure 2 : Raman spectra of ECM-chitosan and dried chitosan solution at pH2

The Raman spectrum of electrochemically deposited material also showed various new peaks as compared to the air-dried chitosan: a peak at 3345 cm^{-1} from the N-H stretching vibration in C=N-H groups while the peak at 1860 cm^{-1} is expected to be from the C=O stretching vibration. The peak at 1589 cm^{-1} can be attributed to N-H bending in amine and the peak at 1525 cm^{-1} is consistent with the asymmetric NO$_2$ stretching vibration. The peak at 901 cm^{-1} is consistent with the C-N stretching vibration in nitro groups, and the new peak at 491 cm^{-1} is expected to result from the NO$_2$ rocking vibration. Vibrational spectroscopy also indicates oxidation of a fraction of amine to imine and nitro groups as well as oxidation of some of the hydroxyl groups to aromatic ketone.

To confirm this oxidation, XPS of the underside of a removed film layer (as described previously) was carried out. The resulting XPS spectrum of electrodeposited chitosan is shown in Figure 3. Two peaks are present for N1s; with charge correction, they correspond to binding energies of 398.6 eV (NH$_2$) and a large, broad peak at 407 eV. The peak at 407eV confirms oxidation of nitrogen in chitosan, falling in the range of nitrate in cellulose nitrate (408.1eV) and C-NO$_2$ nitro compounds (406.3 eV).[15]

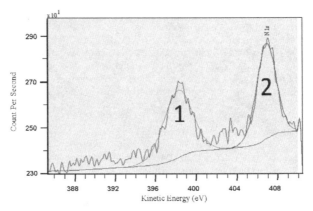

Figure 3: XPS of electrodeposited chitosan . Peak 1: primarily amine (-NH$_2$), Peak 2: similar to nitrate, possible nitronyl (N=O).

Hence, vibrational and X-Ray spectroscopy provide clear evidence oxidation of at least a portion of the amine groups in chitosan to nitro groups and some hydroxyl groups to aromatic ketone. The estimated structure of this novel molecule is shown in Figure 4. The oxidation is expected to have taken place by reaction of amine and hydroxyl group with OH Ions formed by electrolysis of entrapped water molecules in the hydrogel structure as shown in equations below:

-NH$_2$ + OH$^-$ → =NH + H$_2$O + e$^-$ equation (1)
=NH + 5OH$^-$ → =NO$_2$ + 3H$_2$O + 5e$^-$ equation (2)
-OH + OH$^-$ → =O + H$_2$O + e$^-$ equation (3)

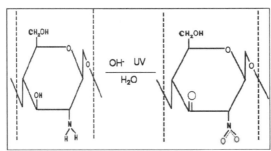

Figure 4: Structures of chitosan and nitro-chitosan synthesized by electrochemical deposition and photochemical oxidation.

3.3 Effects of UV exposure
Exposure of hydrogel to ultraviolet light was found to increase the intensity of the oxidized nitrogen peak in X-ray photoelectron spectroscopic spectrum (not shown) indicating oxidation of

higher fraction of amine. Enhanced oxidation of nitrogen by hydroxyl ion in presence of UV light has also been observed by other researchers.[16]

Synchrotron FTIR microspectroscopy was used to compare electrochemically-deposited chitosan coatings, with and without post-deposition UV irradiation. As can be noted in Figure 5, UV post-processing irradiation leads to decrease in intensity from free NH bending in amine, an increase in the signal from the -OH stretch, and a decrease in intensity of the hydrocarbon region of the FTIR spectra. In addition, the relative intensity of free amino groups (1590 cm^{-1}) versus amide I (1660 cm^{-1}) decreased with UV irradiation, and significant changes may be noted in the C-N region with UV-irradiation (not shown). These changes are indicative of cross-linking, in particular at the amino ligands of the chitosan molecule.

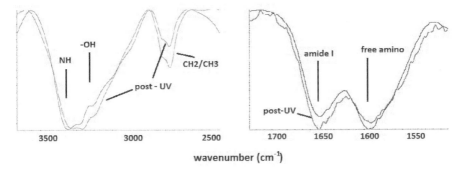

Figure 5: Synchrotron FTIR spectroscopy from as-deposited and UV-irradiated chitosan layers deposited on 304 stainless steel

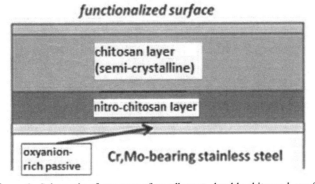

Figure 6: Schematic of structure of an adherent, durable chitosan layer formed electrochemically on a stainless steel cathode

Figure 6 is a schematic showing the proposed structure of the electrodeposited chitosan layer formed on the stainless steel cathode. X-ray diffraction of the deposited layer (removed and crushed for powder diffraction) provides a broad two-line diffraction pattern, indicative of a semi-crystalline state.[17] The functionalized surface shown in the diagram may be that created through UV irradiation (resulting in additional oxidation and cross-linking) or may be that which results from further chemical processing, such as the synthesis and formation of nanoparticles (as noted below).

3.4 Use as substrate for silver metal nanoparticle formation and deposition

We have recently developed a patent-pending process for noble metal nanoparticles formation in and on an electrochemically-deposited chitosan layer, formed as described above. Through exposing an air dried electrochemically deposited chitosan layer on stainless steel to an aqueous solution containing 0.001 to 0.1 M concentration of silver nitrate salts while maintaining a negative potential on the substrate, we have observed the formation of metallic nanoparticles in a size range of 10 to 100 nm. The details of this process are provided elsewhere.[18]

An example of silver nanoparticles deposited from a 0.01 M solution of silver nitrate at -0.5 V (vs Ag/AgCl) is shown in Figure 7. UV-vis spectroscopy (in reflection mode), SEM-EDAX and X-ray absorption near edge synchrotron (XANES) spectroscopy confirmed the metallic nature of the resulting nanoparticles. It was also found that the distribution and average size of nanoparticles can be controlled by varying the time and voltage applied during synthesis, with longer times and more negative voltages resulting in growth of nanoparticles and aggregation into larger dendritic structures. Well dispersed nanocomposite surfaces similar to those shown in the micrograph are possible with less than one minute of deposition, and the surfaces formed have been shown to remain stable for at least six months of storage under atmospheric conditions. Initial studies have also shown some antimicrobial and catalytic properties (for the oxygen reduction reaction) by the nanocomposites surfaces formed in this manner, though this work is quite preliminary and requires much refinement.

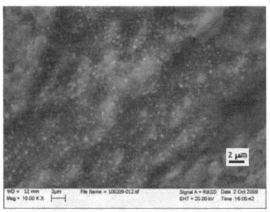

Figure 7: Ag nanoparticles formed on a chitosan layer deposited on 304 stainless steel

Metal nanoparticles have been identified as an extremely important class of catalysts for a variety of chemical processing needs, including for the conversion of organic compounds in energy generation[19], use in polymer membranes in hydrogen fuel cells, chemical synthesis (e.g. carbon-carbon bond formation and oxidation reactions[20]) and a number of other reactions. Most reported fabrications of catalytic metal nanoparticles (e.g. Ag, Au, Pt and Pd) involve a three stage process requiring: (a) a metal salt in solution; (b) a "shaping" or encapsulation agent, which is usually an organic molecule, such as chitosan[21]; and (c) a strong reducing agent to reduce the metal ions for the formation of nanoparticles. It is this third highly reactive component which is often an environmental or health hazard, and which limits the sustainability of the chemical process. Our process uses a room temperature, rapid technique taking advantage of electrochemical reduction to develop a nanoparticle/chitosan layer composite. Our method not only incorporates the principles of "green" chemistry but also is an important step toward design of a sustainable manufacturing technology for metal nanoparticle composites that is rapid, safe and inexpensive.

While the role of nitro-type groups formed in the chitosan layer in nanoparticle formation and encapsulation is not clear, the strong adhesion of the coatings to the substrate helps make the subsequent processing steps possible and hence forms a critical stage in further chemical processing for nanoparticle formation, catalysis and other applications.

4. CONCLUSIONS

While the chitosan coatings and Ag-chitosan nanocomposites materials described have been developed for distinct applications, they share a unique manufacturing process which limits the use of hazardous and toxic materials, occurs at room temperature and at high speed. Hence the value of the process lies not only in the unique materials created, but also in the safe and inexpensive synthesis process. The chitosan layers formed on stainless steel are strongly adherent, likely due to bonding at the passive layer-deposited chitosan interface, taking advantage of the formation of oxidized nitrogen groups in the polymer. Through a benign electrochemical process, metallic silver nanoparticles have been shown to readily form on these substrates, with a range of particle sizes and spatial distribution. Additional studies are being conducted to provide (a) the ability to improve control over deposition rate and particle distribution, (b) a better understanding of the impact of substrate preparation and surface chemistry on interfacial and mechanical properties, and (c) further chemical and structural analysis to enhance understanding of the structure/properties/processing relationship. While initial results are highly encouraging, further work will provide the necessary knowledge to make manufacturing possible and expand uses in biomedical, electronic, structural and research applications.

ACKNOWLEDGEMENTS:

The process for Ag-chitosan composite formation is Pat. Pending: PCT/US11/26075, WO/2011/106526. We gratefully acknowledge the support of Stony Brook University, and an NSF-REU site grant, #1062806, which supported the work of James Ging and Kweku Acquah.

REFERENCES:

[1] K.D. Yao, T. Peng, Y.J. Yin, M.X. Xu, and M.F.A. Goosen, Microcapsules/Microspheres Related to Chitosan, *J. Macromol. Sci.-Rev. Macromol. Chem. Phys.*, **C35**, 155–180 (1995).
[2] In-Yong Kim, Seog-Jin Seo, Hyun-Seuk Moon, Mi-Kyong Yoo, In-Young Park, Bom-Chol Kim and Chong-Su Cho, Chitosan and its Derivatives for Tissue Engineering Applications, *Biotechnology Advances*, **26 (1)**, 1-21 (2008).
[3] E. Guibal, Heterogeneous Catalysis on Chitosan-Based Materials: A Review, *Prog. Polym. Sci.*, **30**, 71-109 (2005).
[4] J. Lumelsky, M. S. Shur, and S. Wagner, Sensitive Skin, *IEEE Sensors J.*, **1**, 1, 41–51 (2001).
[5] Stephanie P. Lacour, Joyelle Jones, Sigurd Wagner, Teng Li, and Zhigang Suo, Stretchable Interconnects for Elastic Electronic Surfaces, *Proc. of the IEEE*, **93** (8), 1459–1467 (2005).

[6] Gregory F. Payne and Srinivasa R. Raghavan, Chitosan: A Soft Interconnect for Hierarchal Assembly of Nano-scale Components, *Soft Matter*, **3**, 521-527 (2007).

[7] J.D. Bumgardner, R. Wiser, S.H. Elder, R. Jouett, Y. Yang and J.L. Ong, Contact Angle, Protein Adsorption and Osteoblast Precursor Cell Attachment to Chitosan Coatings Bonded to Titanium, *Journal of Biomaterials Science, Polymer Edition*, **14** (12), 1401-1409 (2003).

[8] Alex H. Greene, Joel D. Bumgardner, Yunzhi Yang, Jon Moseley and Warren O. Haggard, Chitosan-Coated Stainless Steel Screws for Fixation in Contaminated Fractures, *Clin. Orthop. Relat. Res.*, **466**, 1699-1704 (2008).

[9] H. Dahms and I.M. Croll, The Anomalous Codeposition of Nickel-Iron Alloys, *J. Electrochem Soc.*, **112**, 771-775 (1965).

[10] V.K. Moury and N.N. Inamdar, Chitosan-Modifications and Applications: Opportunities Galore. *Reactive and Functional Polymers*, **68**, 1013-1051 (2008).

[11] A. Sionkowska, H. Kaczmarek, M. Wisniewski, J. Skopinska, S. Lazare and V. Tokarev, The Influence of UV Irradiation on the Surface of Chitosan Films, *Surface Science*, vol. 600, pp. 3775-3779 (2006).

[12] Lionel T. Keene, Gary P. Halada and Clive R. Clayton, Failure of navy Coatings Systems 1: Chemical Depth Profiling of Artificially and Naturally Weathered High-Solids Aliphatic Poly(ester-urethane) Military Coating Systems, *Progress in Organic Coatings*, **52**, 173-186 (2005).

[13] G.P. Halada, P. Jha, K. Nelson, W. Zhao, C.S. Korach, A. Neiman and S.J. Lee, Formation and Characterization of Chitosan-Based Coatings on Stainless Steel, *Biomaterials*, ACS symposium series, **8**, 159-171 (2010).

[14] R. Fernendes, L.Q. Wu, T. Chen, H. Yi, G.W. Rubloff, R. Ghodssi, W.E. Bentley and G.F. Payne, Electrochemically Induced Deposition of a Polysaccharide Hydrogel onto a Patterned Surface, *Langmuir*, **19**, 4058-4062 (2003).

[15] B. Beard, Cellulose Nitrate as a Binding Energy Reference in N(1s) XPS Studies of Nitrogen-Containing Organic Molecules, *Applied Surface Science*, **45**(3), 221-227 (1990).

[16] P. Berger, N.K. Vel Leitner, M. Dore and B. Legube, Ozone and Hydroxyl Radicals Induced Oxidation of Glycine, *Water Research*, **33**(2), 433-441 (1999).

[17] K.V. Harish Prashanth, F.S. Kittur and R.N. Tharanathan, Solid State Structure of Chitosan Prepared Under Different N-Deacetylating Conditions, *Carbohydrate Polymers*, **50**(1), 27-33 (2002).

[18] G.Halada, P. Jha, J. Ging, M. Cuiffo, Submitted to *Nano Research*

[19] Catalysis for Energy: Fundamental Science and Long-Term Impacts of the U.S. Department of Energy Basic Energy Science Catalysis Science Program, Board on Chemical Sciences and Technology, National Academies press (2009)

[20] Astruc, D. ed., Nanoparticles and Catalysis, Wiley-VCH, 2008, isbn 978-3-527-31572-7

[21] M. Adlim, M.A. Bakar, K.Y. Liew and J. Ismail, Synthesis of chitosan-stabilized platinum and palladium nanoparticles and their hydrogenation activity, *Journal of Molecular Catalysis A: Chemical*, **212**, 141-149 (2004).

SYNTHESIS AND CHARACTERIZATION OF NANOCRYSTALLINE NICKEL/ZINC OXIDE PARTICLES BY ULTRASONIC SPRAY PYROLYSIS

Ilayda Koc, Burcak Ebin, Sebahattin Gürmen

Istanbul Technical University, Department of Metallurgical & Materials Engineering, Istanbul-Turkey

ABSTRACT

Nano-sized particles research has become very important field in material science. Among the various nanomaterials, oxide nanoparticles have attracted increasing technological and industrial interest. This interest has mainly to do with their properties (magnetic, electrical, and catalytic properties). In this study, nanocrystalline Ni/ZnO particles were synthesized by ultrasonic spray pyrolysis and hydrogen reduction method. Stochiometric amount of nickel nitrate and zinc nitrate were used to prepare the corresponding solution. Particles were synthesized by hydrogen reduction of the aerosol droplets of the precursor solution under constant H_2 flow rate at 700°C, 800°C and 900°C. Thermodynamics of the composite particle formation by hydrogen reduction/decomposition of metal nitrates were investigated by HSC software. Morphology, size, chemical composition and crystal structure of the particles were investigated by scanning electron microscope-energy dispersive spectroscopy (SEM-EDS) and X-ray diffraction (XRD). We have explored catalytic properties of nanocrystalline Ni/ZnO particles.

INTRODUCTION

The design, characterization, production and application of materials , devices and systems by controlling shape and size of the nano-scale can be a simple definition of nanotechnology.[1] Nanomaterials are used in space and aircraft, automotive, information technology, environmental, textile, and catalytic applications.[2] Nano-scale particles research has recently become a very significant field in materials science. Such metal nanoparticles often exhibit very interesting electronic, magnetic, optical, and chemical properties. The reactivity of nanoparticles depends on its size, shape, surface composition, and surface atomic arrangement.[3] There are two basic strategies are used to produce nanoparticles: "top-down" and "bottom-up". The term "top-down" refers to the mechanical crushing of source material using a milling process and mechanical-physical particle production processes based on principles of microsystem technology and it includes high- energy mechanical milling, electrodeposition and lithography/etching processes.[4-6] In the "bottom-up (Chemo- physical production process)" strategy, structures are built up by chemical processes and this methods, are based on physicochemical principles of molecular or atomic self-organization. This approach produces selected, more complex structures from atoms or molecules, better control sizes, shapes and size ranges and it includes aerosol processes, precipitation reactions and sol-gel processes. The selection of the respective process depends on the chemical composition and the desired features specified for the nanoparticles.[7] Ultrasonic spray pyrolysis (USP) is a continuous flow process that operates at ambient pressure with involving four steps. The first step is aerosol generation in the untrasonic atomizer. Aerosols are obtained from aqueous solution of metal salts and transferred into the reaction furniture. Then, shrinkage of the aerosol droplets comes due to the evaporation. Next step is the chemical reaction in the furnace. Final step is the formation of solid particle. Moreover, this method allow to produce spherical, homogenous, non-agglomerated particles.[8]

Zinc oxide materials have a wide band gap (3.37 eV), large exciton binding energy (60 meV) and semiconductor properties. Thus, they have great importance to several applications such as UV light emitters, gas sensors and phototheraphy agents. Changes in optical, electrical, and magnetic properties could occur when impurities were added into a wide gap semiconductor so a certain element into ZnO has become an important route to optimize its optical, electrical, and magnetic performance. At this point, nickel starts to become significant as a transition and magnetic metarial.[8]

In this study, Ni/ZnO nanocomposites were produced with USP method and particles were characterized with X-ray diffraction (XRD), scanning electron microscopy – energy dispersive spectroscopy (SEM-EDS).

EXPERIMENTAL PROCEDURE

In this research, nanoparticles were synthesized by ultrasonic spray pyrolysis process Fig. 1 shows the laboratory equipments of USP system.

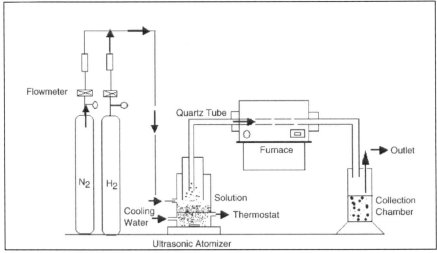

Fig. 1. Laboratory equipments of USP system for the synthesis of nanoparticles from Ni/ZnO nanoparticles.

Within the scope of experimental study, high purity nickel nitrate hexahydrate, {Ni(NO$_3$)$_2$.6H$_2$O} and zinc nitrate hexahydrate {Zn(NO$_3$)$_2$. 6H$_2$O} are used and Ni^{2+} and Zn^{2+} including an aqueous solution were prepared to be stoichiometric Ni:Zn ratio to 1. To atomize the appropriate solution aerosol, 1.3 MHz ultrasonic generator was used in this method. To transport aerosol, N$_2$ was selected as a carrier gas and H$_2$ was selected as a reductant gas (performed into the furnace with 500 mL/min constant flow rate) at the 700 °C, 800 °C and 900°C temperatures.

Table 1 contains data for concentration of the precursor solution, temperatures, H$_2$ flow rate and volume of the solution.

Table 1 Concentration of the precursor solution, temperatures, H_2 flow rate, volume of the solution.

Concentration of $Ni(NO_3)_2$ (M)	Concentration of $Zn(NO_3)_2$ (M)	Temperatures (°C)	H_2 flow rate (mL/min)	Volume of the solution (mL)
0.1	0.1	700	500	300
0.1	0.1	800	500	300
0.1	0.1	900	500	300

HSC Chemistry software was used to obtain possibility of reactions in system. HSC Chemistry is the world's favorite thermochemical software with a versatile flowsheet simulation module. It is designed for various kinds of chemical reactions and equilibria calculations as well as process simulation[9]. X-ray diffraction (XRD) and scanning electron microscope-energy dispersive spectroscopy (SEM-EDS) were used to characterize synthesized nano particles. SEM is used to analyse morphological and chemical analysis of materials at high magnification by using back-scattered electron image and secondary electron image systems. XRD analysis was perform with RIGAKU MiniFlex X-ray difractometer and JEOL JSM 7000F field emission scanning electron microscope were used to obtain scanning electron micrographs and elemental analysis.

RESULTS AND DISCUSSION
Some possible reactions in this system were investigated thermodynamically.
Reaction equations are as follows:

$$Ni(NO_2)_2 + 2H_2\,(g) \rightarrow Ni + 2NO_2\,(g) + 2H_2O \tag{1}$$

$$Zn(NO_3)_2 + H_2\,(g) \rightarrow ZnO + 2NO_2\,(g) + H_2O \tag{2}$$

$$Zn(NO_3)_2 + 2H_2\,(g) \rightarrow Zn + 2NO_2\,(g) + 2H_2O \tag{3}$$

For the reactions (1), (2), and (3), Gibbs free energy change and reaction equilibrium constant were calculated by HSC. Gibbs free energy change with respect to temperature is given in Fig. 2.

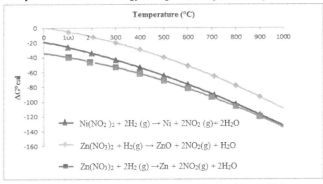

Fig. 2. Gibbs free energy changes calculated from HSC.

It is seen that zinc have negative Gibbs free energy changes so zinc can join nanocomposite structure. Moreover, the realization rate of the reaction (2) is more than reaction (3) because ZnO have more negative Gibbs free energy than Zinc.

Synthesized nano particles chemical composition and crystal structure characterize by the XRD. The pattern of particles at 700°C, 800°C and 900°C and Ni (JCPDS Card no: 00-001-1260) peaks are shown in Fig. 3. Ni, and ZnO were obtained in composite structure. It was seen that Ni peaks were observed at 44.62°, 51.96° and 76.34°, as well as ZnO peaks appeared with increasing temperature. Peaks were shifted according to XRD analysis beacause Ni may be intersititial atom in zinc structure and Ni_xZn_y alloy was occured. Crystallite size of the particles were calculated with Debye-Scherrer formula using (111) peak ($2\theta = 44.6°$) of Ni and (101) peak of ZnO ($2\theta = 36.5°$). Crystallite size of Ni is 61.3 nm whereas for ZnO is 17.6 nm for particles prepared at 800°C.

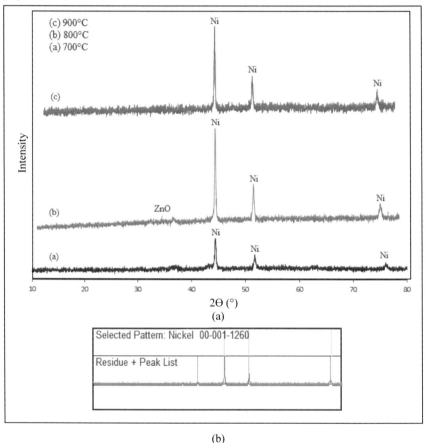

Fig. 3. (a) XRD pattern of Ni/ZnO and (b) Ni (JCPDS Card no: 00-001-1260) peaks.

Presence of atomic percentages of each element is given in Table 2 and EDS analysis of Ni/ZnO nanoparticles is seen Fig. 4. Although, atomic ratio of zinc and oxygen is 1:1 for ZnO structure, zinc and oxygen have different percentages in obtained particles and amount of oxygen is lesser than zinc because excess zinc may be occupied in nickel structure as an interstitial atom and formed Ni_xZn_y alloy.

Table 2 Presence of atomic percentages of each element.

Elements (%)	700°C	800°C	900°C
O	9.34	6.88	8.65
Ni	75.49	82.58	73.08
Zn	15.17	10.54	18.27

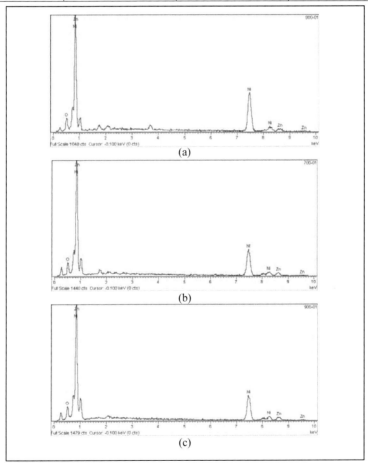

Fig. 4. EDS analysis of Ni/ZnO particles, (a) 700 °C, (b) 800 °C and (c) 900 °C.

The morphology of produced particles was investigated by SEM. Micrographs of particles prepared at 700°C, 800°C and 900°C are shown in Fig. 5. All samples have spherical shape morphology and their sizes are ranging nearly from 150 nm to 750 nm. The average particle size of the samples slightly increasing from 280 nm to 450 nm by elevating reaction temperature. It is clearly observed that aggregation of nanoparticles (primary particles) formed secondary submicron particles. Elevating temperature increased the aggregation, which caused denser secondary particles and smoother surface morphology. In addition, primary particle sizes of Ni/ZnO nanocomposites observed at high magnified SEM images were consistent with the crystallite size of samples calculated from XRD results.

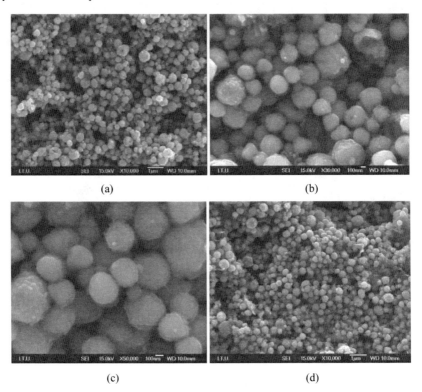

(a)

(b)

(c)

(d)

Fig. 5. Scanning Electron Micrographs of Ni/ZnO (a) 700°C, x10K, (b) 700°C, x30K, (c) 700°C, x50K, (d) 800°C, x10K, (e) 800°C, x30K, (f) 800°C, x50K, (g) 900°C, x10K, (c) 800°C, x30K, (c) 700°C, x50K.

CONCLUSIONS

Spherical Ni/ZnO nanocomposite particles were produced by ultrasonic spray pyrolysis (USP) method with 0.1M precursor solution at 700°C, 800°C, and 900°C reaction temperatures. Decomposition and reduction reactions was also discussed. Accordingly to XRD, SEM-EDS analyzes were performed on Ni/ZnO nano composites and these nanocomposites contains only Ni, Zn and O element. Composite particles prepared at 800°C consist of Ni and ZnO crystallites which sizes are 61.3 and 17.6 nm, respectively. Particle morphology were analyzed via SEM and spherical particle morphology was determined. It was obtained that particle sizes are ranging from 150 nm to 750 nm and average particle sizes are between 280 and 450 nm. Results indicate that USP method is a suitable process for nanocrystalline Ni/ZnO particle preparation and it is expected that Ni/ZnO particles have effective catalytic properties.

ACKNOWLEDGEMENT
This work has been supported by the ITU Research fund.

REFERENCES
[1] J. Ramsden, *Essentials of Nanotechnology,* Jeremy Ramsden & Ventus Publishing (2009).
[2] J. Ramsden, *Applied Nanotechnology,* Elsevier, Burlington : MA, (2009).
[3] R.W. Siegel, *Annual Review of Material Science.* 21 559–578, (1991).
[4] S F. L. Zhang, C. Y. Wang and M. Zhu, *Nanostructured WC/Co composite powder prepared by high energy ball milling,* Scripta Materialia, vol. 49, 1123-1128, (2003).
[5] W. Zhan, J. Alvarez and R. M. Crooks, *Electrochemical Sensing in Microfluidic Systems Using Electrogenerated Chemiluminescence as a Photonic Reporter of Redox Reactions,* Journal of American Chemical Society, vol. 124, 13265-13270, (2002).
[6] C.L. Haynes, A.J. Haes and R.P. Van Duyne, *Nanosphere Lithography: Synthesis and Application of Nanoparticles with Inherently Anisotropic Structures and Surface Chemistry,* MRS Proceedings 1-6, (2000).
[7] C. Raab, M. Simkó, U. Fiedeler, M. Nentwich, A. Gazsó, *Production of Nanoparticles and Nanomaterials,* Nano Trust-Dossier No. 006, (2011).
[8] Y.Wang, X. Liao, Z. Hauang, G. Yin, J. Gu, and Y. Yao, *Preparation and Characterization of Ni-doped ZnO Particles via a Bioassisted Process,* Colloids and Surfaces A: Physicochemical and Engineering Aspects, vol. 372, 165-171, (2010).
[9] HSC Chemistry for Thermodynamic Calculations, [cited 10/01/2013], Available from: http://www.chemsw.com/Software-and-Solutions/Laboratory-Software/HSC-Chemistry.aspx

IRON–NICKEL-COBALT (Fe–Ni-Co) ALLOY PARTICLES PREPARED BY ULTRASONIC SPRAY PYROLYSIS AND HYDROGEN REDUCTION (USP-HR) METHOD

Cigdem Toparli, Burcak Ebin, and Sebahattin Gürmen
Istanbul Technical University, Department of Metallurgical & Materials Engineering
34469, Istanbul-Turkey.

ABSTRACT
    Nanocrystalline FeNiCo ternary alloy particles were synthesized by ultrasonic spray pyrolysis and hydrogen reduction method. $Fe(NO_3)_3.9H_2O$, $Ni(NO_3)_2.6H_2O$ and $Co(NO_3)_2.6H_2O$ were used as a starting material to synthesize the nanocrystalline spherical ternary alloy particles. The experiments were performed at 600, 800 and $1000^{\circ}C$ using iron-nickel-cobalt nitrate aqueous solution. Shape, morphology, size, chemical composition and crystal structure of the particles were investigated. X-ray diffraction (XRD), scanning electron microscope (FEG-SEM) and energy dispersive spectroscopy (EDS) studies were carried out to characterize the ternary alloys particles. Vibrating sample magnetometer (VSM) was used to characterize the magnetic properties. Spherical nanocrystalline FeNiCo ternary alloy particles were prepared in the size range of 65 and 500 nm. Average crystallite size of the particles is 50 nm.

1. INTRODUCTION

    Developing new methods for preparation of nano sized materials have attracted scientific attention due to their interesting size-dependent electronic, magnetic, optical and chemical properties compared to bulk forms [1, 2]. Their unique features depend on their size, shape, surface composition and surface atomic arrangement.

    Recently, nanometer-sized magnetic structured materials are of great interest of scientific research because of the unique magnetic properties which differ from those of bulk materials [3-5]. These materials have some both important technological applications and research area.
Binary and ternary alloy ferromagnetic nanoparticles have gained great attention because of their modified magnetic and catalytic properties comparing with monometallic nanoparticles as a result; many kinds of compositions including Fe, Ni, and Co such as FeNi and FeCo, CoNi and FeNiCo have been investigated during recent years. FeNiCo ternary alloys have had many important applications not only due to their remarkable magnetic properties, but also because of their suitable catalytic properties. Thus, latest scientific research has focused on the preparation and characterization of the FeNiCo nanoparticles [6-10].

    FeNiCo nanostructured materials are prepared by various methods including sol-gel, mechanical alloying, electro deposition, hydrogen plasma-metal reaction, solvothermal process, micro-wave assisted synthesis [11, 12]. Azizi et. al. [13] and his colleaques were studied the effect of the hydrogen reduction on magnetic properties of FeNiCo nano-powder. They prepared Fe–16.5Ni–16.5Co nano-powder by two major steps including by mechanochemical method and hydrogen reduction and they report that after hydrogen reduction coercivity was decreased. Li and Takahashi [11] studied the synthesis and magnetic properties of Fe-Co-Ni nanoparticles by hydrogen plasma-metal reaction. They prepared the FeNiCo nanoparticles in six compositions and worked on the effect of compostion change on the magnetic properties and found that the coercive force is increased as the cobalt content increased. Wu et.al [14] investigated the magnetic properties of FeNiCo samples prepared into two stages including hydrothermal method and carbon coating the particles obtained. They reported that carbon coated particles have smaller coercivity and saturation magnetization than bulk form.

It is obviously seen that several methods are applied to obtain FeNiCo nanostructured materials. However, these methods need either special equipments or reducing agents. Therefore, the production of FeNiCo ternary alloy nanoparticles in a simple and single step process is a significant field to investigate the magnetic properties. Among these production techniques, ultrasonic spray pyrolysis (USP) is an advantageous method that allows one-step production. Spherical, non-agglomerated submicron particles of complex composition and controlled phase content of high technology sintered nano powders can be easily prepared by USP.

In our previous work [15, 16] the production of Fe-Ni and Fe-Co binary alloy particles by USP method were investigated and reported that the nanostructured particles of these alloys can be prepared with uniform particles size, spherical and desired composition.

In this study, production of nanoparticles of FeNiCo ternary alloys by ultrasonic spray pyrolysis and hydrogen reduction technique by using ıron (III), nickel (II) and cobalt (II) nitrate solutions were reported. Effect of reduction temperature on particle size and in parallel with magnetic properties was studied under 1.3 MHz ultrasonic frequency and 1.0 L/min H_2 flow rate conditions and discussed respectively.

2. EXPERIMENTAL

2.1. Materials
All the reagents (Co $(NO_3)_2 \cdot 6H_2O$, Fe $(NO_3)_3 \cdot 9H_2O$ and Ni $(NO_3)_2 \cdot 6H_2O$) were analytical grade and used without further purification. The nitrate salts were dissolved in deionized water and stirred by magnetic stirrer for 30 minutes. The concentration of the precursor solution was 0.1 m/L.

2.2. Experimental Procedure
The nanocrystalline particles of ternary alloy of FeNiCo were synthesized by USP method using aqueous solution of iron, cobalt and nickel nitrate under hydrogen (H_2) gas flow at temperatures 600, 800 and 1000°C. The experimental set up consists of ultrasonic spray generator, heating zone and collection chamber. Nitrogen with 1 L/min flow rate was used to create an inert atmosphere prior to and after the reduction process due to the safety regulations. The aerosol droplets of the corresponding solution were generated by high frequency ultrasonic atomizer (1.3 MHz). Then the mist was carried into a preheated zone by H_2 gas flow. H_2 was used without mixing any inert gas in the experiments as a carrier/reducing agent in 1 L/min gas flow rate. Table I shows the experimental parameters for the production FeNiCo ternary alloy particles.

Table I. Experimental parameters

Process temperature (°C)	Ni$(NO_3)_2$.6H$_2$O (mol/L)	Co$(NO_3)_2$.6H$_2$O (mol/L)	Fe$(NO_3)_3$.9H$_2$O (mol/L)	H_2 flow rate (L/min)	Frequency (MHz)
600	0.1	0.1	0.1	1	1.3
800	0.1	0.1	0.1	1	1.3
1000	0.1	0.1	0.1	1	1.3

The crystallite sizes were calculated by Scherrer formula using XRD data. The chemical compositions of particles were analyzed by energy dispersive spectroscopy (EDS) instrument. Particle size and morphology of the samples were investigated by field emission scanning electron microscopy (FE-SEM, JEOL JSM 700F). The magnetic properties of samples were measured by Vibrating Sample Magnetometer. The magnetic hysteresis loop measurements were carried at room temperature with an applied magnetic field up to 20 KOe.

3. RESULTS AND DISCUSSION

XRD patterns of the particles produced at 600, 800 and 1000°C reaction temperatures by using 0.1M concentrated solution are given in Fig. 1. FeNiCo ternary alloy particles shows three main peak at 43.92, 51.1, 75.22 °. The X-ray analyses of the powders reveal that FeNiCo ternary alloys particles present the body centered cubic crystalline structure with Space Group: Fm-3m.

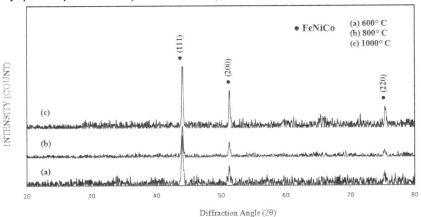

Figure 1. XRD analysis of the particles produced at (a) 600°C, (b) 800°C and (c) 1000°C

Crystallite sizes of the particles were determined by Scherrer equation using (111) and (220) diffraction line. Instrumental broadening was taken into account to obtain accurate crystallite size in the calculation. Average crystallite size of the particles calculated by using scherrer equation and shown at Table II. Crystallite size of the particles range slightly between 45.6- 61.2 nm.

Table II. Crystallite size of the FeNiCo ternary alloy particles.

Process Temperature (°C)	Average Crystallite Size (nm)
600	61.2
800	45.6
1000	45.6

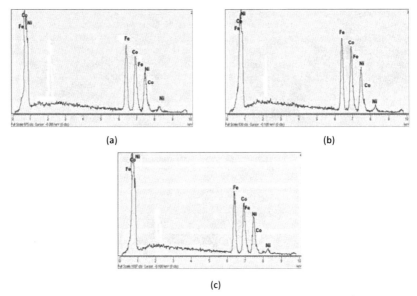

(a)　　　　　　　　　　　　　　　(b)

(c)

Figure 2. EDS analysis of the particles produced at (a) 600°C, (b) 800°C and (c) 1000°C

     EDS results of particles are shown at Fig 2. EDS results indicate that particles obtained different temperature contain iron, nickel and cobalt. Besides, any impurities which possibly contaminate from the waste product of the reduction/decomposition reactions, such as N were not detected. Table III represents the EDS results in atomic percentage of particles obtained at different temperatures.

Table III. EDS results in atomic percentage of FeNiCo ternary alloy nanoparticles

Elements	Atomic% (Theoretical)	Atomic% (Exp. at 600°C)	Atomic% (Exp. at 800°C)	Atomic% (Exp. at 1000°C)
Fe	33.33	32.08	33.44	31.95
Ni	33.33	35.33	35.38	35.83
Co	33.33	32.59	31.16	32.24

The stoichiometric atomic ratio of Fe:Ni:Co in the precursor is 1:1:1 and EDS analysis denoted that atomic Fe:Ni:Co ratio of the particles produced by USP method are nearly same for all the samples. EDS results indicate ternary alloy particles can produce in desired elemental and homogeneous composition by USP method.

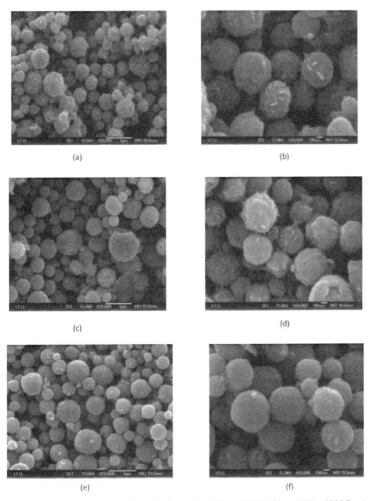

Figure 3. SEM images of the particles obtained a) x 20K., 600°C b) x 50K., 600°C c) x 20K., 800°C d) x 50K.,800°C e) x 20K., 1000°C and f) x 50K., 1000°C

Fig. 3 exhibited the SEM images of the FeNiCo ternary alloy particles. All samples shows nearly spherical shape morhology. FeNiCo particles prepared at 600°C, 800°C and 1000°C are in the size range of 65-500 nm. Average particle size of samples prepared at 600°C, 800°C and 1000°C are 300, 350 and 400 nm, respectively. Agglomeration of the primary particles is obviously seen in Fig. 3d, e and f. increasing the reaction temperature increases the agglomeration and sphericity. However, as the temperature elevated average particle size is increased and due to the sintering of the particles denser nanocyrstalline particles obtained.

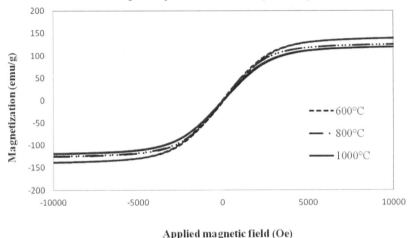

Applied magnetic field (Oe)

Figure 4. Magnetic analysis of particles produced at 600, 800 and 1000°C

Magnetic properties were characterized by vibrating sample magnetometer (VSM). The coercive force of the samples was derived from the magnetic hysteresis loop. Changes in magnetic features such as Hc depending on the production temperature are shown in the Fig. 4. It is shown that all FeNiCo nanocrystalline particles prepared with USP method exhibit ferromagnetic behavior. Their special coercivity (Hc) and saturation magnetization are listed in Table IV Magnetic properties of the samples prepared at different temperature.

Table IV. Magnetization and coercivity of the samples

Sample	Hc (Oe)	Ms(emu/g)
FeNiCo produced at 600°C	48.24	138.26
FeNiCo produced at 800°C	22.72	127.48
FeNiCo produced at 1000°C	11.15	121.997

It is found that Hc and Ms values are decreased with increasing process temperature. The variations of magnetic properties can be attributed to the changes of not only particle sizes but also particle morphology. It is assumed that as the crystalline size decreased Hc is decreased. However, Ms value of the particles is decreased as the particles size is increased.

Magnetic measurements reveal that the synthesized materials all exhibit ferromagnetic behaviour at room temperature. Hc values usually increase with the increasing of Co content or decreasing the Ni content and is affected by crystallite size, packing density and morphology of the particles [11]. Nanocrystalline FeNiCo particles were obtained at different reduction temperature using same corresponding solution to investigation the dependence of magnetic properties such as Hc on the particle production temperature in USP method. Results exhibit that coercivity and saturation magnetization of the samples is affected by the crystallite size and morphology in parallel with process parameter. We assume that the reason for reduction in Hc due to the decreasing of the crystallite size.

In all cases, FeNiCo ternary alloy nanocrystalline particles prepared with USP method show almost spherical morphology, homogeneous size distribution and smooth surface and ferromagnetic behavior. Futher, nanocyrstalline of FeNiCo ternary alloy powders can be applied as soft magnetic materials for high density magnetic recording media.

4. CONCLUSION

Ultrasonic spray pyrolysis was used for the synthesis of nanocrystalline spherical FeNiCo ternary alloy particles from nitrate-based solution. The effect of the reduction temperature on the particle size and magnetic behavior was investigated under the conditions of 2.5 h running time, 600, 800 and 1000°C reaction temperature and 1.0 L/min H_2 volumetric flow rate. The X-ray analysis of the particles indicated the presence of body centered cubic crystalline structure of nanocrystalline of FeNiCo ternary alloy. It is found that average crystallite size of the particles 50 nm. SEM images exhibited that spherical nanocrystalline Fe-Ni-Co ternary alloy particles were prepared in the size range of 65 and 500 nm. Average particle size of samples prepared at 600, 800 and 1000°C are 300, 350 and 400 nm, respectively. The coercivity values of particles are 48.24, 22.72, and 11.15 Oe for nanocrystalline particles obtained at 600, 800 and 1000°C, respectively. Magnetic analysis reveal that saturation magnetization value of particles at 600, 800 and 1000°C were 138.26, 127.48 and 121.997 emu/g, respectively.

5. ACKNOWLEDGEMENT

This work has been supported by the Istanbul Technical University Research fund. Authors thanks to Prof.Dr. Hakan Köckar and Dr. Oznur Karaagac for their help in magnetic characterization study.

6. REFERENCES

[1]M.G. Lines, Nanomaterials for practical functional uses, J. Alloy. Compd. 449 (1-2), 242-245 (2008).
[2]D.S. Jung, S.K. Hong, J.S. Cho, Y.C. Kang, Nano-sized barium titanate powders with tetragonal crystal structure prepared by flame spray pyrolysis, J. Eur. Ceram. Soc. 28(19), 109-115 (2008).

[3]Ajay Kumar Guptaa, Mona Gupta, Synthesis and surface engineering of iron oxide nanoparticles for biomedical applications, Biomaterials, 26, 3995–4021(2005).

[4]S. Gurmen, B. Ebin, Production and characterization of the nanostructured hollow iron oxide spheres and nanoparticles by aerosol route, J. Alloy. Compd. 492, 585–589 (2010).

[5]D.Yang, X. Ni, D. Zhang, H. Zheng, J. Cheng, and P. Li, Prepation and characterization of hcp Co-coated Fe nanoparticles, J. Cryst. growth, 286, 152-155(2006).

[6]J. S. Lee, S. S. Im, C. W. Lee, J. H. Yu, Y. H. Choa, and S.T Oh., Hollow nanoparticles of ß iron oxide synthesized by chemical vapor condensation , J. Nanopart. Res.6, 627-631 (2004).

[7]P.M. Paulus, F. Luis, M. KroK ll, G. Schmid, L.J. de Jongh, Low-temperature study of the magnetization reversal and magnetic anisotropy of Fe, Ni, and Co nanowires, J. Magn. Magn. Mater. 224, 180-196 (2001).

[8]Y.M. Kim, D. Choi, S.R. Kim, K.H. Kim, J. Kim, S.H. Han, H.J. Kim, Magnetic properties of as-sputtered Co-Ni-Fe alloy films, J. Magn. Magn. Mater. 226-230, 1507-1509 (2001).

[9]E.V. Khomenko, E.E. Shalyguina, N.G. Chechenina, Magnetic properties of thin Co–Fe–Ni films, J. Magn. Magn. Mater.316, 451–453 (2007).

[10]R.H. Kodama, Magnetic nanoparticles, J. Magn. Magn. Mater. 200, 359-372(1999).

[11]Xingguo Li, Seiki Takahashi, Synthesis and magnetic properties of Fe-Co-Ni nanoparticles by hydrogen plasma-metal reaction, J. Magn. Magn. Mater. 214, 195-203 (2000).

[12]E. Jartych, On the magnetic properties of mechanosynthesized Co–Fe–Ni ternary alloys, J. Magn. Magn. Mater. 323, 209–216, (2011).

[13]A.Azizi , H.Yoozbashizadeh , S.K.Sadrnezhaad , Effect of hydrogen reduction on microstructure and magnetic properties of mechanochemically synthesized Fe–16.5Ni–16.5Co nano-powder, J. Magn. Magn. Mater. 321, 2729–2732, (2009).

[14]Aibing Wu, Xuwei Yang, Hua Yang, Magnetic properties of carbon coated Fe, Co and Ni nanoparticles, Journal of Alloys and Compounds 513, 193– 201 (2012).

15S. Gurmen, A. Guven, B. Ebin, S.Stopic, B. Friedrich, Synthesis of nano-crystalline spherical cobalt–iron (Co–Fe) alloy particles by ultrasonic spray pyrolysis and hydrogen reduction, J. Alloy. Compd. 481, 600–604 (2009).

[16]S. Gurmen, B.Ebin, S.Stopic, B. Friedrich, Nanocrystalline spherical iron–nickel (Fe–Ni) alloy particles prepared by ultrasonic spray pyrolysis and hydrogen reduction (USP-HR), J. Alloy. Compd. 480, 529–533 (2009).

Electronic and
Functional Ceramics

SYNTHESIS AND CHARACTERIZATION OF POLYVINILIDENE FLUORIDE (PVDF) CERIUM DOPED

*Evaristo Alexandre Falcão[1], Laís Weber Aguiar[1], Eriton Rodrigo Botero[1], Anderson Rodrigues Lima Caires[1], Nelson Luis Domingues[1], Cláudio Teodoro de Carvalho[1], Andrelson Wellinghton Rinaldi[2]

[1]Federal University of Grande Dourados (UFGD), Faculty of Science and Technology (FACET) - Rodovia Itahum, km 12,
CEP 79.804-970 - Dourados, MS, Brazil.
[2]State University of Maringa, Department of Chemistry - Av. Colombo 5790, Jd. Universitário, CEP 87020-900 - Maringá, PR, Brazil

Abstract

In this work cerium compound was synthesized with trans-3,4-(Methylenedioxy)cinnamate as ligand. The compound characterization was carried out using simultaneous thermogravimetry and differential thermal analysis (TG-DTA), Fourier transform infrared spectroscopy (FT-IR). Cerium dehydrated compounds was added in the polymeric PVDF matrix in different concentrations. PVDF sample doped with cerium compound characterized by FT-IR measurements and fluorescence. The FT-IR measurements showed that the Ce compound was incorporated in the polymer matrix. In addition, a decreasing of the percentages of β phase in relation of α phase was observed as a function of Ce addition in the PVDF matrix. From the optical measurements in the PVDF/Ce doped, were observed a broadening of absorption and fluorescence spectra as a function of the Ce addition and a displacement of maximum fluorescence of PVDF from green to the orange wavelength region. In summary, the experimental results revealed that the PVDF/Ce is a potential candidate to optical and photonic applications.

Keywords: PVDF; cerium; optical characterization.

*Corresponding author. Tel/FAX +55-67-3410-2088
E-mail address: evaristofalcao@ufgd.edu.br.

Introduction

The polymeric materials have gained a special attention as they present good ferroelectric properties as well as good flexibility. Among the polymeric materials with ferroelectric properties can be cited the poly (vinylidene fluoride) (PVDF) which has been used as coating to prevent corrosive attack, transducer electro-mechanical and electro-thermal, speakers, sensors and to excite the growth of bone tissue in animals[1]. The organic material possesses structural flexibility, low production cost combined with excellent optical and electric properties. On the other hand, the inorganic metals provide improvement in the electrical, magnetical, mechanical and thermal properties of materials[2]. Thus, because of these excellent physical and chemical properties presented by organic and inorganic materials in the last years the hybrid materials have received great attention from many researchers. Some authors have reported that the processing of ceramics and polymer composites (eg, PZT / polymer composite) has the advantage of low density, high piezoelectric response and high electromechanical coupling factor[3-5]. Recently, PVDF ferroelectric polymers have also attracted interest in the area of photonics to be used as hosts for rare-earth ions, which enable its use in optical devices[6]. Among them rare earth ions the cerium coumponds have been largely studied because they present excellent thermal stability, nontoxicity and biocompatibility[2-7]. Cerium compound, in particular Ce^{3+}, has also received much attention to present a broad emission band in the visible region of electromagnetic spectra which permits white light (WL) generation for light sources used in display devices[8]. In this context, the present study aimed to synthesized and characterized the PVDF/$Ce(C_{10}H_7O_4)_3$ doped material to investigated its optical properties. The preliminary results revealed that the PVDF/$Ce(C_{10}H_7O_4)_3$ present a great potential to be used in photonic devices.

Experimental

Compound preparations

The trans-3,4-(Methylenedioxy)cinnamic acid (HL) with 99 % purity was obtained from Sigma. Aqueous solution of NaL 0.1 mol L^{-1} was prepared from aqueous HL suspension by treatment with sodium hydroxide solution 0.1 mol L^{-1}.

Cerium (III) was used as its nitrate and ca. 0.1 mol L^{-1} aqueous solution of this ion was prepared by direct weighing of the salt. The solid state compound was prepared by adding slowly, with continuous stirring, the aqueous solutions of the ($NaC_{10}H_7O_4$) to the respective metal chloride solutions until total precipitation of the metal ions. The precipitate was washed with distilled water until elimination of nitrate ions, filtered through and dried on Whatman no. 42 filter paper and kept in desiccator over anhydrous calcium chloride.

Compositions Ce-Complex-doped PVDF matrix were prepared from the dissolution of PVDF of dimethylformamide under agitation and heating. An amount of stock solution of the $Ce(C_{10}H_7O_4)_3$ dehydrated was prepared by dissolving it in dimethylacetamide and placed in an ultrasonic bath. Then, PVDF was added in six Petry dishes, in which was added the solution of $Ce(C_{10}H_7O_4)_3$ to obtain concentrations of 0.0, 0.2, 0.4, 0.6, 0.8 and 1.0 wt% of $Ce(C_{10}H_7O_4)_3$. For a better comprehension, the $Ce(C_{10}H_7O_4)_3$ will be referred from this point on only as Ce. After stirring the mixture, they were taken for drying the solvent in a furnace.

Instrumental measurements

The measurements of infrared absorption spectroscopy by Fourier transform (FTIR-ATR) were done by using the equipment JASCO FT/IR – 4100. The data were collected by performing 150 scans with a resolution of 2 cm^{-1} in the 400–4000 cm^{-1} range. The fluorescence measurements were performed with a portable Spectrofluorimeter composed of two laser beams operating at 405 nm and 532 nm, a monochromator (USB 2000 FL / Ocean Optics), a fiber-type "Y" and a laptop. In the fluorescence measurements were used as the laser excitation with 405 nm wavelength.

Simultaneous TG-DTA curves were obtained with the thermal analysis system, model SDT 2960 from TA Instruments. The purge gas was air with a flow rate of 100 mL min^{-1}, heating rate of 20 °C min^{-1} and with samples mass of about 5 mg. Alumina crucibles were used for TG-DTA.

Results and Discussion

Cerium compound characterization

Cerium compound with initial mass of 5.17 mg is shown in the TG-DTA curves, as shown in Fig. 1. The first mass loss (TG) 30-150 °C, corresponding to an endothermic peak at 116 °C (DTA) is due to dehydration with loss of $3H_2O$ (TG = 7.10%). The anhydrous compound is stable up to 250 °C and above this temperature the two following steps of thermal decomposition occurs in overlapping steps between 250-294 and 294-311 in the TG curve, corresponding to an exotherm between 261-313 °C in the DTA curve attributed to the decarboxylation and oxidation of the organic matter.

For the fourth step the thermal decomposition occurs between 311-379 °C, corresponding to the intense exothermic peak at 372 °C attributed to the oxidation reaction of Ce(III) to Ce(IV), together with the oxidation and/or combustion of the organic matter, as already observed for other cerium compounds[9]. The mass loss up to 379 °C (TG) attributed to ligand is 70.17 % with formation of CeO_2 (22.73 %), as final residue. These results have permitted to establish the stoichiometry of this compound, which is in agreement with the general formula: $Ce(C_{10}H_7O_4)_3\cdot3H_2O$.

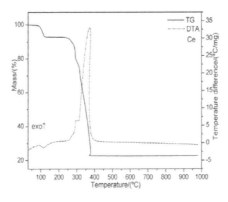

Figure 1. TG-DTA curves of the compound: $Ce(L)_3\cdot3H_2O$ where L = 3,4-(methylenedioxy)cinnamate.

FT-IR

The attenuate total reflectance infrared spectroscopic (ATR) data on 3,4-(methylenedioxy)cinnamate (sodium salt) and its compounds with the metal ions considered in this paper are shown in Table 1. The investigation was focused mainly in the 1700-1300cm^{-1} range, because this region is potentially the most informative in attempting to assign coordination sites[10].

Table I. Spectroscopic data for sodium and cerium 3,4-(methylenedioxy)cinnamate.
IR spectra cm^{-1}.

Compound	v_{as} (COO$^-$) cm^{-1}	v_{sym} (COO$^-$) cm^{-1}	Δv (v_{as} - v_{sym})
NaL	1547vs	1412s	135
Ce(L)$_3$.3H$_2$O	1500vs	1425s	75

vs = very strong; s = strong; v_{as} (coo-) and v_{sym} (coo$^-$); anti-symmetrical and symmetrical vibrations of the coo$^-$ group, respectively.

In NaC$_{10}$H$_7$O$_4$ (sodium salt), high intensity bands located at 1547 cm^{-1} and 1412 cm^{-1} are attributed to the anti-symmetrical and symmetrical frequencies of the carboxylate groups, respectively[11,12]. For cerium compound the band assigned to the anti-symmetrical stretching frequencies of the carboxylate are shifted to lower values and the symmetrical ones to higher, relative to the corresponding frequencies in the sodium salt. The Δv ($\Delta v_{asym} - \Delta v_{sym}$) for this compound is indicative of the coordination of Ce(III) to the carboxylate group by a chelating binding, according to Figure 2. This behaviour is in agreement with the observed in the compounds of light trivalent lanthanides with succinate ion[9].

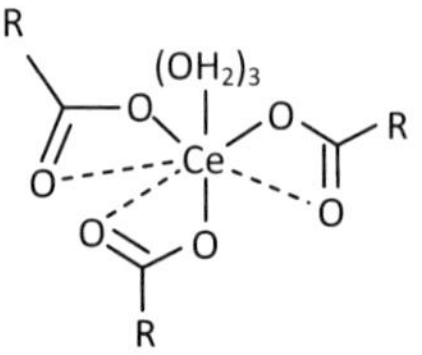

Figure 2. Coordination sites of the metal-ligand, where R: (C$_9$H$_6$O$_2$).

Ce-Complex-doped PVDF matrix

Different phases of PVDF compositions can be obtained, the beta phase has attracted attention of researches for, technological applications, because in this phase the PVDF presents electroactive properties[1]. It is possible to distinguish the many possible phases of the

PVDF, as well as determine the dopant incorporation by using the FTIR-ATR spectra. Figure 3 shows the FTIR-ATR spectra of the PVDF samples with 0.0, 0.2, 0.4, 0.6, 0.8 and 1.0 wt % of Ce added. The symbols represented by, υ-CF_2 refers to stretching modes[13], α refers to alpha phase, β refers to beta phase, γ refers to gamma phase[14]. Vibrational bands at 615 cm^{-1} and 763 cm^{-1} (CF_2 bending and skeletal bending) and 795 cm^{-1} (CH_2 rocking) refer to α-phase. Vibrational band at 840 cm^{-1} (CH_2 rocking) correspond to β-phase[15]. As it can be seen, the cerium compound addition (Fig. 3 a) changes the absorption regions at 840 and 763 cm^{-1}. From the FTIR-ATR data is possible estimated the percentage of β phase in relation of α phase, following the method proposed by Mohammadi[16]. A decreasing of the percentage of β phase in relation of α phase was observed, as a function of Ce addition in the PVDF matrix from 62 % (represented by *PVDF/Ce 0.0* sample) to 40 % (represented by *PVDF/Ce 1.0* sample).

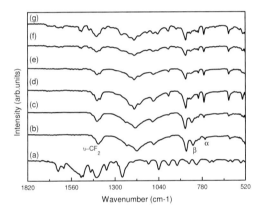

Figure 3. FT-IR measurements of PVDF/Ce as a function of Ce concentration.

Figure 4 shows the behavior of optical absorption and the emission intensity of the PVDF and Ce. As it can be seen, the PVDF polymer presented a maximum absorption at near 350 nm, while the Ce compound showed a maximum absorption at 389 nm. The maximum fluorescence emission was observed at 514 and 553 nm for PVDF and Ce, respectively, when excited at 405 nm. It is well known that the Ce^{3+} present a broad band emission centered at near 550 nm, associated to the allowed $5d \rightarrow 4f$ electronic transition[8].

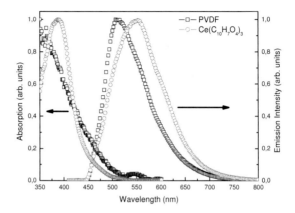

Figure 4. Normalized absorption and emission spectra for PVDF and Ce samples.

Fig. 5 shows the normalized absorption and fluorescence spectra as a function of Ce added in the PVDF matrix. As it can be observed, the addition of Ce in the polymer matrix leads to an absorption band broadening associated with a red shift, from a maximum absorption at 350 nm to 517 nm. A similar broadening and shift behavior is observed in the fluorescence spectra with the addition of Ce in the polymeric matrix. In fact, the PVDF polymer presents a natural fluorescence with a maximum centered at 514 nm which is broadened when Ce is added, emitting between 489 (green wavelength region) and 616 nm (orange region). Andrade and co-workers observed a broad emission, centered at 540 nm (yellow), for cerium doped low-silica-calcium-alumino-silicate (LSCAS) glass when excited at 405 nm, and according to the authors the yellow broad emission is characteristic of Ce^{3+}-doped garnet crystals.

The observed gap (as shown in Fig. 4) between the maximum of the first absorption band and the maximum of fluorescence is called Stokes shift $\left(\Delta \bar{v} = \bar{v}_a - \bar{v}_f \right)$, where $\bar{v}_a$ is the maximum of the first absorption band and $\bar{v}_f$ is the maximum of fluorescence, expressed in wavenumbers[17]. As shown in Table 2, the Stokes shift decrease approximately 39 % with the increase of Ce content.

From the point of view of technological application, the results of this work show that PVDF/Ce is a potential candidate to be used in optical and photonic applications.

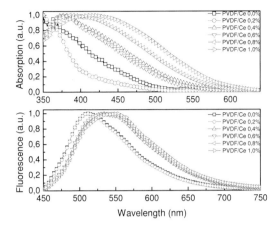

Figure 5. Normalized absorption and emission spectra of PVDF as a function of Ce addition in the polymeric matrix.

Table II. Maximum wavenumber of absorption and fluorescence spectra of the PVDF/Ce doped and Calculus of Stokes shift.

	Ce^{3+} Compound	PVDF $Ce^{3+}0.2\%$	PVDF $Ce^{3+}0.2\%$	PVDF $Ce^{3+}0.4\%$	PVDF $Ce^{3+}0.6\%$	PVDF $Ce^{3+}0.8\%$	PVDF $Ce^{3+}1.0\%$
Absorption cm^{-1}	25707	28571	28571	26455	26667	25907	23866
Emission cm^{-1}	18149	19531	19531	18692	18692	18450	18349
$\Delta\bar{\nu}$ cm^{-1}	7558	9040	9040	7763	7975	7457	5517

Conclusion

In this work the FT-IR measurements showed that the Ce compound was incorporated in the polymer matrix and a decreasing of the percentages of β phase in relation of α phase, as a function of Ce addition in the PVDF matrix. The optical measurements showed that incorporation of Ce in the PVDF matrix induced a broadening of absorption and fluorescence spectra of the compound. In addition, the increase of Ce percentage induced a red shift in the absorption and fluorescence bands, moving the maximum fluorescence band from green to orange wavelength region. It was also observed a decreasing of Stokes shift with the increasing of Ce content in the polymer matrix. In summary, our results showed that the optical behavior of PVDF/Ce may be used in optical and photonic application, revealing that PVDF/Ce compound has a great potential to be applied to develop optical and photonic devices.

Acknowledgements

The authors thank CNPq Foundation (Process N° 483683/2010-8) for financial support.

References

[1] Nalwa, H. S., *Ferroelectric Polymers: Chemistry, Physics and Applications*, (Marcel Dekker, New York, 1995), 183.

[2] Anees A. Ansari1, M. A. M. Khan, M. Naziruddin Khan, Salman A. Alrokayan1, M. Alhoshan, M. S. Alsalhi, Optical and electrical properties of electrochemically deposited polyaniline/CeO2 hybrid nanocomposite film, *Journal of Semiconductors* Vol. 32, No. 4, (2011), 043001-1-043001-6.

[3] Moulson A. J., Herbert J. M., *Electroceramics, Materials-Properties-Applications*, (Champman and Hall: London, UK, 1990) 310.

[4] Janas V. F., Safári A., "Overview of Fine-Scale Piezoelectric Ceramic/Polymer Composite Processing", *J. Am. Ceram. Soc.* 78 (11), (1998), 2945-2955.

[5] Smith W. A., Shaulov A. A., "Composite piezoelectrics: basic research to a practical device", *Ferroelectrics* 87, (1998) 309-320.

[6] Chunhua Xu, Runping Jia, Chunfa Ouyang, Xia Wang, and Guoying Yao, "Preparation and optical properties of poly(vinylidene difluoride)/(Y0.97Eu0.03)2O3 rare-earth nanocomposite," *Chin. Opt. Lett.* 6, (2008), 763-766.

[7] W. C. Choi, H. N. Lee, E. K. Kim, Y. Kim,C. Y. Park, H. S. Kim, J. Y. Lee, "Violet/blue light-emitting cerium silicates", *App. Phys. Letters* 75 [16], 2389 (1999).

[8] L.H.C. Andrade, S.M. Lima, M.L. Baesso, A. Novatski, J.H. Rohling, Y. Guyot, G. Boulon, "Tunable light emission and similarities with garnet structure of Ce-doped LSCAS glass for white-light devices", *Journal of Alloys and Compounds* 510, (2012), 54-59.

[9] Lima L. S., Caires F. J., Carvalho C. T., Siqueira A. B. and Ionashiro M., "Synthesis, characterization and thermal behaviour of solid-state compounds of light trivalent lanthanide succinates", *Thermochim Acta*. 2010: 501: 50-4.

[10] Deacon G. B., Phillips R. J., "Relationships between the carbon-oxygen stretching frequencies of carboxylato complexes and the type of carboxylate coordination". *Coord. Chem. Rev.* 33 (1980) 227-50.

[11] Socrates G., *Infrared Characteristic Group Frequencies*, (2nd ed., Wiley, New York, 1994), 91 and 236-37.

[12] Silverstein R. M., Webster F. X., *Spectrometric Identification of Organic Compounds*, (6th ed., Wiley, New York; 1998), 92-3 and 96-7.

[13] Silverstein, R. M., Webster, F. X., Kiemle, D. J., *Spectrometric Identification of Organic Compounds*, (7th ed.John Wiley and Sons: Hoboken, USA (2005), 107.

[14] Boccaccio, T., Bottino, A., Camannelli, G., Piaggio, P., "Characterization of PVDF Membranes by Vibrational Spectroscopy", *Journal of Membrane Science* 210, (2002), 315-317

[15] A. Salimi, A.A. Yousefi, "Analysis Method: FTIR studies of β-phase crystal formation in stretched PVDF films", *Polymer Testing* 22, (2003), 699-704.

[16] Behzad M., A. A. Yousefi, S. M. Bellah, "Effect of tensile strain rate and elongation on crystalline structure and piezoelectric properties of PVDF thin films", *Polymer Testing* 26, (2007), 42-50.

[17] Valeur B., *Molecular Fluorescence: Principles and Applications*", (Wiley – VCH Verlag GmbH, 2001), 54.

EFFECT OF POLING FIELD ON ELASTIC CONSTANTS IN PIEZOELECTRIC CERAMICS

Toshio Ogawa, Keisuke Ishii, Tsubasa Matsumoto and Takayuki Nishina
Department of Electrical and Electronic Engineering, Shizuoka Institute of Science and Technology,
2200-2 Toyosawa, Fukuroi, Shizuoka 437-8555, Japan

ABSTRACT
Elastic constants were measured in piezoelectric ceramics composed of lead zirconate titanate, lead titanate, alkali niobate and alkali bismuth titanate. While ferroelectric domain switching and domain rotation can be explained by DC poling field dependence of dielectric and piezoelectric properties, Young's modulus and Poisson's ratio toward poling field were investigated by measuring longitudinal and transverse wave velocities to clarify the relationships between high piezoelectricity, domain clamping and the values. Basically, increasing domain alignment by applying DC poling field, longitudinal wave velocity increased and transverse wave velocity decreased. As a result, Young's modulus decreased in lead zirconate titanate and alkali niobate, and moreover, increased in lead titanate and alkali bismuth titanate. Poisson's ratio increased despite the ceramic compositions.

INTRODUCTION
The behavior of ferroelectric domains toward a DC field was investigated on the basis of the DC poling field dependence of dielectric and piezoelectric properties in various types of piezoelectric ceramics[1-7] and single crystals[8,9]. In addition, we have developed a method to evaluate elastic constants, such as Young's modulus and Passion's ratio, by measuring longitudinal and transverse wave velocities using an ultrasonic thickness gauge with high-frequency pulse oscillation[10] in comparison with a conventional method[11-14]. The acoustic wave velocities can be measured by this method in the cases of ceramics and single crystals despite the degree of the DC poling, including as-fired ceramics and as-grown single crystals[10].

Therefore, in order to clarify the relationships between elastic constants and electrical properties vs DC poling fields, we studied the poling field dependence of acoustic wave velocities and dielectric and piezoelectric properties in ceramics. Here, we report the relationships between DC poling fields, acoustic wave velocities, Young's modulus, and Passion's ratio to realize a high piezoelectricity, especially in lead-free ceramics.

EXPERIMENTAL PROCEDURE
The piezoelectric ceramic compositions measured were as follows: $0.05Pb(Sn_{0.5}Sb_{0.5})O_3$-$(0.95-x)PbTiO_3$-$xPbZrO_3$ (x=0.33, 0.45, 0.66) with (hard PZT) and without 0.4 wt% MnO_2 (soft PZT)[3]; $0.90PbTiO_3$-$0.10La_{2/3}TiO_3$(PLT) and $0.975PbTiO_3$-$0.025La_{2/3}TiO_3$(PT)[6]; $(1-x)(Na,K,Li,Ba)(Nb_{0.9}Ta_{0.1})O_3$-$xSrZrO_3$ (SZ) (x=0.00, 0.05, 0.07)[15,16]; $(1-x)(Na_{0.5}Bi_{0.5})TiO_3$(NBT)-$x(K_{0.5}Bi_{0.5})TiO_3$ (KBT) (x=0.08, 0.18) and $0.79NBT$-$0.20KBT$-$0.01Bi(Fe_{0.5}Ti_{0.5})O_3$(BFT) (x=0.20)[17]; and $(1-x)NBT$-$xBaTiO_3$(BT) (x=0.03, 0.07, 0.11)[17].

DC poling was conducted for 30 min at the most suitable poling temperature depending on the Curie points of the ceramic materials when the poling field (E) was varied from $0\rightarrow+0.25\rightarrow+0.5\rightarrow+0.75\rightarrow+1.0\rightarrow$ --- $+E_{max}\rightarrow0\rightarrow-E_{max}\rightarrow0$ to $+E_{max}$ kV/mm. $\pm E_{max}$ depended on the coercive fields of the piezoelectric ceramics. After each poling process, the dielectric and piezoelectric properties were measured at room temperature using an LCR meter (HP4263A), a precision impedance analyzer (Agilent 4294A), and a piezo d_{33} meter (Academia Sinica ZJ-3D). Furthermore, the acoustic wave velocities were measured using an ultrasonic precision thickness gauge (Olympus 35DL), which has PZT transducers with 30 MHz for longitudinal-wave generation and 20 MHz for transverse-wave generation[10]. The acoustic wave velocities were evaluated on the basis of the propagation time between the second-pulse echoes in the thickness of ceramic disks parallel to the poling field with dimensions of 14 mm diameter and 0.5-1.5 mm thickness. The sample thickness was measured using a precision micrometer (Mitutoyo MDE-25PJ). Moreover, Young's modulus in the thickness direction of ceramic disks (Y_{33}^E) and Passion's ratio (σ) were calculated on the basis of the longitudinal (V_L) and transverse (V_S) wave velocities as shown in the following equations of (1) and (2):

$$Y_{33}^E = 3\rho V_S^2 \frac{V_L^2 - \frac{4}{3}V_S^2}{V_L^2 - V_S^2} \quad ----- \quad (1)$$

and

$$\sigma = \frac{1}{2}\left\{1 - \frac{1}{\left(\frac{V_L}{V_S}\right)^2 - 1}\right\} \quad ----- \quad (2),$$

where ρ is the density of ceramic disks.

RESULTS AND DISCUSSION
Poling field dependence in PZT ceramics
         Figures 1-3 show the poling field dependence of longitudinal (V_L) and transverse (V_S) wave velocities, Young's modulus (Y_{33}^E), and Passion's ratio (σ) in $0.05Pb(Sn_{0.5}Sb_{0.5})O_3$-$(0.95-x)PbTiO_3$-$xPbZrO_3$ (x=0.33, 0.45, 0.66) with (hard PZT) and without 0.4 wt% MnO_2 (soft PZT) ceramics at a poling temperature (T_P) of 80 ℃ (hard PZT in Fig. 1 and soft PZT in Fig. 2), $0.90PbTiO_3$-$0.10La_{2/3}TiO_3$ (abbreviate to PLT/T_p=80 ℃) and $0.975PbTiO_3$-$0.025La_{2/3}TiO_3$ (abbreviate to PT /T_p= 200 ℃) ceramics (Fig. 3), respectively. While the poling field dependence of V_L has almost same tendency in spite of hard and soft PZT, the one of V_S at x=0.45, which corresponds to MPB[18], abruptly decreases in the both cases of hard and soft PZT (Figs. 1, 2). Since the highest coupling factor in PZT is obtained at the MPB (x=0.45), the origin of the highest piezoelectricity is due to the decrease in V_S with increasing the domain alignment by DC poling field. Furthermore, the lowest Y_{33}^E and the highest σ are realized at the MPB. The change in V_L, V_S, Y_{33}^E and σ vs E of soft PZT is smaller than the one of hard PZT because of the softness of the materials. As mentioned details to the next session, it was indicated that minimum V_L, σ and maximum V_S, Y_{33}^E were obtained at the domain clamping such as the domain alignment canceled each other (↑ ↓), at which the lowest piezoelectricity is realized[3]. On the other hand, the values of V_L, V_S, Y_{33}^E and σ vs E in PT are smaller than the ones of PLT. Moreover, at the domain clamping fields, minimum V_L, σ, Y_{33}^E and maximum V_S were obtained in both the PLT and PT. The reason of minimum Y_{33}^E at the DC field of the domain clamping will be discuss in the next session. In addition, the change in Y_{33}^E of PLT and PT while applying ±E is smaller than the one of PZT, and higher Y_{33}^E and lower σ appear in comparison with the ones of PZT, because these come from the hardness of PLT and PT ceramics.

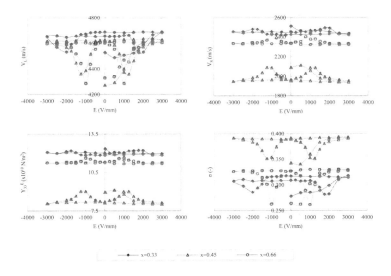

Fig. 1 Poling field dependence of longitudinal (V_L) and transverse (V_S) wave velocities, Young's modulus (Y_{33}^E), and Poisson's ratio (σ) in $0.05Pb(Sn_{0.5}Sb_{0.5})O_3$-$(0.95-x)PbTiO_3$-$xPbZrO_3$ (x=0.33, 0.45, 0.66) with 0.4 wt% MnO_2 (hard PZT) ceramics.

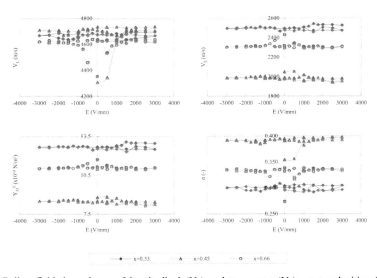

Fig. 2 Poling field dependence of longitudinal (V_L) and transverse (V_S) wave velocities, Young's modulus (Y_{33}^E), and Poisson's ratio (σ) in $0.05Pb(Sn_{0.5}Sb_{0.5})O_3$-$(0.95-x)PbTiO_3$-$xPbZrO_3$ (x=0.33, 0.45, 0.66) without 0.4 wt% MnO_2 (soft PZT) ceramics.

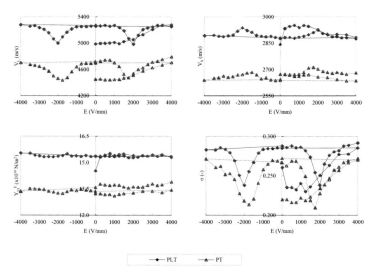

Fig. 3 Poling field dependence of longitudinal (V_L) and transverse (V_S) wave velocities, Young's modulus (Y_{33}^E), and Poisson's ratio (σ) in $0.90PbTiO_3$-$0.10La_{2/3}TiO_3$ (PLT) and $0.975PbTiO_3$-$0.025La_{2/3}TiO_3$ (PT) ceramics.

Poling field dependence in lead-free ceramics

Figures 4-6 show the poling field dependence of longitudinal (V_L) and transverse (V_S) wave velocities, Young's modulus (Y_{33}^E), and Passion's ratio (σ) in $(1-x)(Na,K,Li,Ba)(Nb_{0.9}Ta_{0.1})O_3$-$xSrZrO_3$(SZ) (x=0.00, 0.05, 0.07) ceramics at a poling temperature (T_P) of 150 ℃ (Fig. 4), in $(1-x)(Na_{0.5}Bi_{0.5})TiO_3$(NBT)-$x(K_{0.5}Bi_{0.5})TiO_3$(KBT) (x=0.08, 0.18/ T_p=70 ℃) and 0.79NBT-0.20KBT-$0.01Bi(Fe_{0.5}Ti_{10.5})O_3$(BFT) (x=0.20/ T_p=70 ℃) ceramics (Fig. 5), and in $(1-x)NBT-xBaTiO_3$(BT) (x=0.03, 0.07, 0.11/ T_p=70 ℃) ceramics (Fig. 6), respectively. With the enhancement of domain alignment with an increase in poling field from E=0 to +E_{max}, V_L increased and V_S decreased independently of the ceramic composition. From the composition dependence of Y_{33}^E and σ, high piezoelectricity [high planar coupling factor (k_p)] compositions show lower Y_{33}^E and higher σ values at $0.95(Na,K,Li,Ba)(Nb_{0.9}Ta_{0.1})O_3$-$0.05SZ$ (k_p=46%) (Fig. 4) and 0.82NBT-0.18KBT (k_p=27%) (Fig. 5) as well as the ones at $0.05Pb(Sn_{0.5}Sb_{0.5})O_3$-$0.47PbTiO_3$-$0.48PbZrO_3$ (k_p=52% in hard ceramics and k_p=65% in soft ceramics) than the other compositions. Although morphotropic phase boundaries (MPBs) were observed in the NBT-KBT[19,20] and PZT[3,18] ceramics, there was no evidence of the existence of MPBs in the $(Na,K,Li,Ba)(Nb_{0.9}Ta_{0.1})O_3$-SZ ceramics[15,16]. The effects of 0.01BFT modification in NBT-KBT on Y_{33}^E and σ were as follows: the composition of 0.79NBT-0.20KBT-0.01BFT (k_p=22%; x=0.20 in Fig. 5) showed the highest piezoelectric d_{33} constant of 150 pC/N in NBT-KBT and higher Y_{33}^E and lower σ values than that of 0.82NBT-0.18KBT (k_p=27%; x=0.18 in Fig. 5) because the relative dielectric constant increased from 800 (0.82NBT-0.18KBT) to 1250 (0.79NBT-0.20KBT-0.01BFT)[17]. Moreover, a high coupling factor composition at 0.93NBT- 0.07BT (k_p=16%) existed in a MPB[21] with a low Y_{33}^E (Fig. 6). Basically, research on piezoelectric ceramics with high planar coupling factors, especially in lead-free ceramics, has been focused on determining the MPB composition because many different polarization axes are generated in MPB.

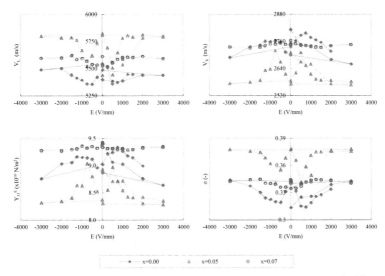

Fig. 4 Poling field dependence of longitudinal (V_L) and transverse (V_S) wave velocities, Young's modulus (Y_{33}^E), and Poisson's ratio (σ) in $(1-x)(Na,K,Li,Ba)(Nb_{0.9}Ta_{0.1})O_3$-$xSrZrO_3$ ($x=0.00$, 0.05, 0.07) ceramics.

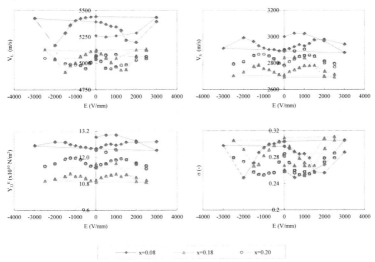

Fig. 5 Poling field dependence of longitudinal (V_L) and transverse (V_S) wave velocities, Young's modulus (Y_{33}^E), and Poisson's ratio (σ) in $(1-x)NBT$-$xKBT$ ($x=0.08$, 0.18) and $0.79NBT$-$0.2KBT$-$0.01BFT$ ($x=0.20$) ceramics.

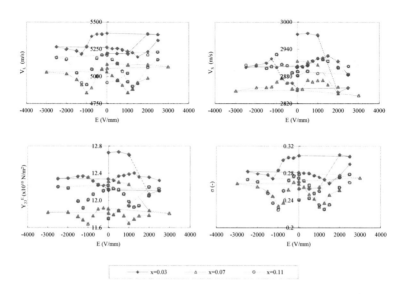

Fig. 6 Poling field dependence of longitudinal (V_L) and transverse (V_S) wave velocities, Young's modulus (Y_{33}^E), and Poisson's ratio (σ) in (1-x)NBT-xBT (x=0.03, 0.07, 0.11) ceramics.

Through our study on acoustic wave measurement, it was found that there was an important factor for obtaining a high piezoelectricity regarding Y_{33}^E and σ; lower Y_{33}^E and higher σ values in the cases of PZT, alkali bismuth titanate [NBT-KBT], and alkali bismuth barium titanate [NBT-BT] with MPB, and in the case of alkali niobate [(Na,K,Li,Ba)(Nb$_{0.9}$Ta$_{0.1}$)O$_3$-SZ] without MPB. Therefore, as mentioned later, we need a new concept in addition to conventional research on MPB compositions for developing piezoelectric ceramics with high coupling factors.

The DC poling fields under domain clamping (E_d) can be estimated on the basis of E values to realize the minimum V_L and maximum V_S in Figs. 4-6. Although the maximum Y_{33}^E and minimum σ were obtained at E_d, which corresponds to the E of the minimum coupling factor, in (Na,K,Li,Ba)(Nb$_{0.9}$Ta$_{0.1}$)O$_3$-SZ (Fig. 4), NBT-KBT and 0.79NBT-0.20KBT-0.01BFT (Fig. 5), the minimum Y_{33}^E and minimum σ were obtained at E_d in NBT-BT (Fig. 6). It was clarified that the minimum Y_{33}^E in NBT-BT was realized in the cases of $\Delta V_S/\Delta V_L < 1/4$ at E_d, where ΔV_S and ΔV_L denote the variations in V_S and V_L at E_d, respectively. This may be due to the poor domain alignment (lower Poisson's ratio) perpendicular to the poling field (radial direction in disk ceramics) in comparison with the domain alignment parallel to the poling field (thickness direction in disk ceramic). Furthermore, the increase in coupling factor corresponds to the increase in σ at all compositions including lead-containing and lead-free ceramic compositions.

Materials road maps in piezoelectric ceramics on elastic constants

Figure 7 shows the relationships between longitudinal (V_L) and transverse (V_S) wave velocities, Young's modulus (Y_{33}^E), and Passion's (σ) ratio vs planar coupling factors (k_p) in (1-x)(Na,K,Li,Ba)(Nb$_{0.9}$Ta$_{0.1}$)O$_3$-xSZ (abbreviated to "SZ"), (1-x)NBT-xKBT ("KBT"), 0.79NBT-0.20KBT-0.01BFT ("KBT"), and (1-x)NBT-xBT ("BT") lead-free ceramics compared with 0.05Pb(Sn$_{0.5}$Sb$_{0.5}$)O$_3$-

$(0.95-x)$PbTiO$_3$-xPbZrO$_3$ ceramics with ("hard PZT") and without 0.4 wt% MnO$_2$ ("soft PZT"), and with 0.90PbTiO$_3$-0.10La$_{2/3}$TiO$_3$ ("PLT") and 0.975PbTiO$_3$-0.025La$_{2/3}$TiO$_3$ ("PT") lead-containing ceramics after full DC poling. Although the V_L values of the PZT ceramics were almost constant at approximately 4,600-4,800 m/s independently of the composition x, their V_S values linearly decreased from 2,500 to 1,600 m/s with increasing k_p from 20 to 65% (solid lines). In addition, the V_L and V_S values of the PZT ceramics were smaller than those of the lead-free ceramics (V_L=5,000-5,800 m/s and V_S=2,600-3,000 m/s; dashed and dotted lines). Although the V_L values of the PT ceramics were almost the same (4,800 m/s) as those of the PZT ceramics, the V_S values of the PT ceramics were approximately 2,700 m/s. On the other hand, the V_L values of the SZ ceramics were relatively high (5,500-5,800 m/s); furthermore, the V_S values of the SZ ceramics also increased (2,600-2,700 m/s) and linearly decreased with increasing k_p from 25 to 50% (dashed lines), the behavior of which was almost the same as that of the V_S values of the PZT ceramics. The V_L values of the KBT, BT, and PLT ceramics (5,000-5,400 m/s) were between those of the PZT, PT, and SZ ceramics. However, the V_S values of the KBT, BT, and PLT ceramics (2,800-3,000 m/s) were the highest. Therefore, it was possible to divide V_L and V_S into three material groups, namely, PZT and PT/ KBT, BT (alkali bismuth titanate), and PLT/ SZ (alkali niobate). In addition, k_p lineally increased from 4 to 65% with decreasing Y_{33}^E from 15×10^{10} to 6×10^{10} N/m^2 and lineally increased with increasing σ from 0.25 to 0.43. It was clarified that higher k_p values can be realized at lower Y_{33}^E and higher σ values.

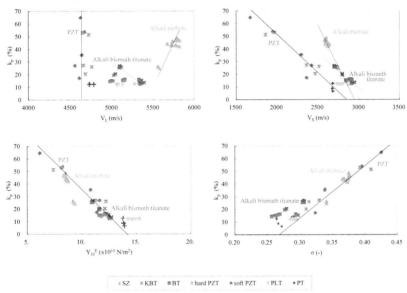

Fig. 7 Relationships between longitudinal (V_L) and transverse (V_S) wave velocities, Young's modulus (Y_{33}^E), and Passion ratio (σ) vs planar coupling factors (k_p) in $(1-x)$(Na,K,Li,Ba)(Nb$_{0.9}$Ta$_{0.1}$)O$_3$-xSZ (abbreviated to "SZ"), $(1-x)$NBT-xKBT ("KBT"), 0.79NBT-0.20KBT-0.01BFT ("KBT"), and $(1-x)$NBT-xBT ("BT") lead-free ceramics compared with 0.05Pb(Sn$_{0.5}$Sb$_{0.5}$)O$_3$-$(0.95-x)$PbTiO$_3$-xPbZrO$_3$ ceramics with ("hard PZT") and without 0.4 wt% MnO$_2$ ("soft PZT"), and with 0.90PbTiO$_3$-0.10 La$_{2/3}$TiO$_3$ ("PLT") and 0.975PbTiO$_3$-0.025La$_{2/3}$TiO$_3$ ("PT") lead-containing ceramics after full DC poling.

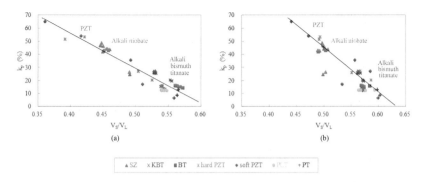

(a)   (b)

Fig. 8 Ratio of longitudinal wave velocity (V_L) to transverse wave velocity (V_S): V_S/V_L vs k_p in lead-containing ceramics [$0.05Pb(Sn_{0.5}Sb_{0.5})O_3$-$(0.95$-$x)PbTiO_3$-$xPbZrO_3$ with ("hard PZT") and without 0.4 wt% MnO_2 ("soft PZT"), $0.90PbTiO_3$-$0.10La_{2/3}TiO_3$ ("PLT"), and $0.975PbTiO_3$-0.025 $La_{2/3}TiO_3$ ("PT")] and in lead-free ceramics [$(1$-$x)(Na,K,Li,Ba)(Nb_{0.9}Ta_{0.1})O_3$-$xSZ$ (abbreviated to "SZ"), $(1$-$x)NBT$-$xKBT$ ("KBT"), $0.79NBT$-$0.20KBT$-$0.01BFT$ ("KBT"), and $(1$-$x)NBT$-xBT ("BT")] (a) after DC full poling and (b) before poling.

Figures 8 and 9 show the ratio of V_L to V_S vs k_p and V_S/V_L vs piezo strain constant of d_{33} in lead-containing and lead-free ceramics (a) after DC full poling and (b) before poling, respectively. There are linear relationships between V_S/V_L vs k_p and d_{33} in spite of DC poling treatment; with decreasing V_S/V_L from 0.58 to 0.36, k_p increased from 4 to 65% (correlation coefficient r=-0.96 after poling and r=-0.89 before poling) and d_{33} also increased from 52 to 463 μC/N (r=-0.87 in both after and before poling). This means that it is possible to estimate the degree of piezoelectricity (the values of k_p and d_{33}) from V_S/V_L, even though the ceramics were without DC poling treatment (as-fired).

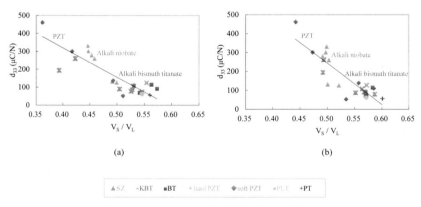

(a)   (b)

Fig. 9 Ratio of longitudinal wave velocity (V_L) to transverse wave velocity (V_S): V_S/V_L vs piezo d_{33} constant in lead-containing ceramics ["hard PZT", "soft PZT", "PLT" and "PT"] and in lead-free ceramics ["SZ", "KBT" and "BT"] (a) after DC full poling and (b) before poling.

Furthermore, we can mention that it is significant to evaluate the both values of k_p and d_{33}, which are not directly corresponding to each other because of $d=k (\varepsilon \cdot s)^{1/2}$ (d: piezo strain constant, k: coupling factor, ε: dielectric constant, s: compliance). Therefore, our developed method for acoustic wave velocity measurement, especially V_S/V_L, is useful for materials research on piezoelectric ceramics including lead-free ceramics with higher coupling factors and higher piezo d_{33} constants in the cases of as-fired (before poling) [Figs. 8(b) and 9(b)] as well as after DC fully polarized [Figs. 8(a) and 9(a)] ceramics.

In this study, it was described the road maps in piezoelectric ceramics regarding the relationships between longitudinal and transvers wave velocities, Young's modulus and Poisson's ratio to research and develop new piezoelectric ceramic materials, especially lead-free ceramics with high piezoelectricity.

CONCLUSIONS

Longitudinal and transverse wave velocities of PZT, lead titanate and lead-free ceramics were measured by an ultrasonic precision thickness gauge with high-frequency pulse oscillation to calculate elastic constants such as Young's modulus and Poisson's ratio. The DC poling field dependence of the longitudinal and transverse wave velocities of PZT, lead titanate, and lead-free ceramics was investigated to evaluate the elastic constants vs poling field. The effect of the poling field on domain alignment, such as in the cases of DC full poling and domain clamping, could be explained by the relationships between acoustic wave velocities, Young's modulus, and Poisson's ratio vs poling field. The directions of the research and development of piezoelectric ceramics including lead-free ceramics with high coupling factors could be proposed on the basis of the findings of this study.

ACKNOWLEDGMENTS

This work was partially supported by a Grant-in-Aid for Scientific Research C (No. 21560340) and a Grant of Strategic Research Foundation Grant-aided Project for Private Universities 2010-2014 (No. S1001032) from the Ministry of Education, Culture, Sports, Science and Technology, Japan (MEXT).

REFERENCES

[1]T. Ogawa, A. Yamada, YK. Chung, and DI. Chun, Effect of Domain Structures on Electrical Properties in Tetragonal PZT Ceramics, *J. Korean Phys. Soc.*, **32**, S724-S726 (1998).
[2]T. Ogawa and K. Nakamura, Poling Field Dependence of Ferroelectric Properties and Crystal Orientation in Rhombohedral Lead Zirconate Titanate Ceramics, *Jpn. J. Appl. Phys.*, **37**, 5241-5245 (1998).
[3]T. Ogawa and K. Nakamura, Effect of Domain Switching and Rotation on Dielectric and Piezo-electric Properties in Lead Zirconate Titanate Ceramics, *Jpn. J. Appl. Phys.*, **38**, 5465-5469 (1999).
[4]T. Ogawa, Domain Switching and Rotation in Lead Zirconate Titanate Ceramics by Poling Fields, *Ferroelectrics*, **240**, 75-82 (2000).
[5]T. Ogawa, Domain Structure of Ferroelectric Ceramics, *Ceram. Int.*, **25**, 383-390 (2000).
[6]T. Ogawa, Poling Field Dependence of Crystal Orientation and Ferroelectric Properties in Lead Titanate Ceramics, *Jpn. J. Appl. Phys.*, **39**, 5538-5541 (2000).
[7]T. Ogawa, Poling Field Dependence of Ferroelectric Properties in Barium Titanate Ceramics, *Jpn. J. Appl. Phys.*, **40**, 5630-5633 (2001).
[8]T. Ogawa, Poling Field Dependence of Ferroelectric Properties in Piezoelectric Ceramics and Single Crystal, *Ferroelectrics*, **273**, 371-376 (2002).
[9]R. Kato and T. Ogawa, Chemical Composition Dependence of Giant Piezoelectricity on k_{31} Mode in $Pb(Mg_{1/3}Nb_{2/3})O_3$-$PbTiO_3$ Single Crystals, *Jpn. J. Appl. Phys.*, **45**, 7418-7421 (2006).

[10]T. Ogawa and T. Nishina, Acoustic Wave Velocities Measurement on Piezoelectric Ceramics to Evaluate Young's Modulus and Poisson's Ratio for Realization of High Piezoelectricity, *Ceramic Transactions, Advances and Applications in Electroceramics II* (John Wiley Publication, Hoboken, 2012), **235**, p.105-112.

[11]K. Shibayama et al., *Danseiha Soshi Gijyutsu Handbook* (Technical Handbook of Acoustic Wave Devices) (Ohmsha, Tokyo,1991), p.29 [in Japanese].

[12]WP. Mason et al., *Physical Acoustics I*, **Part A**. (Academic Press, New York, 1964), p.182.

[13]K. Takagi et al., *Cyouonpa Binran* (Ultrasonic Handbook) (Maruzen, Tokyo, 1999), p.395 [in Japanese].

[14]K. Negishi and K. Takagi, *Cyouonpa Gijutsu* (Ultrasonic Technology) (Univ. of Tokyo Press, Tokyo, 1984), p.109 [in Japanese].

[15]M. Furukawa, T. Tsukada, D. Tanaka, and N. Sakamoto, Alkaline Niobate-based Lead-Free Piezoelectric Ceramics, *Proc. 24th Int. Japan-Korea Semin. Ceramics*, 2007, p.339-342.

[16]T. Ogawa, M. Furukawa, and T. Tsukada, Poling Field Dependence of Piezoelectric Properties and Hysteresis Loops of Polarization versus Electric Field in Alkali Niobate Ceramics, *Jpn. J. Appl. Phys.*, **48**, 709KD07-1-5 (2009).

[17]T. Ogawa, T. Nishina, M. Furukawa, and T. Tsukada, Poling Field Dependence of Ferroelectric Properties in Alkali Bismuth Titanate Lead-Free Ceramics, *Jpn. J. Appl. Phys.*, **49**, 09MD07-1-4 (2010).

[18]T. Ogawa, Highly Functional and High-Performance Piezoelectric Ceramics, *Ceramic Bulletin*, **70**, 1042-1049 (1991).

[19]W. Zhao, H. Zhou, Y. Yan, and D. Liu, Morphotropic Phase Boundary Study of the BNT-BKT Lead-Free Piezoelectric Ceramics, *Key Eng. Mater.*, **368-372**, 1908-1910 (2008).

[20]Z. Yang, B. Liu, L. Wei, and Y. Hou, Structure and Electrical Properties of $(1-x)Bi_{0.5}Na_{0.5}TiO_3$-$xBi_{0.5}K_{0.5}TiO_3$ Ceramics near Morphotropic Phase Boundary, *Mater. Res. Bull.*, **43**, 81-89 (2008).

[21]Y. Dai, J. Pan, and X. Zhang, Composition Range of Morphotropic Phase Boundary and Electrical Properties of NBT-BT System., *Key Eng. Mater.*, **336-338**, 206-209 (2007).

ENERGY HARVESTING UTILIZED RESONANCE PHENOMENA OF PIEZOELECTRIC UNIMORPH

Toshio Ogawa[1], Hiroshi Aoshima[1], Masahito Hikida[1] and Hiroshi Akaishi[2]
[1]Department of Electrical and Electronic Engineering, Shizuoka Institute of Science and Technology, 2200-2 Toyosawa, Fukuroi, Shizuoka 437-8555, Japan
[2]Plus Comfort Co., Ltd., 3-11-3 Heiwa, Aoi-Ku, Shizuoka 420-0876, Japan

ABSTRACT

Piezoelectric buzzer for energy harvesting was investigated. While external force was added to a buzzer, a PZT unimorph in the buzzer, the ceramic disc dimensions and capacitance of which were 14 mm in diameter, 0.2 mm in thickness and 10 nF, generated resonance vibration. As a result, alternating voltages with the value of around 30 V and the frequency of 5 kHz were observed. When the generated voltages were applied to a LED lamp, new devices such as "night-view footwear" and "piezo-walker" were developed. It was confirmed that the energy harvesting by piezo-buzzer utilized resonance phenomena is one of an effective tool to obtain clean energy.

INTRODUCTION

Recently, energy harvesting for energy conversion between electrical energy and mechanical energy in piezoelectric ceramics was focused on to realize one of clean energy. There are practical uses such as a floor for electrical generation: "Electricity-Generation-Floor" at ticket gates in Tokyo terminal of Japan Railway East Company (JR-East) and LED light illumination for the Metropolitan Expressway generated electricity by deformation of the bridge piers passing through automobiles.

We newly develop high-efficiency piezoelectric energy harvesting utilized resonance phenomena of piezoelectric ceramics. While applying piezo-buzzer to new devices, "night-view footwear" and "piezo-walker", could be developed.

EXPERIMENTAL PROCEDURE

Generation voltages were measured by a digital oscilloscope (Agilent DSO1012) in the cases of (1) adding external force to unimorph and piezo-buzzer and (2) connecting to a rectifying circuit. Furthermore, how can be added external force to piezo-buzzers effectively was studied. The piezo-buzzers and LED lamp were used produced by Kyocera (KBS-20DB-5A)[1], by Murata (PKM17EWH4000)[2] and by Linkman (LFTLED-R301)[3], respectively. External force was added by dropping a sphere-weight of 6.15 g to unimorph and piezo-buzzer from 10 cm height (Fig.1). Here, the

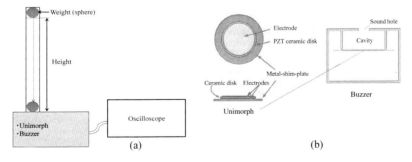

Fig.1 (a) Adding external force by sphere-weight to (b) unimorph and piezo-buzzer.

piezoelectric materials used in the unimorphs were composed of soft PZT's with morphotropic phase boundaries (MPB's), which can realize high electromechanical coupling factors, high relative dielectric constants and high piezoelectric strain constants of d_{31} and d_{33}[4].

RESULTS AND DISCUSSION
Generation Voltage by Piezo-Buzzer

Figure 2 shows the generated voltages vs reaction times by PZT unimorph (dimensions: 14 mm in diameter and 0.2 mm in thickness) on (a) a sponge and on (b) a desk while adding external force from 10 cm height. On the sponge, the added force was absorbed by the sponge, and after that vibrated the unimorph again. Therefore, two peaks of voltage were observed in the one (+) direction. On the other hand, the unimorph on the desk generated one peak of + voltage because of non-repulsion for the unimorph. The maximum generated voltage reached to about +50 V in the both cases. We call this procedure of the energy harvesting "non-resonant system" which is the same mechanism as the "Electricity-Generation-Floor" of JR-East.

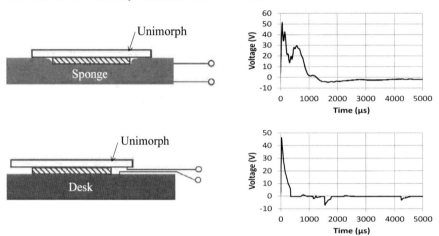

Fig. 2 Generated voltage-waves from PZT unimorphs on sponge and on desk; this procedure corresponds to "non-resonance system" as same mechanism as "Electricity-Generation-Floor" of JR-East.

Figures 3 shows the generated voltages vs reaction times by piezo-buzzer (KBS-20DB-5A) while adding external force from 10 cm height (a) without and (b) with rectifying circuit of a diode bridge. Adding the external force, the unimorph in the piezo-busser resonated and generated alternating voltages with a period of 200 μ sec (5 kHz), which corresponds to the sound frequency of piezo-buzzer, as shown in Fig. 3 (a). We call this procedure "resonance system". In the "resonant system", the generated voltage decreases from +50 V ("non-resonant system") to ±30 V ("resonant system") even through adding the same external force. The phenomena reducing the generated voltage are due to the decrease in the added force in the case of the piezo-buzzer. However, it is confirm that the durability of devices against the external force was improved in the cases of piezo-buzzer in comparison with unimorph itself. Moreover, it was found that the electric charges accompanied with alternating voltages were taken out efficiently using a rectifying circuit [Fig. 3 (b)].

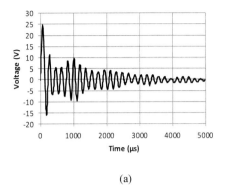

(a)

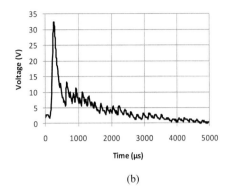

(b)

Fig. 3 Generated voltage-waves by piezoelectric buzzer while adding external force (a) without and (b) with rectifying circuit.

Application of Energy Harvesting by Piezo-Buzzer to Devices

Figure 4 shows the output voltages vs reaction times of piezo-buzzer (KBS-20DB-5A) in the case of "night-view footwear". Walking one footstep, there was additional force by collisions between the buzzer and its surroundings in addition to the resonance of unimorph itself up to 5,000 μ sec in Fig. 4. The generated voltages occurred by the additional force between 7,000 and 10,000 μ sec. We call these additional force "secondary resonance force". Therefore, it was confirmed that the voltages generated continuously during the one footstep. Since the generated charges per footstep were 1.5 μC, it was possible that the charges could be put on LED lamps.

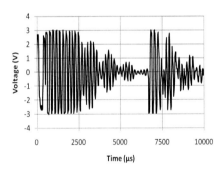

Fig. 4 Generated voltage-waves in the case of "night-view footwear" while walking one footstep; there were two kinds of added force during one footstep: below 5,000 μ sec and between 7,000 and 10,000 μ sec.

Figure 5 shows devices of (a) "night-view footwear" and (b) "piezo-walker" utilizing the energy harvesting in the case of "resonant system" of PZT unimorph in piezo-buzzer[5]. In the "night-view footwear", while walking, force generated from each footstep pushes the piezo-buzzer, and charges are supplied to LED lamp. Therefore, people can walk for traffic road safety at night. In the "piezo-walker", while walking, force generated from actions of legs and hands attached with the "piezo-walker" moves a sphere-weight and the weight runs into the both sides [◀----▶ in Fig. 5 (b)] of piezo-buzzers (PKM17EWH4000). Generated charges produced by resonance of PZT unimorph in piezo-buzzer were supplied to LED lamps. This device is also useful to prevent traffic accident at night; especially in Japan for senior's walking in order to maintain their health well.

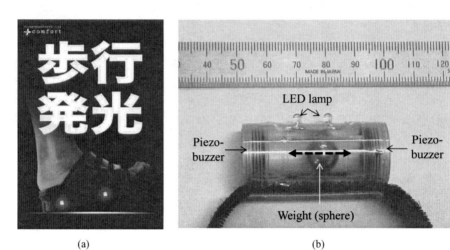

(a)   (b)

Fig. 5 Energy harvesting devices of (a) "night-view footwear" produced by Plus Comfort Co., Ltd.[6], Shizuoka, Japan and (b) "piezo-walker" utilizing "resonant system" of PZT unimorph in piezo-buzzer; Japanese characters of "hokou hatsuko" in Fig. 5 (a) means "walking lights".

Even though energy harvesting utilized "resonant system" of PZT unimorph in piezo-buzzer is relatively small portion, it is possible to realize new devices suitable for it. Our developed procedure of "resonant system" is more practical than "non-resonant system", because the piezo-buzzer possesses more excellent durability against external force than PZT unimorph.

CONCLUSIONS
    New energy harvesting for energy conversion between electrical energy and mechanical energy in piezoelectric ceramics was developed. The "resonant system" of PZT unimorph in piezo-buzzer is preferred to convert energy effectively and to keep reliability of the devices in comparison with the ordinary "non-resonant system". Utilizing "resonant system" of PZT unimorph in piezo-buzzer, the devices such as "night-view footwear" and "piezo-walker" were developed for people, especially seniors in Japan, walking under conditions of traffic safety at night.

ACKNOWLEDGEMENTS
    This work was partially supported by a Grant 2010 of B-nest at Shizuoka City, Japan, a Grant of Strategic Research Foundation Grant-aided Project for Private Universities (No. S1001032) from the Ministry of Education, Culture, Sports, Science and Technology, Japan (MEXT) and a Research

Foundation 2012 between Academy and Industry of Fukuroi City, Shizuoka, Japan.

REFERENCES
[1]www.kyocera.co.jp/.
[2]www.murata.co.jp/.
[3]www.linkman.jp/.
[4]T. Ogawa and K. Nakamura, Effect of Domain Switching and Rotation on Dielectric and Piezo-electric Properties in Lead Zirconate Titanate Ceramics, *Jpn. J. Appl. Phys.*, **38**, 5465-5469 (1999).
[5]T. Ogawa, Piezoelectric Energy Harvesting and Devices for Night-View Footwear and Piezo-Walker Utilized the Energy Harvesting, Japanese Unexamined Patent Application Publication No. 2011-188660 (March 10, 2010).
[6]www.plus-comfort.co.jp/.

INTERNAL STRAIN AND DIELECTRIC LOSSES BY COMPOSITIONAL ORDERING ON THE MICROWAVE DIELECTRICS PSEUDO-TUNGSTENBRONZE $Ba_{6-3x}R_{8+2x}Ti_{18}O_{54}$ (R = RARE EARTH) SOLID SOLUTIONS

Hitoshi Ohsato

Nagoya Industrial Science Research Institute, Nagoya 464-0819, Japan

Nagoya Institute of Technology, Nagoya 466-8555, Japan

ABSTRACT

Microwave dielectrics pseudo-tungstenbronze solid solutions form compositional ordering at $x = 2/3$ on the $Ba_{6-3x}R_{8+2x}Ti_{18}O_{54}$ (R = Sm, Nd, Pr, and La) formula. The Qf value of the $x = 2/3$ composition shows the highest value. The internal strain of the composition is the lowest value based on the full-width at half-maximum (FWHM) of X-ray power diffraction (XRPD) patterns. In the region $x < 2/3$, a part of Ba ions occupy statistically perovskite blocks, so, d-spacing of the unit cell shows fluctuation. As the unit cells without Ba ions in the perovskite blocks are causing tensile stress, dielectric constant and dielectric losses increase by means of the rattling of cations in the octahedral sites.

INTRODUCTION

Microwave dielectrics are generally expected to have high quality factor Q due to the perfect crystal structure without defects and strain, when they have highly symmetric crystal structure and are para-electrics with inversion symmetry i. In this paper, relationship between strain and microwave properties are discussed on the pseud-tungstenbronze type $Ba_{6-3x}R_{8+2x}Ti_{18}O_{54}$ (R = rare earth) solid solutions[1, 2] which have applications as microwave resonators and filters for mobile phone. Up to now, this solid solutions show the highest quality factor Qf at $x = 2/3$ which shows compositional ordering, and the lowest internal strain[3, 4]. Moreover, different species of the R ion yields to different Qf, namely, Sm with intermediate ionic radius perform the highest Qf value which also shows the lowest internal strain[5, 6, 7]. In this paper, the origins of the internal strain are examined by crystallographic considerations, which lead to the result that the highest Qf value occurs at the lowest internal strain.

EXPERIMENTAL

The samples for the measurement of the microwave dielectric properties were prepared by a solid state reaction method. The raw materials were used the high purity ones with over 99.9%. The dielectric properties were measured by the Hakki and Coleman method[8, 9]. Accurate lattice parameters were obtained using the whole-powder-pattern-decomposition method (WPPD) program[10]. The X-ray powder diffraction (XRPD) patterns were obtained by the step scanning technique using Si (99.99%) as an internal standard. The internal strain was obtained from the following equation:

$$\beta\cos\theta = \lambda/t + 2\eta\sin\theta \quad \cdot \ \cdot \ \cdot \ (1)$$

Here, β is the full-width at half-maximum (FWHM) of the XRPD peaks. λ, t, and η are wave length, crystalline size, and internal strain, respectively. The term including crystalline size t was ignored because of sufficiently large grain size. The powder patterns were obtained by multi-detector system (MDS)[11] using synchrotron radiation in the "Photon Factory" of the National Laboratory for High Energy Physics in Tsukuba, Japan. The details of precise measurement of lattice parameters were reported in previous papers[1, 4].

PREVIOUS REPORTS

We presented the relationship between the Qf values and the internal strain on the pseudo tungstenbronze $Ba_{6-3x}R_{8+2x}Ti_{18}O_{54}$ (R = rare earth) solid solutions as shown in Fig. 1(a) and Fig. 2(b)[3,4]. Fig. 1(a) shows the Qf values of the solid solutions with Sm, Nd, Pr, and La as a function of composition x. The Qf properties of the all solid solutions show the highest Qf value at $x = 2/3$, and the solid solutions with Sm show the highest Qf value compared with other R ions with larger size[3]. Fig. 2(b) shows internal strain of the solid solutions as a function of x on the Sm-tungstenbronze solid solutions. The internal strain η at the $x = 2/3$ composition is the smallest in the solid solutions, which was obtained from the linear function depending on the equation (1) as shown in Fig. 2(a). The internal strain η is a slope of the linear function. We clarified that the highest Qf value is generated from the lowest internal strain.

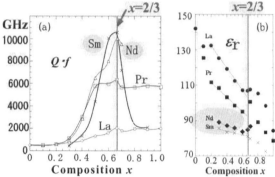

Figure 1. Qf values (a) and ε_r (b) of pseudo-tungstenbronze type solid solutions as a function of x on the $Ba_{6-3x}R_{8+2x}Ti_{18}O_{54}$ solid solutions.

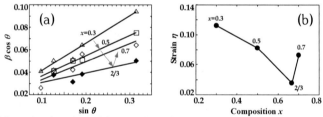

Figure 2. (a) Internal strain η obtained from the slope of equation $b\cos\theta = \lambda/t + 2\eta\sin\theta$. (b) Internal strain η as a function of composition x on the Sm-tungstenbronze solid solutions.

Here, we explain the pseudo-tungstenbronze solid solutions for understanding the origin of the internal strain based on the crystal structure[12, 13, 4]. The solid solutions located on the tie-line with the ratio Ti:O = 1:3 between $BaTiO_3$ and $R_2Ti_3O_9$. The ratio 1:3 shows the crystallographic feature, that is, all TiO_6 octahedra are connected with other octahedra by sharing all apexes as shown in the crystal structures of $BaTiO_3$ and tungstenbronze in Fig. 3(a) and Fig. 3(b), respectively. The tungstenbronze crystal structure has two large cation sites such as A_1 and A_2 with different sizes, compared to the perovskite structure with only one large cation site A. So, the tungstenbronze crystal structure has two sites A_1 with rhombic and A_2 with pentagonal sites. Mainly, R ions occupy the rhombic A_1 site with small size, and Ba ions occupy the pentagonal A_2 site with large size. On the pseudo-tungstenbronze crystal structure, four rhombic A_1 sites are including in one 2x2 perovskite block as shown in Fig. 4(a). If the two large cations became the same ionic size, the structure would change to the perovskite structure.

As this solid solutions are substituted 3 Ba for 2 R, the structural formula is $[R_{8+2x}Ba_{2-3x}]_{A1}$ $[Ba_4]_{A2}Ti_{18}O_{54}$ which R and Ba ions locate in A_1 site, and 4 Ba ions locate in A_2 site. When the number of Ba ions in A_1 site become zero, namely at $x = 2/3$, compositional ordering will appear as shown in Fig. 4(b). Moreover, on the tungstenbronze crystal structure, more one different site triangle C locates among the perovskite blocks with pentagonal A_2 sites.

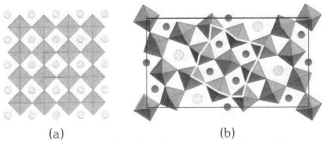

(a)   (b)

Figure 3. Crystal structure of perovskite (a) and pseudo-tungstenbronze (b) composed by TiO$_6$ octahedra connected each other with apexes.

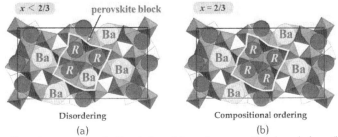

Disordering   Compositional ordering

(a)   (b)

Figure 4. Crystal structure of disordering (a) and compositional ordering (b) on the pseudo-tungstenbronze solid solutions.

RESULTS AND DISCUSSIONS

i) Internal strain and dielectric losses on the pseudo-tungstenbronze solid solutions

1) Origin of the small internal strain at the compositional ordering[3]

FWHM of XRPD peak is depending on the fluctuation of d-spacing of lattice constant. At smaller values of x, namely $x < 2/3$, Ba ions occupy statistically at A_1 sites in perovskite block such as $[R_{8+2x}Ba_{2-3x}V_x]_{A1}$, in which 3 Ba ions are substituted by 2 R ions and one vacancy. Fig. 5 shows schematic lattice image at $x = 2/3$ and $x < 2/3$, in Fig. 5(a) and 5(b), respectively, which is drawn schematically as only one dimensional fluctuation of d-spacing and without consideration of existing vacancies. Fig. 5(a) at $x = 2/3$ shows that all unit cells have the same size without the fluctuation of d-spacing, so the FWHM of the diffraction peaks becomes small. On the other hand, Fig. 5(b) shows fluctuation of the unit cells, when $x < 2/3$, which increases the FWHM. Fig. 6 shows the crystal structure of pseudo-tungstenbronze solid solutions when the unit cell is embedded in a periodic lattice such as Fig. 5(b). When $x < 2/3$ as shown in Fig. 5(b), the fluctuation of d-spacing occurs, because the Ba ions occupy statistically the perovskite blocks. As shown in Fig. 4(b) or 5(a) at $x = 2/3$, however, there is a special point, where all unit cells have the same d-spacing based on the compositional ordering[1,2]. So, FWHM becomes small which is affecting by the internal strain.

2) Microwave dielectric properties depending on the internal strain

In the case of $x < 2/3$, small size cell without Ba ions in the perovskite block is causing the tensile stress by the around large size cells with Ba ions in the block as shown in Fig. 6. The rattling effects of cations in the oxygen polyhedra become large because the volumes of polyhedra especially TiO_6 octahedra become large accompanying with the lattice size by the tensile stress. The rattling yields to increasing dielectric constant ε_r and dielectric losses $\tan\delta$, which are shown in Fig. 1(b), and Fig. 1(a), respectively. Here, Q value is reverse of dielectric losses $\tan\delta$.

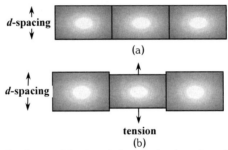

Figure 5. Schematic lattice image of the tungstenbronze structure at $x = 2/3$ (a) and $x < 2/3$ (b), which yield to narrow and broad FWHM, respectively.

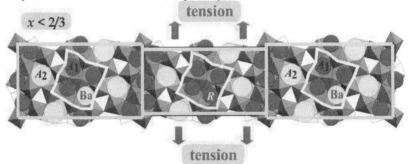

Figure 6. Schematic crystal structure of $x < 2/3$. When there is a perovskite block without Ba, its lattice unit is smaller yielding to tensile stress.

ii) Internal strain and dielectric losses on the solid solutions with different R ions

1) The case of Sm, Nd, Pr, and La pseudo-tungstenbronze solid solutions[5]

As shown in Fig. 1(a), the Qf values decrease in order of increasing of ionic radius of Sm, Nd, Pr, and La ions[3, 4]. On the other hand, the internal strains η increase in the order of the ionic radius as shown in Fig. 7(c), which was obtained from the slope of the linear expression from FWHM of XRPD lines as shown in Fig. 7(b). The precise diffraction line profiles as shown in Fig. 7(a) are obtained using synchrotron radiation, from which 002, 302, 103, and 222 isolated diffraction peaks were selected among more than 1000 other peaks in the 2θ range of 20 to 60°. From the relationship between the Qf values and the internal strain, Sm-tungstenbronze solid solutions with low internal strain show the highest Qf value, and the Qf values of others with large ionic radius decrease in order the ionic radii. Especially, La-tungstenbronze solid solutions have very low Qf, which crystal structure may be unstable depending on the ionic radius of La which is close to that of Ba. Qf values of La-tungstenbronze solid solutions as shown in Fig. 8 deviated from the straight line of other rare earth solid solutions[14]. La-tungstenbronze solid solutions will be unstable, because the difference of ionic radius between Ba and La ions is small. The stability of the tungstenbronze crystal structure is depending on the difference of ionic radius between the two large cations as described above (Fig. 3). The dielectric constants ε_r values are also increasing depending on the ionic radius of R ions (Fig. 1(b)), which are affected by the rattling factor of cations depending on the size of polyhedra, because the lattice constants increase by R ions.

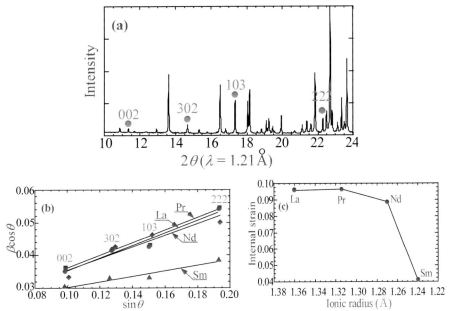

Figure 7. (a) XRPD pattern obtained by synchrotron radiation, (b) equation $b\cos\theta = \lambda/t + 2\eta\sin\theta$, (c) internal strain η as a function of ionic radius.

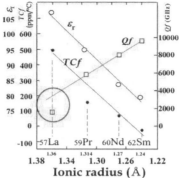

Figure 8. Microwave dielectrics properties as a function of ionic radius of R ion.

2) In the case of Eu, Dy, Ho, Er, and Yb pseudo-tungstenbronze solid solutions[6, 7]

As mentioned above, pseudo-tungstenbronze solid solutions with R ions larger than Sm showed decreasing Qf values. On the other hand, Qf values of Eu-tungstenbronze with small ionic radii than Sm become worth as shown in Fig. 9(a) and (b)[6]. Moreover, Qf of Nd pseudo-tungstenbronze solid solutions substituted by Eu, Dy, Ho, Er, and Yb on the $Ba_4(Nd_{9+1/3-y}R_y)Ti_{18}O_{54}$ ($y = 1$)($R =$ Eu, Dy, Ho, Er, and Yb) decreases in order of small ionic radii as shown in Fig. 10(a)[7]. Fig. 11 shows the unit cell volume of the solid solutions with different R ions as a function of composition y. The volume of the Nd pseudo-tungstenbronze substituted by Eu to Er ions decreases linearly depending of y values. The volume of the solid solutions with the smallest ionic radius Yb has an inflection point at $y = 0.4$. As the ionic radius of Yb ($r = 1.217$ Å for coordination number (CN) = 12) is too small, the site occupied is changed from A_1 rhombic site to C trigonal site, though other R ions of Eu ($r = 1.295$ Å), Dy ($r = 1.254$ Å), Ho ($r = 1.234$ Å), and Er ($r = 1.234$ Å) are occupied A_1 site. This change will bring instability of the crystal structure, internal strain will occur, and Qf value becomes low. The temperature factor TCf is also affected as shown in Fig. 10(b)

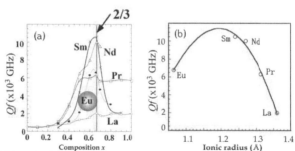

Figure 9. (a) Addition of Qf values of Eu-tungstenbronze to Fig. 1(a). Qf values of each R-compound as a function of ionic radius of R (b).

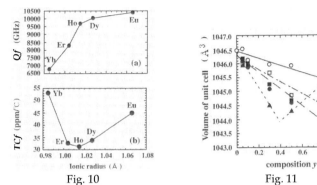

Fig. 10   Fig. 11

Figure 10. Qf and TCf values of Eu to Yb-substituted compounds as a function of ionic radius.
Figure 11. Volumes of unit cells of Eu to Yb-substituted compounds as a function of y on $Ba_4(Nd_{9+1/3-y}R_y)Ti_{18}O_{54}$. That of the Yb-substituted compound has an inflection point at $y = 0.4$.

iii) Improving Qf values in the vicinity of $x = 0$ based on the internal strain origin obtained[15, 16]

Based on the origin of internal strains which are produced by the compositional disordering, low Qf values of ca. 200 GHz on the composition in the vicinity of $x = 0$ as shown in Fig. 1(a) was improved to 6000 GHz by substitution of Sr for Ba as shown in Fig. 12[9,10]. The crystal structural formula of $x = 0$ is $[R_8Ba_2]_{A1}[Ba_4]_{A2}Ti_{18}O_{54}$, in which 2 Ba ions are located in the perovskite block A_1 rhombic sites as shown in Fig.4(a). As the Ba ions bring the d-fluctuation depending on the large ionic radius, the Ba ions were substituted by Sr ions which are smaller than Ba ion as $[R_8Sr_2]_{A1}[Ba_4]_{A2}Ti_{18}O_{54}$ for reducing internal strain. Fig. 12 shows Qf values as a function of α on the $[R_8Ba_{2-\alpha}Sr_\alpha]_{A1}[Ba_4]_{A2}Ti_{18}O_{54}$, which shows improvement of Qf from 200 to 6000 GHz by substitution of Sr for Ba. This substitution of Sr for Ba also introduce a kind of the compositional ordering in $[R_8Sr_2]_{A1}$ and $[Ba_4]_{A2}$ which brings small internal strain. According to our knowledge obtained from crystal structure analysis, Ba or Sr ions can occupy either A_{14} or A_{15}, which are different sites in the Wyckoff notation as presented in a previous paper[17].

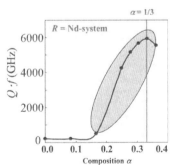

Figure 12. Qf values of the Nd-system with $x = 0$ were improved from 200 to 6000 GHz by substitution of Sr for Ba.

CONCLUSION

i)   Origin of internal strain brought degradation of Qf

1)   In the case of pseudo-tungstenbronze $Ba_{6-3x}R_{8+2x}Ti_{18}O_{54}$ (R = rare earth) solid solutions, the origin of internal strain is the Ba occupation in perovskite blocks which brings fluctuation of d-spacing leading to broadening FWHM. The internal strain yields the degradation of Qf.

2)   In the case of different species of R ions on the pseudo-tungstenbronze compounds, the origin of internal strain is the instability of crystal structure such as mismatch of the R ion compared to the site: the La ion is too large and the Yb ion is too small for A_1 site.

ii)   Application of the origin obtained for improving the Qf values

The internal strain was reduced by substitution of Sr for Ba in perovskite block of the x = 0 on the $Ba_{6-3x}R_{8+2x}Ti_{18}O_{54}$. The Qf value was improved from 200 to 6000 GHz.

ACKNOWLEDGEMENTS

The author would like to express their thanks to Master Engineerings Toshiyuki Ohhashi, Makoto Mizuta, Motoaki Imaeda, Junichi Sugino, Hideyasu Sakashita, Masaki Imaeda, and Yosuke Futamata for study in Nagoya Institute of Technology, and Professors Ken'ichi Kakimoto and Isao Kagomiya for support the experimental and discussions, as well as Professor Wilfried Wunderlich from Tokai University. A part of this work was supported by JSPS KAKENHI Grant Number 22560673.

REFERENCES

[1]H. Ohsato, T. Ohhashi, H. Kato, S. Nishigaki and T. Okuda, "Microwave Dielectric Properties and Structure of the $Ba_{6-3x}R_{8+2x}Ti_{18}O_{54}$ Solid Solutions", *Jpn. J. Appl. Phys.* **34** 187-191 (1995).

[2]H. Ohsato, M. Mizuta, T. Ikoma, Z. Onogi, S. Nishigaki, and T. Okuda, " Microwave Dielectric Properties of Tungsten Bronze-Type $Ba_{6-3x}R_{8+2x}Ti_{18}O_{54}$ (R = La, Pr, Nd and Sm) Solid Slutions", *J. Ceram. Soc. Jpn,* Int. Edition, **106-185**, 184-188 (1998).

[3]H. Ohsato, M. Imaeda, Y. Takagi, A. Komura and T. Okuda, "Microwave quality factor improved by ordering of Ba and rare-earth on the tungstenbronze-type $Ba_{6-3x}R_{8+2x}Ti_{18}O_{54}$ (R = La, Nd and Sm) Solid Solutions", *Proceeding of the XIth IEEE International Symposium on Applications of Ferroelectrics*, 509-512 (1998).

[4]H. Ohsato, "Science of tungstenbronze-type like $Ba_{6-3x}R_{8+2x}Ti_{18}O_{54}$ (R= rare earth) microwave dielectric solid solutions", *J. Euro. Ceram. Soc.* **21**, 2703-2711 (2001).

[5]H. Ohsato, M. Imaeda, H. Sakashita and S. Nishigaki, "The quality factor of the $Ba_{6-3x}R_{8+2x}Ti_{18}O_{54}$ (R = rare earth) solid solutions depend on the ionic size difference between Ba and R", The 9th US-Japan Seminar on Dielectric & Piezoelectric Ceramics (2000, Okinawa).

[6]H. Ohsato, Y. Futamata, K. Kakimoto and S. Nishigaki, "Microwave dielectric properties of $Ba_{6-3x}Eu_{8+2x}Ti_{18}O_{54}$", *Ferroelectrics*, **272**, 249-254 (2002).

[7]H. Ohsato, J. Sugino, A. Komura, S. Nishigaki and T. Okuda, "Microwave Dielectric Properties of $Ba_4(Nd_{28/3-y}R_y)Ti_{18}O_{54}$ (R = Eu, Dy, Ho, Er and Yb) Solid Solutions", *Jpn. J. Appl. Phys.*, **38**, 9B, 5625-5628 (1999).

[8]B. W. Hakki, and P. D. Coleman, "A Dielectric Resonator Method of Measuring Inductive in the Millimeter Range", *IRE Trans.*, **MTT-8**, 402-410 (1960).

[9]Y. Kobayashi, and M. Katoh, "Microwave Measurement of Dielectric Properties of Low-loss Materials by the Dielectric Resonator Method", *IEEE Trans.* **MTT-33**, 586-92 (1985).

[10]H. Toraya, "Whole-Powder-Pattern Fitting without Reference to a Structural Model: Application to X-ray Powder Diffractometer Data," *J. Appl. Cryst.,* **19**, 440-447 (1986).

[11]H. Toraya, H. Hibino, and K. Ohsumi, "A New Powder Diffractometer for Synchrotron Radiation

with Multiple-Detector System," *J. Synchrotron Rad.*, **3**, 75-83 (1996).

[12]H. Ohsato, T. Ohhashi, and T. Okuda, "Structure of $Ba_{6-3x}Sm_{8+2x}Ti_{18}O_{54}$ $(0 < x < 1)$", Ext. Abstr. AsCA '92 Conf., Singapore, Nov., 14U-50 (1990).

[13]H. Okudera, M. Nakamura, H. Toraya, and H. Ohsato, "Tungsten bronze-type solid solutions $Ba_{6-3x}R_{8+2x}Ti_{18}O_{54}$ $(x = 0.3, 0.5, 0.67, 0.71)$ with superstructure", *J. Solid State Chem.*, **142**, 336-343 (1999).

[14]H. Ohsato, M. Mizuta, and T. Okuda, "Crystal Structure and Microwave Dielectric Properties of Tungstenbronze-type $Ba_{6-3x}R_{8+2x}Ti_{18}O_{54}$ $(R = La, Pr, Nd$ and $Sm)$ Solid Solutions", *Applied Crystallography*, 440-447 (1997).

[15]M. Imaeda, K. Ito, M. Mizuta, H. Ohsato, S. Nishigaki, and T. Okuda, "Microwave Dielectric Properties of $Ba_{6-3x}R_{8+2x}Ti_{18}O_{54}$ Solid Solutions Substituted Sr for Ba", *Jpn. J. Appl. Phys.*, **36**(9B), 6012-6015 (1997).

[16]T. Nagatomo, T. Otagiri, M. Suzuki and H. Ohsato, "Microwave dielectric properties and crystal structure of the tungstenbronze-type like $(Ba_{1-\alpha}Sr_{\alpha})_6(Nd_{1-\beta}Y_{\beta})_8Ti_{18}O_{54}$ solid solutions", *J. Eur. Ceram. Soc.*, **26**, 1895-1898 (2006).

[17]H. Ohsato, Y. Futamata, H. Sakashita, N. Araki, K.Kakimoto and S. Nishigaki, "Configuration and coordination number of cation polyhedra of tungstenbronze-type-like $Ba_{6-3x}Sm_{8+2x}Ti_{18}O_{54}$ solid solutions", J. Eur. Ceram. Soc. 23, 2529-2533 (2003).

CHARACTERISTICS OF BaTiO$_3$/(Ba,Sr)TiO$_3$ SUPERLATTICES SYNTHESIZED BY PULSED LASER DEPOSITION

N. Ortega[1], Ashok Kumar[2], J. F. Scott[3] and Ram S. Katiyar[1]

[1]Department of Physics and Institute for Functional Nanomaterials, University of Puerto Rico, San Juan, PR 00931-3343 USA
[2]National Physical Laboratory (CSIR), New Delhi-110012, India
[3]Earth Science Department, University of Cambridge, Cambridge CB2 3EQ, UK

ABSTRACT

BaTiO$_3$/Ba$_{(1-x)}$Sr$_x$TiO$_3$ (BT/BST), (x = 0.3, 0.4, 0.7) thin film superlattices (SLs) were grown by pulsed laser deposition on metallic ferromagnet La$_{0.67}$Sr$_{0.33}$MnO$_3$ (LSMO) coated (001) MgO substrates. X-ray diffraction patterns confirmed the superlattice periodicity of the BT/BST structures. Ferroelectric properties were established in all SLs by the switching and tunability of polarization and dielectric constant respectively under applied external electric field. The slim ferroelectric loops obtained at high applied voltage (80 V ~ 1.3 MV/cm) provide energy storage density ~10 J/cm^3. The observed functional properties may open the possibility of use of this kind of SLs for energy density applications.

INTRODUCTION

Superlattices are artificial structures in which the alternate layers of two or more functional materials are grown in a block called modulation periods (Λ) The thickness of the SL films depends on Λ and the number of repetitions. Superlattices fabricated using oxide materials is a research topic which has gained interest for the possibility of property improvements or the emergence of new properties in the artificially designed structures compared with the single parents layers.

Strain and chemical bonding at the interface are the major factors influencing the physical properties of superlattices structures; lattice strain in the SLs can be modified by the changes in the lattice mismatches between layers in which the interfacial ions are subject to forces different from those in the bulk, these atoms can change positions modifying the properties from single layer oxide thin films. [1] The local stress in the epitaxial films can be controlled by several ways: deposition conditions, substrates, film thickness, thickness of the individual layer i.e. modulation period, and lattice parameters by doping the constituent layers in the artificial SLs.
The influence of strain on the dielectric and ferroelectric properties due to change in thickness of the layers (or modulation period) on the BT/SrTiO$_3$ (ST) SLs are the most studied effects. Das et.al. [2] had measured acoustic phonon dispersions and thereby they had determined the superlattice periodicity. Shimuta et al. [3] investigated the effect of BT/ST composition ratio from 0.2-0.5 in asymmetric SLs, they found the highest remanent polarization (2P$_r$) of 46 μC/cm^2 for 15 unit cell (uc) BT/3 uc ST. Lee et al.[4] have found enhanced dielectric properties in BT/ST with very short stacking (i.e. 2 uc/2uc). Similar studies were carried out in other

systems, such as: BT/BaZrO$_3$ [5], PbZrO$_3$/PbTiO$_3$ [6], PbZrO$_3$/BaZrO$_3$[7]. However few studies have been done to study the effect of manipulation of the degree of strain by changing the composition of one of the constituents of the SLs and investigate the effect on their structural, electrical, and ferroelectric properties. Some related works were done in ferroelectric SLs by Tsurumi et al. [8] using molecular beam epitaxy grown BT/ST SLs where the chemical composition was continuously changed along the film thickness. Hesse's group [9,10] studied the effect of interfacial strain in graded multiferroic SLs of La$_{0.7}$Sr$_{0.3}$MnO$_3$ (LSMO)/PbZr$_x$Ti$_{1-x}$O$_3$ (PZT) grown by pulsed laser deposition. They found magnetic properties of such SLs showing multiple characterizations switching due to modification of the coercive fields of different LSMO layers by strains exerted by the adjacent PZT graded layers.

We have studied the influence of variations in the composition of one of the constituents in the SLs on the structural and ferroelectric properties. We will also present the possible application of these SLs structures as high energy density capacitors.

EXPERIMENTAL METHOD

Superlattices of BT/Ba$_{(1-x)}$Sr$_x$TiO$_3$ (BT/BST) with x = (0.3, 0.4, 0.7) were grown on coated La$_{0.67}$Sr$_{0.33}$MnO$_3$ (LSMO) (001) oriented MgO substrates by pulsed laser deposition (PLD) technique. The thin-film stack was deposited by alternately focusing the beam on stoichiometric BT and BST targets. The film's modulation period of BT$_{\Lambda/2}$/BST$_{\Lambda/2}$ SLs was kept constant (Λ~ 80 - 100 Å). The stacking periodicity $\Lambda/2$ was precisely maintained by controlling the number of laser shots; irrespective of BST concentration, the total thickness of each SL film was ~ 600-700 nm. The SL thin films were prepared with BT as first and final layer. An excimer laser (KrF, 248 nm) with a laser energy density of 1.5 J/cm^2 and pulse repetition rate of 10 Hz was used to deposit the SLs. During deposition the substrate was maintained at 830 °C and oxygen pressure at 200 mTorr. The orientation and phase purity of the films were characterized by XRD using Cu K$_\alpha$ radiation in a Siemens D500 diffractometer. Electrical properties were measures in conventional metal-dielectric-metal (MDM) configuration using LSMO layer as bottom electrode and Pt top square electrodes fabricated by dc sputtering with a area of ~10^{-4} cm^2 utilizing a shadow mask. The dielectric constant vs. electric field was measured using an impedance analyzer HP4294A (Agilent Technology Inc.). The polarization versus electric field (P-E) hysteresis loop of the capacitor was measured using the RT 6000HVS ferroelectric tester (Radiant technology).

RESULTS AND DISCUSSION

Figure 1 (a) shows room temperature X-ray diffraction spectrum of (002) reflection peaks corresponding to BT/Ba$_{0.6}$Sr$_{0.4}$O$_3$ SL structure grown on LSMO/MgO. For comparison, same 2θ-θ scan of the XRD patterns of the LSMO/MgO and MgO substrates are presented in Figure 1(b) and 1(c) respectively. The XRD results revealed that all SLs films are single phase and well oriented, only (00l) series of peaks without evidence of the any additional peaks (i.e. pyrochlore) are observed within detection limit of the instrument. The formation of the superlattice structure in BT/Ba$_{0.6}$Sr$_{0.4}$O$_3$ was corroborated from the appearance of the main peak and the symmetric satellite peaks, labeled as 0, and +1,-1 in Figure 1(a), respectively. Satellite peaks result from the modulated structures indicating the formation of high quality SLs on LSMO/MgO substrates. Dashed lines in Figure 1 confirm that the other peaks which appear in BT/Ba$_{0.6}$Sr$_{0.4}$O$_3$ spectra correspond to LSMO bottom electrode or MgO substrate. Similar XRD results were obtained for the other SLs.

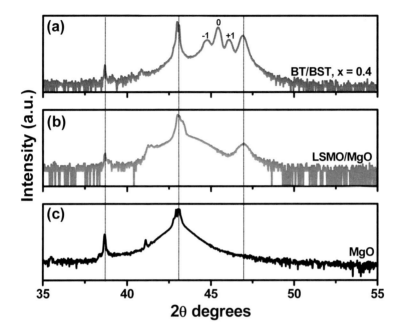

Figure 1. Room temperature X-ray diffraction pattern of (002) reflection peaks for (a) BT/Ba$_{0.6}$Sr$_{0.4}$O$_3$ SLs structures grown on LSMO/MgO, (b) LSMO/MgO and (c) (001) oriented-MgO substrate.

Cross polarized Raman spectra indicate (not showed here) that as the x values (Sr) increase in Ba$_{(1-x)}$Sr$_x$TiO$_3$ layer in BT/BST SLs, the ferroelectric E(1TO) soft mode was found shifted to higher frequencies from 35 cm^{-1} (x = 0) to 115 cm^{-1} (x = 1). It is important to note when x = 0, we have basically BaTiO$_3$ films and when x = 1 we have BT/ST SLs. Namely, the position of the E(TO) soft mode depends on the Ba/Sr ratio in the BST layer, since varying the concentration of ions forming the SLs, it is possible varying the lattice parameters of the layers. These results support upward shift of the soft mode in the SLs which is due to the internal stress induced by the lattice mismatch between the constituent layers, in particular the ferroelectric soft mode which is usually very sensitive to the presence of the strain in thin films. [11,12].

The room temperature ferroelectric character of the $BT/Ba_{(1-x)}Sr_xTiO_3$ SLs was confirmed by capacitance vs. voltage (C-V) measure carried out on the SLs at 8 kHz, the different curves for each SLs structures represent the respective voltage sweep i.e. -40 V to 40 V and vice versa. Figure 2 shows the dielectric constants vs. electric field (ε-E) response, determined from C-V curves, for SLs with x = 0.3, 0.4 and 0.7 values in BST layer. ε-E curves show high nonlinear dielectric response as a function of electric field and a conventional butterfly loop behavior caused by polarization reversal. Maximum dielectric constant at about zero bias was observed for $BT/Ba_{0.3}Sr_{0.7}TiO_3$ and $BT/Ba_{0.7}Sr_{0.3}TiO_3$ (750 and 660 respectively) while a notable decrease was observed for $BT/Ba_{0.4}Sr_{0.6}TiO_3$ (390). These results can be explained in terms of the interlayer strain and properties of the individual layers. Even ones of the requirements for a maximum polarization in SL are: well strained lattice and proper thickness in ferroelectric/paraelectric components of the SLs. However, Lee et al.[13] reported the increase of neighboring Ba layers that is increasing the non-interfacial Ba-surrounded TiO_6 octahedral lead in an enhancement of the polarization.

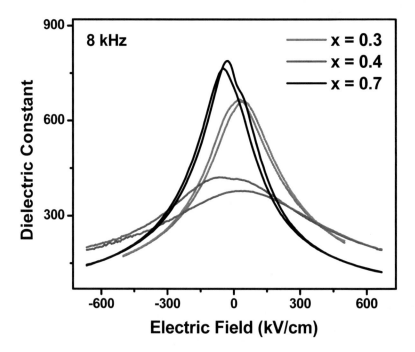

Figure 2. Dielectric constant versus bias electric field curve for BT/ $Ba_{(1-x)}Sr_xTiO_3$ SLs with x = 0.3, 0.4 and 0.7 at room temperature.

The observed highest values in dielectric response of the BT/BST structures with x = 0.3 and 0.7 can be interpreted as follows: while interlayer strain can be minimized by making Ba-rich BST layers, the situation can be reversed (maximal strain) with Sr-rich BST layers. On the other hand, the properties also depend on the composition of the BST layers. The highest values of dielectric properties have been reported for Ba-rich compositions in BST compounds. The ferroelectric nature of the SL films leads to the development of two peaks in ε-E plot, which correspond to the electric coercive field (E$_c$), low E$_c$ values were obtained for x=0.3 and 0.7(~9-10 kV/cm) while ~30 kV/cm was found for x = 0.4. Slim ferroelectric loop (low coercive field) is a typical characteristic for ferroelectric relaxor materials. On the other hand, asymmetry behavior was also observed in the ε-E plot of the SL films, Similar behavior in polarization loops was observed by Lee et al. in case of three-component ferroelectric SLs [13] in BT/ST SLs [8], and BT/Ba$_{0.30}$Sr$_{0.70}$TiO$_3$ [14]; it can be related to: i) the electrode asymmetry, different work function (WF) for the bottom (La$_{0.67}$Sr$_{0.33}$MnO$_3$ –WF~4.96 eV) and top (Pt – WF~ 5.65 eV) electrodes [15] and (iii) compositionally broken inversion symmetry with the creation of new lattices in the SLs [16].

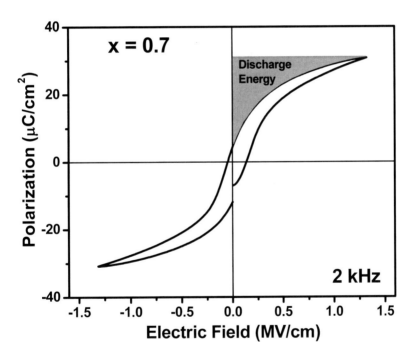

Figure 3. Room temperature ferroelectric hysterisis loops for BT/Ba$_{0.3}$Sr$_{0.7}$TiO$_3$ SL measured at 80 V ~ 1.30 MV/cm. Pink shaded area represent the discharge energy density of the SL capacitor.

With the fast development of power electronics in pulsed power and power conditioning applications, there is a need for next-generation dielectric capacitors with high energy density/low loss. It has been considered that the materials having high bipolar density can be potential candidates for the high power as well as high energy devices. The energy density of either linear or nonlinear dielectric can be obtained from the discharge energy density with applied field, being represented by the integral:

$$\overline{U} = \int E dP \qquad (1)$$

where $\overline{U}$ is energy density, E is the applied electric field and P is the charge density (in case of ferroelectric materials it is polarization or displacement charge (see the pink shaded area in Figure 3). Novel ferroelectric behavior includes relaxor ferroelectric and antiferroelectric-like behavior that are highly desired because of their high storage electric energy density in comparison with the normal ferroelectrics. Figure 3 shows slim asymmetric ferroelectric loop of BT/Ba$_{0.3}$Sr$_{0.7}$TiO$_3$ SL, measured at 80 V ~ 1.30 MV/cm, slim P-E loops is one of the indicators of the presence of relaxor ferroelectrics behavior, more detailed studies were published elsewhere [17]. Equation 1 was used to calculate the discharge energy density, the upper portion of the hysterisis area provides almost ~10 J/cm^3. Similar value of discharge energy density (9 J/cm^3 at 400 MV/cm) were reported for P(VDF-TrFE-CFE) 58.3/43.3/7/5 mol% by Chu et al [18]. These numerical data indicates that SLs may be useful in the fabrication of next generation high energy density device appliances.

CONCLUSIONS

A series of BT/Ba$_{(1-x)}$Sr$_x$TiO$_3$ SLs with different Ba/Sr compositions were prepared by PLD. The XRD patterns demonstrate that the BT/BST thin films grown as superlattice structure. BT/BST SLs exhibited a clear ferroelectric behavior demonstrated by presence of well defined asymmetric hysteresis loop and ε-E plot. The upper area of P-E loop provides high energy density (~10 J/cm^3 at 1.3 MV/cm) which opens the possibility of the application of SLs structure in this area.

ACKNOWLEDGEMENT

This work was partially supported by NSF-EFRI-RPI-1038272 and DOE-DE-FG02-08ER46526 grants.

REFERENCES

1. G. Rijnders and D. H. A. Blank. Materials science: Build your own superlattice. Nature 433 (2005), 369-370.

2. R.R. Das, Yu.I. Yuzuk, P. Bhatacharya, V. Gupta, and R.S. Katiyar, Phys. Rev. 69, 132302 (2004).

3. T. Shimuta et al. Enhancement of remanent polarization in epitaxial BaTiO$_3$/SrTiO$_3$ superlattices with "asymmetric" structure. J. Appl. Phys. 91 (2002), 2290-2294.

4. J. Lee et al. "Dielectric properties of BaTiO$_3$/SrTiO$_3$ ferroelectric thin film artificial lattice" J. Appl. Phys. 100 (2006), 051613 (12pp).

5. M. El Marssi et al. Ferroelectric BaTiO$_3$/BaZrO$_3$ superlattices: X-ray diffraction, Raman spectroscopy, and polarization hysteresis loops. J. Appl. Phys. 108 (2010), 084104 (6pp).

6. T. Choi and J. Lee. Enhancement of dielectric and ferroelectric properties of PbZrO$_3$/PbTiO$_3$ artificial superlattices. J. Korean Phys. Soc. 46 (2005), 116-119.

7. Ch-L Hung et al. Characteristics of constrained ferroelectricity in PbZrO$_3$/BaZrO$_3$ superlattice films. J. Appl. Phys. 97 (2005), 034105 (6pp).

8. T. Tsurumi et al. Preparation and dielectric property of BaTiO$_3$/SrTiO$_3$ artificially modulated structure. Jpn. Appl. Phys. 37 (1998), 5104 – 5107.

9. I. Vrejoiu et al. Interfacial strain effects in epitaxial multiferroic heterostructures of PbZr$_x$Ti$_{1-x}$O$_3$/La$_{0.7}$Sr$_{0.3}$MnO$_3$ grown by pulsed-laser deposition. Appl. Phys. Lett. 92 (2008), 152506 (3pp).

10. M. Ziese et al. Structural, magnetic, and electric properties of La$_{0.7}$Sr$_{0.3}$MnO$_3$/PbZr$_x$Ti$_{1-x}$O$_3$ heterostructures. J. Appl. Phys. 104, (2008) 063908 (9pp).

11. N. Ortega et al. Effect of periodicity and composition in artificial BaTiO$_3$/(Ba,Sr)TiO$_3$ superlattices. Phys. Rev. B. 83 (2011), 144108 (8 pp).

12. O.A. Maslova et al. Raman spectroscopy of BaTiO$_3$/(Ba,Sr)TiO$_3$ superlattices. Phys. Solid State 53 (2011), 1062-1066.

13. H. N. Lee, et al. Strong polarization enhancement in asymmetric three-component ferroelectric superlattices. Nature, 433 (2005), 395-399.

14. Ferroelectric and Dielectric properties of BaTiO$_3$/Ba$_{0.30}$Sr$_{0.70}$TiO$_3$ Superlattices. N. Ortega, Ashok Kumar, Yu. I. Yuzyuk, J. F. Scott and Ram S. Katiyar. Integrated Ferroelectrics 134 (2012),1-7.

15. M. Toda and M. Toda. Asymmetric Hysteresis Loops, Leakage Current and Capacitance Voltage Behaviors in Ferroelectric PZT Films Deposited on a Pt/Al$_2$O$_3$/SiO$_2$/Si Substrate by MOCVD method with a vapor-deposited Gold Top Electrode. International Journal of Applied Physics and Mathematics 1 (2011), 144-148.

16. N. Sai, B. Meyer, and D. Vanderbilt. Compositional Inversion Symmetry Breaking in Ferroelectric Perovskites. Phys. Rev. Lett. 84 (2000), 5636-5639.

17. N. Ortega, et al. R. S. Katiyar. Relaxor-ferroelectric superlattices: A High energy density capacitor. J. Phys.: Condens. Matter. 24 (2012), 445901 (8 pp).

18. B. Chu et al. A Dielectric Polymer with high electric energy density and fast discharge speed. Science 313 (2006), 334-336.

NOVEL DEVICES USING OXIDE SEMICONDUCTORS IN FE-TI-O FAMILY

R. K. Pandey[1,2,x], William A. Stapleton[1], Anup K. Bandyopadhyay[2], Ivan Sutanto[1], and Amanda A. Scantlin[1]

[1] Electrical Engineering Program, Ingram School of Engineering; [2] Department of Physics
Texas State University, San Marcos, TX 78666, USA

[x] Corresponding Author: rkpandey@txstate.edu

Key words: Bipolar Amplifiers, Low Pass Filters, Electronic Switches, Human Auditory Range etc.

ABSTRACT

In this paper we discuss the effects of a biasing voltage and varying frequencies on the varistor properties of two principal members of the iron-titanate family. They are ilmenite-hematite with 45 atomic percent hematite (IHC45) and pseudobrookite, Fe_2TiO_5 (PsB). Both of these are intrinsic n-type semiconductors with high resistivity and dielectric constant greater than 1000. IHC45 is also ferrimagnetic with a well-defined hysteresis loop and the magnetic Curie point at 610 K. PsB is nonmagnetic, or at best feebly magnetic, depending upon the location where this mineral is found or how it is being processed in the laboratory. We identify three major devices based on modified current-voltage characteristics of these two materials: electronic low pass filter, electronic switch, and bipolar signal booster with many possible applications. The low pass filter based on PsB single crystal appears to be well suited for applications in the frequency range of human auditory system (i.e. 2 Hz to 22 kHz) whereas a ceramic-based application can operate up to 2.8 MHz. On the other hand, the same device based on IHC45 ceramic has the bandwidth of about 500 kHz. Both crystal and ceramic substrates can be used for fabrication of variable gain amplifiers but the single crystal PsB substrate appears to be ideally suited for this device because it can produce the largest gain factors among the samples tested. Both IHC45 and PsB can be used for producing electronic switches with switching delays around a few micro-seconds in the mm-scale samples tested. Devices based on these materials have the potential to be suitable for handheld applications, microelectronics, and space electronics and may be tuned by controlling the biasing voltage or frequency. The natural radiation hardness of both IHC45 and PsB make them especially interesting for potential space-based applications. Biocompatibility of IHC45 and PsB make their adoption in bioelectronics a possibility.

INTRODUCTION

This paper deals with the basic characteristics of the varistor diodes based on three different iron titanate substrates, namely, ilmenite-hematite ceramic with 45 atomic percent hematite (IHC45), pseudobrookite (PsB) single crystal, and PsB ceramic. Their current-voltage characteristics show strong frequency dependence between 20 Hz to 4 MHz and undergo major changes by the application of a biasing voltage. The resulting devices are bipolar amplifiers, low pass filters, and electronic switches with applications in microelectronics, handheld devices, space electronics, etc.

Conventionally, one model of an electronic filter consists of two sets of capacitors and resistors; one being in series and the other in parallel configuration. This hybrid circuit then is manipulated by adjusting the capacitance (C) and resistance (R) values to produce the desired output. These filters work solely by selecting appropriate values of C and R as dictated by the design of the desired device. In contrast, our device does not use any external components such as a resistor and a capacitor to work as a filter. It is based completely on the materials properties of the substrates.

Another important aspect of our devices is that they can operate satisfactorily in radiation environment because iron-titanates with band gap > 2.3 eV have been identified to be extremely resistant to radiation damage[3,5,6,7,9]. Raw materials (FeO, Fe_2O_3, and TiO_2) processing of these substrates are readily available from many commercial sources in abundance in high purity grade and rather inexpensively. This would allow the device to be produced in large volumes at an affordable cost. Also, all the three raw materials are compatible to biological systems; and therefore it can be safely assumed that devices built using them can find applications in bio-electronics as well.

PROCESSING, PROPERTIES AND DISCUSSION

I.   IHC45 SUBSTRATE
    Ilmenite ($FeTiO_3$) and hematite (Fe_2O_3) form solid solution (IH) for hematite concentration (x) ranging between 0<x<0.6. Many of its members are ferromagnetic having the magnetic Curie point above 300 K and all are either p-type or n-type semiconductors. Structurally IH exists in layers with sub-lattices of A [xFe^{3+}(1-x) Fe^{2+}] and B [x Fe $^{3+}$(1-x) Ti $^{4+}$]. Its unit cell is hexagonal of the corundum type. Lattice constants of $FeTiO_3$ changes vary little with addition of Fe_2O_3. For example they are for ilmenite (with x= 0): a= 0.504 nm and c = 1.374 nm while for IH (with x=0.6) they are a =0.505 nm and c = 1.385 nm.
    The leading members of the IH series which have been extensively studied recently for spintronics, because of their room temperature magnetism and n-type semiconductor are IHC33 and IHC45. The nonlinear I-V characteristics and their radhard properties are by now well-established [3,5,6]. In fact, neutron radiation enhances the chemical ordering of IHC with different values of hematite concentration and thereby enhances their magnetic properties[7].

High quality chemicals were used in proper proportions to make ceramic samples with the composition of 0.55 $FeTiO_3$·0.45 Fe_2O_3 (hereafter referred as IHC45). The processing parameters followed were identical to those described in our earlier publications[1-3]. After pressing and annealing, pellets of primarily two diameters were obtained: 24-25 mm and 12-13 mm in diameter. The thickness of the pellets varied from a 3-5 mm which was dependent upon the amount of powder of IH used. The pellets were first analyzed for their crystalline structure by Bede d1 X-ray diffractometer (XRD) capable of high angle diffraction, high resolution and glancing incidence diffraction for characterizing the crystallographic orientation. Figure 1 shows the resulting XRD spectrum. The composition was determined by EDAX (Energy dispersion X-ray analysis). Figure 2 shows the EDAX analysis on a comparative scale. Ceramic samples of IHC45 were polished to a high shine and then cut into small rectangular slabs. IHC45 pellets show no evidence of micro cracks and take high polish. Silver point electrodes were placed as three evenly spaced dots on one of the surfaces. Very fine silver wires were used to connect them to a precision parametric analyzer (PA) which is integrated with a PC for data collection.

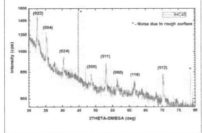

Figure 1. XRD spectrum of IHC 45

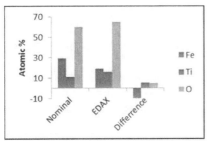

Figure 2. EDAX analysis of IHC45

Figure 1 represents the symmetric high angle diffraction (HAXRD) spectra for IHC45 and after being compared with International Centre for Diffraction data (ICDD) for standard sample, it is found that all peaks represents the diffraction peaks for IHC45 and these are free from the presence of any impurities or secondary phases. From Figure 2 we see that Fe is deficient by about 5 atomic % whereas Ti and O are in excess but by less than 2 atomic % yielding the chemical formula of $Fe_{0.9}Ti_{0.80}O_3$.

In addition to the intrinsic resistance and capacitance of IHC45, it exhibits high dielectric constant, nonlinear I-V characteristics, and magnetism (therefore, inductance). These are all favorable properties for device development. Furthermore, IHC45 is a wide band gap semiconductor of n-type conduction. Its band gap varies between $2.5 < E_g < 3.6$ eV and shows strong dependence on the processing conditions[4].

It is intrinsically ferrimagnetic with well-defined hysteresis loop having the saturated magnetic moment, $M_s \approx 19.4$ emu/g at room temperature and the Curie temperature at 610 K[3]. The strong high temperature magnetic property, n-type semiconductor nature as well as the two point current-voltage characteristics typically found for varistor materials make IHC45 attractive for the development some unique and practical devices some of which are described later in this paper.

II.   PsB SINGLE CRYSTAL SUBSTRATE

Pseudobrookite, Fe_2TiO_5, like ilmenite, $FeTiO_3$, and hematite, Fe_2O_3, is also found widely as a mineral and has been studied extensively for photo-catalysis reactions in search of alternative forms of energy. Besides being environmentally stable and friendly, it is also a wide band gap semiconductor with band gap, $E_g \approx 2.3$ eV, its semiconducting nature is of n-type, and electron mobility is equal to 6.3 $cm^2V^{-1} s^{-1}$ which is relatively high compared to the values for many oxides[7]. It is also exhibits extreme resistance to high levels of radiation[3,9]. The combination of these properties led us to investigate this material for microelectronics applications. For these studies we have used both single crystals and ceramics samples of PsB.

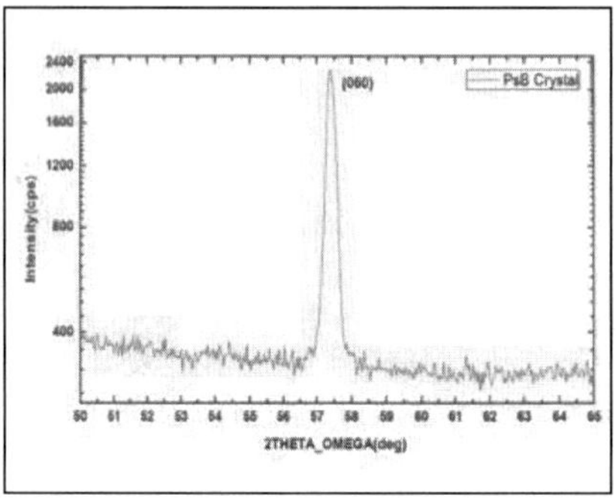

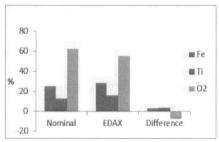

Figure 3 XRD spectrum of PSB single crystal

Figure 4 EDAX analysis of PsB single crystal

Relatively large crystals of PsB were grown using the high temperature solution growth method as described in reference[8]. Its crystal structure is orthorhombic with the lattice constants of: a = 0.979 nm, b= 0.993 nm and c = 0.372 nm. The n-type nature of PsB was confirmed by the negative sign of the Seebeck coefficient which is -35 μV/K at room temperature[8]. Its room temperature dielectric constant is about 1200 at 1 Hz which is a relatively high value[9]. Many of its physical properties including the radhard property have been reported in references[8,9]. Figures 3 and 4 show the HAXRD spectrum and EDAX analyses, respectively. The presence of the single XRD peak at (060) confirms the single phase nature and TEM micrograph, as shown in reference[8], of the single crystal PsB sample. From Figure 4 we conclude that O is deficient and Fe and Ti are slightly in excess. The chemical formula based on this analysis then corresponds to $Fe_{2.27}Ti_{1.3}O_{4.22}$.

III.   PsB Ceramic Substrate

The desired amount of high purity grade powder of pseudobrookite, Fe_2TiO_5, purchased from Alfa Aesar, was mixed with 2.5% PVA by weight and left to dry at room temperature. Once dried completely the mixture was ball milled for approximately 15 minutes using a SFM-3 Desk-Top High Speed Vibrating Ball Miller.

Approximately 4.14 g of finely ball milled powder, ranging in size from 100 nm to 150 nm, was put in a stainless steel die of approximately 13 mm diameter. The die was then heated to about 175°C under a pressure of 1500 lbs. for one hour using a 3891 4NE18 Carver press and then slowly increased to 10,000 lbs. It was left there for 3 hours after which it was released. The pellet was then recovered from the die. The density of the green sample was 3.24g/cm³. Subsequently, the pellet was annealed in air at 1150°C for 10 hours and then cooled to room temperature slowly to prevent cracks from developing. The sample was later annealed for 2 hours in flowing argon at the peak temperature of 1100°C. This produced a very dense and solid pellet which then could be cut into pieces and polished for XRD and SEM analyses and electrical characterizations. Air annealed PsB was highly insulating and its surface resistance was of the order of a few MΩ. Ar annealing makes it slightly less insulating facilitating the output signal to be less noisy in electrical measurements. Figures 5 and 6 show its HAXRD spectrum and SEM micrograph, respectively. For better comparison study the two XRD spectra for PsB and MnPsB ceramic samples were superimposed. The disappearance of some of the PsB peaks after being doped with Mn clearly established the role of Mn doping on PsB Ceramic. It is to be noted that Mn doping made PsB ferrimagnetic; however, the results of MnPSB studies are not covered in this paper. Three spots shown in Figure 6 were used to determine the composition of the sample. It was found to be slightly off from the nominal values; giving the composition to be $Fe_{1.94}Ti_{1.04}O_{5.02}$.

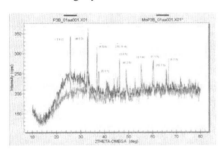

Figure 5 XRD of PsB ceramic (blue) superimposed on the MnPsB ceramic spectrum (red)

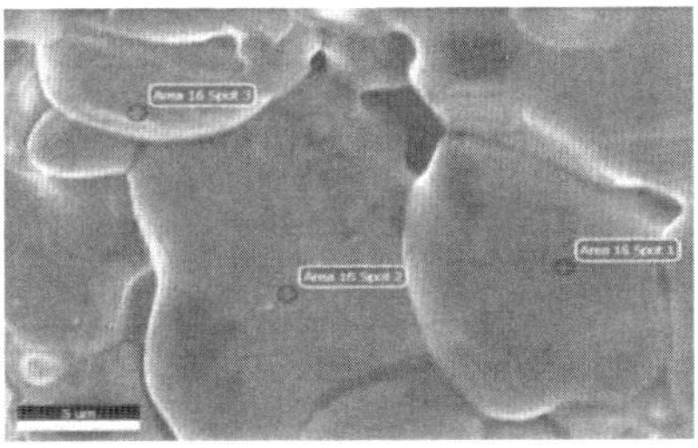

Figure 6 SEM micrograph of PsB ceramic

DETERMINATION OF I-V

I. DC MODE WITH BIAS

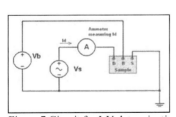

Figure 7 Circuit for I-V determination with a bias and ac of varying frequency; S =source, B = bias and D =drain.

The I-V characteristics were determined first for the dc mode with and without a bias voltage applied to the sample using the standard circuit similar to the one shown in except that first no ac signal was applied between the source and the drain. In Figure 8 we see the varistor nature of the I-V for IHC45 ceramic sample as well as the effect of a bias voltage. Varistors, as is well known, are widely used as circuit protection devices and may be modeled as two Schottky diodes configured back-to-back. The unique feature of a varistor device is that at low

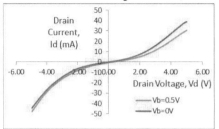

Figure 8. I-V characteristics of IHC 45 with and without a bias voltage

voltage it has high impedance and at high voltage it has low impedance; and therefore it is also called voltage-dependent-resistor or VDR. This feature makes a varistor or a VDR uniquely suitable for circuit protection. Further in figure 8, the effect on the varistor response of introducing a bias voltage is shown. For clarity, figure 9, shows on a relative scale the contributions to the output current of each of two conditions of the bias voltage i.e., when $V_b = 0$ and when $V_b = 0.5$ V.

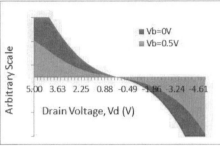

Figure 9. Current –Voltage characteristics of IHC45 with bias and no bias on arbitrary scale

II. AC MODE WITH BIAS: FREQUENCY DEPENDENCE

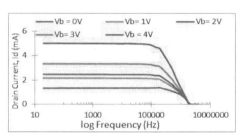

Figure 10 Output current, Id vs. log frequency with variable bias voltage with Vd= 1 Vpp for IHC 45

For an ac signal with varying frequency, the I-V was determined, using again the circuit shown in Figure 7. The range of frequency tested was from 20 Hz to 2 MHz which covers completely also the audio range of 20 Hz to 20 kHz. As seen from Figure 7 the bias voltage applied between point B and ground was in dc mode whereas the voltage between source S and drain D was in ac mode. This allowed us to study the dependence of the drain current (I_d) as a function of the drain voltage (V_d) in ac mode varying the input frequency in a wide range.

Figure 10 shows the relationship between the drain current, Id, and the frequency of the input signal while varying the bias voltage from 0V to 4 V. Plotting the frequency in log scale enables us to ascertain that the characteristics are typical to the curves found for a

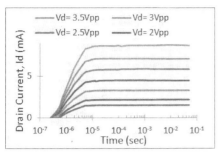

Figure 11. Drain current vs. time for IHC45

low pass filter. From 20 Hz to up to 100 kHz the output signal remains constant and then reduces to effectively zero above 1 MHz. The 3 dB point, that is where the output signal reduces to 70.7 % of its maximum value, is found to be around 420 kHz. The 3 dB point has an important engineering significance. This determines the bandwidth of the filter for satisfactory operation. In other words, this filter can operate satisfactorily upto about 420 kHz.

Figure 11 shows the output signal I_d as a function of time (inverse of the frequency). This feature could be important for using this device for switching purposes. The transition speed for the IHC45 device is such that the output reaches its final value in less than 10 µs regardless of the driving voltage. Switching speeds of this magnitude are sufficient for frequencies less than the 100 kHz bandwidth limit of an audio-range low pass filter.

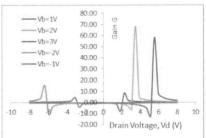

Figure 12. Gain factor, G vs. drain voltage Vd of IHC45

Figure 12 represents the bipolar nature of the signal gain factor G as a function of the bias voltage, V_b, The gain varies from -10 to almost 70 depending upon the chosen bias. The maximum amplification is obtained for Vb = 2 V and 3V, for which the peaks are sharp and well defined. This feature would allow for the fabrication of a simple amplifier using a ceramic substrate of IHC45.

We define the gain factor as follows:

$$G = \frac{I_{d,x} - I_{d,0}}{I_{b,x}} = \frac{\Delta I_{d,x}}{I_{b,x}} \qquad (1)$$

where, $I_{d,x}$ is the drain current at bias voltage $V_b = x$, $I_{b,x}$ = current at the bias when $V_b = x$ and $I_{d,0}$ = drain current for $V_b = 0$.

Figures 13 and 14 show the I-V characteristics for the PsB single crystal substrate for $V_b = 0, \pm 1$, and $\pm 3V$. The effect of a bias voltage on the output signal on relative scale is shown in Figrure 14 for each value of the bias voltage. From these two figures we conclude that the output signal can be modified by using the appropriate bias voltage and this gives us an avenue to build some novel devices using this substrate.

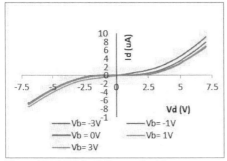
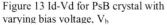
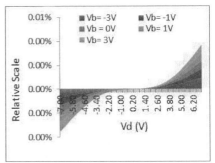

Figure 13 Id-Vd for PsB crystal with varying bias voltage, V_b

Figure 14 Id-Vd for PsB crystal with varying bias voltage, V_b on relative scale

When the drain voltage is kept constant at 3Vpp and the output signal is studied as a function of frequency while varying the bias voltage, V_b, Figure 15 emerges. Here we see, as in Figure 10 for the IHC45 substrate, that the drain current, I_d, shows an increasing trend with increasing bias. Above 10 kHz it rapidly increases reaching its peak at about 20 kHz. The corresponding 3 dB point lies at about 140 kHz, making the bandwidth to be 140 kHz. This device shows an appropriate characteristic for the fabrication of a tunable filter for human audio range which covers the frequency from 2 to 22 kHz.

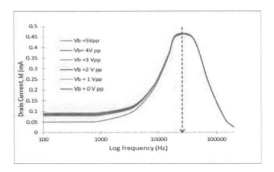

Figure 15 Drain current, Id vs. log frequency
with variable bias V_b at drain voltage, V_d, constant
at 3 Vpp for PsB crystal

The signal gain factor calculated according to the equation (1) shows remarkable trend for the single crystal sample. They are shown in figures 16 and 17. The gain factor, G, varies from -800 to +400 with respect to the bias as shown in the figure. On the other hand, it varies in thousands for $V_b = \pm 2V$ and ± 4 V. This remarkable increase in G appears to be associated with the band gap of the substrate which as mentioned previously is about 2.3 eV. Such anomalies have been observed in different experiments both for IHC45 and PsB. The exact nature of such a change and its mechanisms need to be studied further. However, from the data as presented here leads us to believe that a good bipolar amplifier could be built using the single crystal substrate.

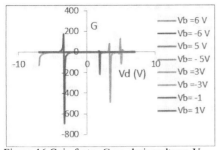

Figure 16 Gain factor G vs. drain voltage, V_d with bias $V_b = \pm 1, \pm 3, \pm 5$ and ± 6 V for PsB crystal

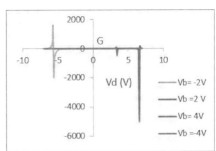

Figure 17 Gain factor G vs. drain voltage, Vd with bias $V_b = \pm 2$, and ± 4 V for PsB crystal

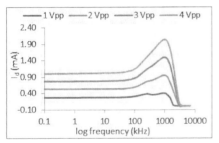

Figure 18 Drain current vs. log frequency with variable drain voltage at constant bias = 3 V for PsB ceramic

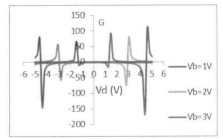

Figure 19 Gain factor G vs. drain voltage with variable bias for PsB ceramic

PsB ceramic substrate also exhibits filter and gain characteristics as shown in Figures 18 and 19, respectively. The bandwidth here is much larger than found for the single crystal device. It shows a peak at about 1 MHz with 3 dB point corresponding to 2.8 MHz. The resulting bandwidth covers the range from 20 Hz to almost 2.8 MHz. Like IHC 45, it can also be used as an electrical switch with on-state at ≈ 0.4 μs and off-state below that.

As for IHC45 we notice here too that G is at least one order of magnitude smaller than for the values obtained when a single crystal substrate is used. It is possible that large number of grains and grain boundaries present in a ceramic adversely affects this property.

CONCLUSION

In this paper we have discussed the processing, properties and devices of two principal members of the Fe-titanate family, namely, IHC45 (0.55 Ilmenite+0.45 Hematite) and PsB (Fe$_2$TiO$_5$). The current-voltage (I-V) characteristics of the substrates of these two materials have been used for the determination of the effects of a bias voltage and frequency on the basic nature of the nonlinear I-V behavior (typically found for a varistor) exhibited by these materials. Ceramic substrates of ilmenite-hematite (IHC45) have been used whereas both single crystal and ceramic substrates of pseudobrookite (PsB) have used for these investigations. Based on the response of the I-V characteristics under the influence of a biasing voltage and varying frequencies many devices with multiple potential applications have been identified[10]. The devices with their key features are summarized in Figure 20. For these substrates Fe$_2$O$_3$, FeO and TiO$_2$ are the raw materials and they are commercially available in high quality inexpensively And therefore are cost effective for mass production. They also are bio compatible.

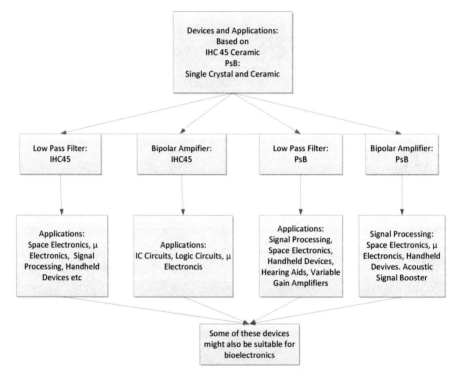

Figure 20. Summary of devices and their applications

ACKNOWLEDGMENT

We gratefully acknowledge the support of the National Science Foundation (NSF), Grant # ECCS-1025395, and the support of the Office of the Associate Vice President for Research and Ingram School of Engineering at Texas State University-San Marcos for this research.

REFERENCES

[1] Feng Zhou, <u>Sushma Kotru</u> and Raghvendra K. Pandey, "Nonlinear current-voltage characteristics of ilmenite–hematite ceramic" *Materials Letters*, **57**, (2003, 2104,

[2] L. Navarrete, J. Dou, D.M. Allen, R. Schad, P. Padmini, P. Kale and R.K. Pandey, " Magnetization and Curie Temperature of Ilmenite-Hematite Solid Solution Ceramics", *J. Am. Ceram. Soc., 89(5),* (2006). 1601.

[3] R. K. Pandey, P. Padmini, R. Schad, J. Dou, H. Stern, R. Wilkins, R. Dwivedi W. J. Geerts and C. O' Brien, " Novel Magnetic-Semiconductors in Modified $FeTiO_3$ for Radhard Electronics", *J. Electro-Ceramic*, DOI 10.1007/s10832-007-9390-1 (2008) and (2009), **22,**334. (2009)

[4] J. Dou, L. Navarrete, R. Schad, P. Padmini, R. K. Pandey, H. Guo and A. Gupta, "Magnetic properties of ilmenite-hematite films and bulk samples", *J. Appl. Phys*, 103, 07D117, (2008).

[5] P. Padmini, M. Pulikkathara, R. Wilkins and R. K. Pandey, " Neutron radiation effects on the nonlinear current-voltage characteristics of ilmenite-hematite ceramics," *Applied Physics Letters,* **82(4),** (2003), 586.

[6] P. Padmini, S. Ardalan, F. Tompkins, P. Kale, R. Wilkins and R. K. Pandey, "Influence of proton radiation on the current-voltage characteristics of ilmenite-hematite ceramics", *J. Elec. Mats,***34(2),** (2005), 1095.

[7] D. M. Allen, L. Navarrete, J. Dou, R. Schad, P. Padmini, P. Kale and R. K. Pandey, S. Shojah-Ardalan and R. Wilkins, " Chemical ordering in ilmenite-hematite bulk ceramics through proton irradiation", *Appl. Phys. Lett.* **85,**(2004), 5902.

[8] R.K. Pandey, P. Padmini, L.F. Deravi, N.N. Patil, P. Kale, J. Zhong, J.Dou, L. Navarette, R. Schad and M. Shamzuzhoa, "Magnetic Semiconductors in Fe-Ti-Oxide Series and their Potential Applications", IEEE Proceedings of the 8th International Conference on Solid State and Integrated Circuit Technology, ICSICT (2006), Part 2, , ISBN: 1-4244- 0160-5, (2006), 992.

[9] R. K. Pandey, P. Padmini, P. Kale, J. Dou, C. Lohn, R. Schad, R. Wilkins and W. Geerts, " Multifunctional nature of modified Iron Titanates and their potential applications" ,*Ceramic Transactions*, **226**, 61, (2011); A John Wiley Publication.

[10] International Patent on "Varistor-Transistor Hybrid Devices"; Inventors: Raghvendra K. Pandey, William. A. Stapleton, Ivan Sutanto and Amanda A. Scantlin (pending).

FABRICATION AND IMPROVEMENT OF THE PROPERTIES OF Mn-DOPED BISMUTH FERRITE–BARIUM TITANATE THIN FILMS

Yuya Ito, Makoto Moriya, Wataru Sakamoto and Toshinobu Yogo
Division of Nanomaterials Science, EcoTopia Science Institute, Nagoya University
Furo-cho, Chikusa-ku, Nagoya 464-8603, Japan

ABSTRACT
    Bismuth ferrite ($BiFeO_3$)−barium titanate ($BaTiO_3$) solid-solution thin films that simultaneously exhibit ferroelectricity and ferromagnetism at room temperature were fabricated using chemical solution deposition. Perovskite-structured, single-phase $0.7BiFeO_3$–$0.3BaTiO_3$ thin films were successfully synthesized at 700°C on $Pt/TiO_x/SiO_2/Si$ substrates. Although the $0.7BiFeO_3$–$0.3BaTiO_3$ thin films exhibited poor polarization (P)−electric field (E) hysteresis loops owing to their low insulating resistance, the leakage current under high applied fields was effectively reduced through Mn doping at the Fe sites of the $0.7BiFeO_3$–$0.3BaTiO_3$ thin films, which led to improved ferroelectric properties. The 5 mol% Mn-doped $0.7BiFeO_3$–$0.3BaTiO_3$ thin films simultaneously exhibited ferroelectric polarization and ferromagnetic hysteresis loops at room temperature.

INTRODUCTION

    Multiferroic bismuth ferrite ($BiFeO_3$) thin films, which simultaneously exhibit ferroelectricity and ferromagnetism have been receiving great attention as promising materials for the development of novel electronic devices.[1] However, the synthesis of pure $BiFeO_3$ without undesirable second phases is usually difficult because of the low structural stability of perovskite $BiFeO_3$. In the case of $BiFeO_3$ thin films, the crystallization of the $BiFeO_3$ phase on a substrate often results in the formation of a bismuth-deficient second phase, such as $Bi_2Fe_4O_9$, which degrades the surface morphology as a result of exaggerated grain growth, thereby leading to poor electrical properties.[2] Moreover, the evaluation of the ferroelectric properties of $BiFeO_3$ at ambient temperature is difficult owing to its low insulating resistance. In addition, $BiFeO_3$ is known to exhibit very weak ferromagnetism with relatively low magnetization.[3] Therefore, the modification of its electrical and magnetic properties is necessary.
    The formation of $BiFeO_3$–ABO_3 solid-solution systems is an effective method for improving the structural stability and other properties of $BiFeO_3$ thin films. Among various ABO_3 compounds, barium titanate ($BaTiO_3$) is a suitable counterpart because it is a stable ferroelectric perovskite oxide. Thus, a solid solution of $BiFeO_3$ and $BaTiO_3$ is expected to achieve the desired structural stabilization and electrical and magnetic properties. Furthermore, the doping of $BiFeO_3$ thin films with functional elements, such as Mn, to improve their electrical and magnetic properties has been intensively studied.[4,5] However, the multiferroic properties of $BiFeO_3$−$BaTiO_3$ (BF–BT) thin films reported in previous studies were not satisfactory.[6]
    In this study, the synthesis of room-temperature multiferroic BF–BT thin films on Si-based substrates using chemical solution deposition has been examined. M. M. Kumar et al. have reported that perovskite BF–BT maintains a rhombohedrally distorted perovskite structure when the content of $BaTiO_3$ is less than 33 mol%.[7] Therefore, the $0.7BiFeO_3$–$0.3BaTiO_3$ composition was selected in this study to adequately stabilize the perovskite structure and to improve the multiferroic properties. The effect of Mn doping on the electrical and magnetic properties of the BF–BT thin films was also investigated.

313

EXPERIMENTAL DETAILS

$Bi(O^tAm)_3$, $Fe(OEt)_3$, $Ba(OEt)_2$, $Ti(O^iPr)_4$, and $Mn(O^iPr)_2$ were selected as starting materials for the preparation of precursor solutions. The desired amounts of starting metal alkoxides that corresponded to a composition of $0.7Bi(Fe_{1-x}Mn_x)O_3$–$0.3BaTiO_3$ (BFM_x-30BT; x = 0, 0.03, 0.05) with a 3 mol% excess of Bi were dissolved in absolute 2-methoxyethanol. The mixed solution was subsequently heated at 90°C for 18 h to yield a 0.2 M precursor solution. The entire procedure was conducted under a dry nitrogen atmosphere.

Thin films were fabricated using the BFM_x-30BT precursor solution by spin coating onto $Pt/TiO_x/SiO_2/Si$ substrates. The precursor films were dried at 150°C for 5 min and then calcined at 400°C for 1 h (heating rate: 5°C/min) under flowing oxygen (100 ml/min). After the films were subjected to five drying and calcining cycles, they were crystallized at 600–700°C for 30 min in an oxygen flow using rapid thermal annealing (180°C/min). The coating and calcining cycles were repeated 10 times with two heat treatments for crystallization, which resulted in a film thickness of approximately 500 nm.

The crystallographic phases of the crystallized thin films were identified by X-ray diffraction (XRD) analysis using monochromated Cu Kα radiation. The surface morphology of the films was observed using atomic force microscopy (AFM). Platinum top electrodes with a diameter of 0.2 mm were deposited via DC sputtering onto the surface of the BFM_x-30BT films, which were subsequently annealed at 400°C for 1 h. The ferroelectric properties of the films were measured using a ferroelectric test system under vacuum (1.0 Pa). The current density–electric field characteristics were also evaluated using an electrometer/high-resistance meter. The magnetization behavior of the prepared thin films at 300 K was evaluated using a superconducting quantum interference device (SQUID) magnetometer.

RESULTS AND DISCUSSION

Figure 1 shows the XRD patterns of $0.7BiFeO_3$–$0.3BaTiO_3$ (BF-30BT), $0.7BiFe_{0.97}Mn_{0.03}O_3$–$0.3BaTiO_3$ ($BFM_{0.03}$-30BT) and $0.7BiFe_{0.95}Mn_{0.05}O_3$–$0.3BaTiO_3$ ($BFM_{0.05}$-30BT) thin films fabricated on $Pt/TiO_x/SiO_2/Si$ substrates after the films were heat-treated at 700°C. All the BFM_x-30BT thin films crystallized as single-phase perovskite $BiFeO_3$–$BaTiO_3$ (BF–BT) with a random orientation. Crystallization in the perovskite single phase is achieved by the stabilization of the perovskite phase by the formation of a solid solution with $BaTiO_3$. The formation of the $Bi_2Fe_4O_9$ second phase in pure $BiFeO_3$ films has been attributed to the low structural stability of perovskite $BiFeO_3$ and to the volatility of Bi ions at high temperatures (>600°C).[2] The temperature range of crystallization in single-phase perovskite BF–BT was enlarged by the formation of a $0.7BiFeO_3$–$0.3BaTiO_3$ solid solution. In addition, Mn doping of the BF-30BT thin films did not influence the crystallization in perovskite BF-30BT.

Figure 2 shows the AFM images of BF-30BT and $BFM_{0.05}$-30BT thin films prepared at 700°C on $Pt/TiO_x/SiO_2/Si$ substrates. Both films exhibited homogeneous surface morphologies, which can be attributed to the suppression of the formation of a second phase, such as $Bi_2Fe_4O_9$. Grain growth was enhanced through Mn doping, and the grain size of the films increased from 30–60 nm (Mn0%) to 80–160 nm (Mn5%). The root-mean-square (RMS) roughness values of the BF-30BT and $BFM_{0.05}$-30PT thin films were 6.3 and 3.8 nm, respectively. Based on Fig. 2, Mn-doped thin films exhibited larger grain sizes and smaller surface roughnesses, which indicates that Mn enhanced the grain growth and improved the surface morphology.

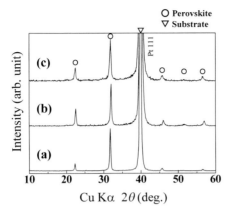

Figure 1. XRD profiles of BFM$_x$-30BT thin films crystallized at 700°C:
(a) $x = 0$, (b) $x = 0.03$, and (c) $x = 0.05$.

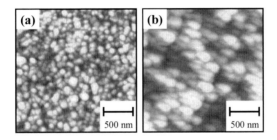

Figure 2. AFM images of BFM$_x$-30BT thin films crystallized at 700°C:
(a) $x = 0$ and (b) $x = 0.05$.

We performed $P–E$ hysteresis measurements to characterize the ferroelectric properties for BF-30BT and Mn-doped BF-30BT thin films crystallized at 700°C. Figure 3(a) shows the $P–E$ hysteresis loops of a BF-30BT thin film measured at room temperature and at −190°C. Although the BF-30BT films exhibited a ferroelectric hysteresis loop, a saturated $P–E$ loop could not be obtained because a sufficiently high electric field could not be applied at room temperature. This inability to apply a high electric field is related to the result of leakage current measurements described later. Because the electrical resistivity was supposed to be sufficiently high to characterize the net ferroelectricity of the synthesized thin films, a ferroelectric measurement at −190°C was also performed. In this case, the BF30BT thin film exhibited relatively saturated ferroelectric hysteresis loops compared with those at room temperature. The remnant polarization (P_r) and coercive field (E_c) of the BF-30BT thin film were approximately 30 μC/cm^2

and 170 kV/cm^2, respectively.

The BF-30BT thin films exhibited relatively poor electrical resistivity. Therefore, in order to improve their insulating properties, Mn-modified BF–BT thin films were prepared and examined in this study. Improvement in the electrical properties of perovskite $BiFeO_3$–$PbTiO_3$ (BF–PT) films, especially their insulating resistance, as a result of Mn doping has been reported by Sakamoto et al.[8] Figure 3(b) shows the P–E hysteresis loops of the BF-30BT, BFM$_{0.03}$-30BT, and BFM$_{0.05}$-30BT thin films measured at room temperature. Among them, the BFM$_{0.05}$-30BT thin film exhibited a well-shaped hysteresis loop with a small leakage component. This result suggests that 5 mol% Mn doping of the BF-30BT thin films is an appropriate level for improving polarization properties. The BFM$_{0.05}$-30BT thin film exhibited a remanent polarization of 27 μC/cm^2. This value is consistent with that for the BF-30BT thin film at $-190°C$ and is larger than that of 0.67BF–0.33BT ceramics that contain 0.1 wt% MnO_2.[9]

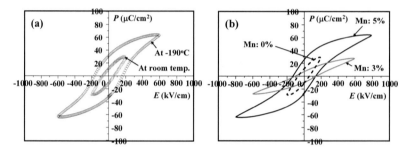

Figure 3. P–E hysteresis loops of BFM$_x$-30BT thin films crystallized at 700°C:
(a) $x = 0$ (measured at room temperature and at $-190°C$) and
(b) $x = 0$, 0.03, and 0.05 (measured at room temperature).

Figure 4 shows the leakage current properties of the BF-30BT, BFM$_{0.03}$-30BT and BFM$_{0.05}$-30BT thin films measured at room temperature. Curves of the leakage current density as a function of the applied electric field for the films with and without Mn doping clearly exhibited different behaviors. The leakage current density of the nondoped BF-30BT films exhibited an abrupt increase with an applied field of approximately 30 kV/cm, whereas, in the low-electric-field region, the BF-30BT thin films exhibited a lower current density than did the BFM$_{0.03}$-30BT and BFM$_{0.05}$-30BT thin films. The doping of the BF-30BT films with Mn clearly decreased the leakage current at high applied fields, as shown in Fig. 4. The results represented in Fig. 4 are consistent with those reported for $BiFeO_3$ and Mn-doped $BiFeO_3$ thin films.[4]

The leakage current properties of the BF-30BT films were markedly improved after being doped with Mn as previously described. Therefore, the ln J–ln E plots of the BF-30BT and Mn-doped BF-30BT thin films in the positive applied field (from the bottom to the upper electrode) area of Fig. 4 were characterized to examine the difference in the leakage current mechanism, as demonstrated in Fig. 5. From the ln J–ln E plot in Fig. 5, ohmic conduction, which is confirmed from the slope of the ln J–ln E plot ($n \approx 1$), was observed over a wide range of applied fields (< 200 kV/cm) for the $BFM_{0.05}$-30BT films, whereas the nondoped BF–BT thin films exhibited an abrupt increase in leakage current in fields stronger than 30 kV/cm. Doped Mn was found to effectively suppress the abrupt increase in leakage current under strong applied fields and to maintain the ohmic conduction with a low leakage current under a strong electric field. Similar to the case of previously reported Mn-doped $BiFeO_3$–$Bi_{0.5}Na_{0.5}TiO_3$ (BF–BNT) thin films,[10] Mn, when used as a dopant, might play an important role as an acceptor for trapping free carriers in the BF-30BT thin films. However, Mn substitutes at the Fe^{3+} site as Mn^{2+}, which is accompanied by the formation of an oxygen vacancy. This might have a negative influence on the films under an increasing concentration of the Mn dopant. Thus, the optimum Mn and $BaTiO_3$ concentrations should be determined in future work.

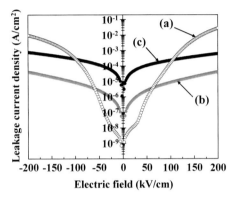

Figure 4. Leakage current density of BFM_x-30BT thin films as a function of the applied electric field: (a) $x = 0$, (b) $x = 0.03$, and (c) $x = 0.05$.

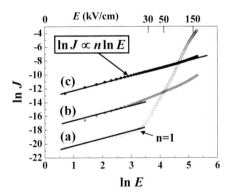

Figure 5. Plots of $\ln J$ vs. $\ln E$ for BFM_x-30BT thin films: (a) $x = 0$, (b) $x = 0.03$, and (c) $x = 0.05$.

The best electrical properties were obtained with BFM$_{0.05}$-30BT thin films in this study. The magnetic properties of perovskite BFM$_{0.05}$-30BT and BiFeO$_3$ films were also evaluated using a SQUID magnetometer. Figure 6 shows the magnetization (M)–magnetic field (H) hysteresis loops of pure BiFeO$_3$ and BFM$_{0.05}$-30BT thin films on Pt/TiO$_x$/SiO$_2$/Si substrates after the films were heat-treated at 600°C and 700°C, respectively. These measurements were performed at 300 K, and the thickness of the films was 1 μm (20 coatings). The data was corrected for the diamagnetism component from the substrate and sample holder. The BFM$_{0.05}$-30BT sample exhibited typical ferromagnetic characteristics, because the typical $M–H$ hysteresis loop of ferromagnetism shown in Fig. 6(a) shows a magnetization at an applied field of 30 kOe, remanent magnetization, and coercive force of approximately 16 emu/cm^3, 7.5 emu/cm^3, and 4.0 kOe, respectively. Kumar et al. reported on (1–x)BiFeO$_3$–xBaTiO$_3$ ceramics with ferromagnetic $M–H$ hysteresis loops.[11] Although the XRD data in Fig. 1 indicates that the films in this study are pseudocubic perovskite, their weak ferromagnetism depends on the structural distortion that occurred in BF–BT. Similar magnetic behavior, including the presence of ferromagnetism, has been observed in a previous study of magnetization in BF–PT thin films.[12] The BFM$_{0.05}$-30BT thin film is judged to be multiferroic at room temperature on the basis of the results shown in Fig. 3(b).

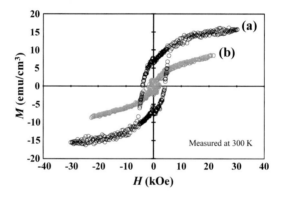

Figure 6. $M–H$ curves of (a) BFM$_{0.05}$-30BT (●) and (b) BiFeO$_3$ (○) thin films.

A BiFeO$_3$ single crystal was reported to exhibit a near-linear $M–H$ relationship.[3] The usual polycrystalline BiFeO$_3$ samples also exhibited very small magnetization and $M–H$ hysteresis loops without spontaneous magnetization.[13] Moreover, for the current BFM$_{0.05}$-30BT thin-film sample, the observed induced magnetization was greater than that observed for the BiFeO$_3$ thin films. The pure BiFeO$_3$ thin films prepared in this study showed smaller magnetization (approximately 8.9 emu/cm^3) at 20 kOe, with remanent magnetization of 1.2 emu/cm^3 and a coercive force of 1.3 kOe, as shown in Fig. 6(b). The enhanced magnetization of the BFM$_{0.05}$-30BT thin films, which is similar to that reported by other authors[5,14], is considered to be due to the magnetic structural change from the spiral to collinear G-type antiferromagnetic ordering of $BiFeO_3$ upon Mn doping. Furthermore, the spatial confinement[15] such as the size and surface effects associated with the small grain size (approximately 100 nm, as shown in Fig. 2), the existence of Ti ions[15], and the defect structure of the films (especially the valence states of Mn and Fe ions in the B sites of the perovskite $BiFeO_3$ structure, given the superexchange interaction through the oxide ions in BF–BT) also affect the magnetic properties, i.e., the magnetization behavior and coercive force. The reason for these effects is not clear and is under investigation.

CONCLUSIONS

Room temperature multiferroic $0.7BiFeO_3$–$0.3BaTiO_3$ (BF-30BT) thin films were successfully fabricated using chemical solution deposition. Through the formation of a solid solution with $BaTiO_3$, the BF-30BT films directly crystallized as a perovskite single phase, and the process window for perovskite BF–BT films with a homogeneous surface microstructure was enlarged. Although the electrical resistivity of the non-doped BF-30BT thin films was not sufficiently high, the leakage current density was effectively reduced under high applied fields through Mn doping. In particular, the 5 mol% Mn-doped BF-30BT thin films simultaneously exhibited improved ferroelectric properties and typical ferromagnetic M–H hysteresis loops at ambient temperatures. Thus, Mn doping was found to be highly effective in improving the electrical properties and in enhancing the magnetization of the multiferroic $BiFeO_3$–$BaTiO_3$ thin films.

ACKNOWLEDGMENTS

This work was partly supported by a Grant-in-Aid for Scientific Research (B), No. 23360284, from the Ministry of Education, Culture, Sports, Science and Technology, Japan.

REFERENCES

[1] J. Wang, J. B. Neaton, H. Zheng, V. Nagarajan, S. B. Ogale, B. Liu, D. Viehland, V. Vaithyanathan, D. G. Schlom, U. V. Waghmare, N. A. Spaldin, K. M. Rabe, M. Wuttig, and R. Ramesh, Epitaxial $BiFeO_3$ Multiferroic Thin Film Heterostructures, *Science*, **299**, 1719-22 (2003).

[2] F. Tyholdt, S. Jorgensen, H. Fjellvag, and A. E. Gunnaes, Synthesis of Oriented $BiFeO_3$ Thin Films by Chemical Solution Deposition: Phase, Texture, and Microstructural Development, *J. Mater. Res.*, **20**, 2127-39 (2005).

[3] F. Bai, J. Wang, M. Wittig, J.-F. Li, N. Wang, A. P. Pyatakov, A. K. Zvezdin, L. E. Cross, and D. Viehland, Destruction of Spin Cycloid in $(111)_c$-Oriented $BiFeO_3$ Thin Films by Epitiaxial Constraint: Enhanced Polarization and Release of Latent Magnetization, *Appl. Phys. Lett.*, **86**, 032511 (2005).

[4] S. K. Singh, H. Ishiwara, and K. Maruyama, Room Temperature Ferroelectric Properties of Mn-Substituted $BiFeO_3$ Thin Films Deposited on Pt Electrodes Using Chemical Solution Deposition, *Appl. Phys. Lett.*, **88**, 262908 (2006).

CORRELATING THE CRYSTAL STRUCTURE AND THE PHASE TRANSITIONS WITH THE DIELECTRIC PROPERTIES OF $K_xBa_{1-x}Ga_{2-x}Ge_{2+x}O_8$ SOLID SOLUTIONS

Marjeta Maček Kržmanc[1], Qin Ni[2], and Danilo Suvorov[1]

[1]Advanced Materials Department, Jožef Stefan Institute, Jamova 39, 1000 Ljubljana, Slovenia

[2]State Key Laboratory of Optoelectronic Materials and Technologies, School of Physics and Engineering, SunYat-Sen University, Guangzhou 510275, P. R. China

ABSTRACT

$K_xBa_{1-x}Ga_{2-x}Ge_{2+x}O_8$ ($0 \leq x \leq 1$) solid solutions exist in two structural modifications: a low-temperature-stable phase with a pseudo-orthorhombic $P2_1/a$ symmetry and a high-temperature-stable phase with a monoclinic $C2/m$ symmetry. The $P2_1/a$ phase is necessary for high Qxf values. Therefore, the synthesis of the pure $P2_1/a$-structure for $K_xBa_{1-x}Ga_{2-x}Ge_{2+x}O_8$ ceramics is essential for microwave dielectric applications. In this work, the $K_xBa_{1-x}Ga_{2-x}Ge_{2+x}O_8$ solid solutions were studied in terms of their crystal structure, phase transitions, sintering behavior and dielectric properties. The solid-state reaction between the reagents resulted in the $P2_1/a$ phase for the compositions $0 \leq x \leq 0.67$ and in the $C2/m$ phase for the compositions $0.9 \leq x \leq 1$. However, the $C2/m$ powders undergo a $C2/m \Rightarrow P2_1/a$ phase transformation after prolonged annealing at a temperature that is 50–80°C below the lowest $P2_1/a \Rightarrow C2/m$ transition temperature of the particular composition. The sintering of the $P2_1/a$ powders was also restricted to a critical temperature range in order to avoid the formation of the $C2/m$ phase. The thermal history and the processing conditions were found to influence the $C2/m \Rightarrow P2_1/a$ phase transformation by complex mechanisms. The well-sintered $K_xBa_{1-x}Ga_{2-x}Ge_{2+x}O_8$ (x=0.9 and 0.67) ceramics with the $P2_1/a$ structure showed a good combination of a low permittivity (ε_r=5.6-6), high Qxf-values (96700-134000 GHz), a small temperature coefficient of resonant frequency (~-20 ppm/°C) and a low sintering temperature (910-970°C).

INTRODUCTION

Rapid developments in various wireless telecommunications and radar systems have been observed in the past 15 years. During the course of this progress, the utilizable region of the communication frequencies has expanded to frequencies higher than 10 GHz[1,2] High permittivities, which for reasons of miniaturization of the electronic component are required for lower frequencies (<10 GHz), are not appropriate for high frequencies, because their small size requires very accurate processing. In addition, a high permittivity means a low signal-propagation velocity. Low-permittivity materials have several advantages for the high-frequency range. Firstly, the size is in accordance with the accuracy of the processing and, secondly, low-permittivity materials also allow a high signal-propagation velocity.

Feldspar-based ceramics are typical low-permittivity dielectrics and good candidates for several applications in the field of electroceramics. Recrystallized feldspars have been used as the constituents of commercial, low-temperature, cofired ceramic (LTCC) dielectrics.[3] However, due to the high sintering temperature of crystallized feldspars, LTCC sintering was only achieved with the addition of a low-melting-point glass. Since the remains of the glass after recrystallization increased the dielectric losses, the Qxf value of such a system is lower compared to that of glass-free feldspars. Our previous studies of aluminosilicate feldspars revealed that these materials exhibited low permittivity (ε=6-9) and Qxf values, which varied over a wide range from 6000 to 100000 GHz. Composition, crystal structure, sintering temperature, time and cooling rate were found to be the key factors influencing the Qxf values.[4,5,6,7] Among the $MeAl_2Si_2O_8$ (Me=Ca, Sr, Ba) feldspars, $CaAl_2Si_2O_8$ anorthite showed the lowest Qxf values, which varied between 5500 and 8500 GHz and from 7100 to 11000 GHz for the $I\bar{1}$ and $P\bar{1}$ structures, respectively. In addition to prolonging the annealing at high temperature (1400°C), slow cooling, that enables a displacive $I\bar{1} \leftrightarrow P\bar{1}$ phase transition, contributed to an improvement in the anorthite's Qxf-values.[6] Considerably higher Qxf values of 42000–93000 GHz were measured for the $BaAl_2Si_2O_8$ and $Sr_xBa_{1-x}Al_2Si_2O_8$ (0≤x<0.1). Similar to anorthite, the Qxf values were improved by prolonged annealing at a high temperature of 1400°C. A decrease in the concentration of crystal-structure defects, such as antiphase boundaries, was the main structural change that accompanied this improvement in the Qxf values.[7] The reason for the sluggish, high-temperature ordering behavior of the aluminosilicate feldspars was attributed to the strong Si-O bond. Based on the weaker Ge-O bond, Ge-substituted feldspars exhibit faster tetrahedral ordering kinetics and sinter at lower temperatures.[8,9] In our previous studies we reported the existence of the low-temperature-stable $P2_1/a$ and the high-temperature-stable C2/m crystal modifications of $K_xBa_{1-x}Ga_{2-x}Ge_{2+x}O_8$ solid solutions.[10,11,12] The former exhibited much higher Qxf values than the latter.

In the present article we are focused on the preparation and sintering of the high-Q $P2_1/a$ modification of $K_xBa_{1-x}Ga_{2-x}Ge_{2+x}O_8$ ceramics. The dielectric properties of the $P2_1/a$ and C2/m phases were compared and correlated with the characteristics of their crystal structure.

EXPERIMENTAL

The $K_xBa_{1-x}Ga_{2-x}Ge_{2+x}O_8$ (0 ≤ x ≤ 1) solid solutions were synthesized using the solid-state reaction technique from reagent-grade Ga_2O_3, (Aldrich, 99.99%,), GeO_2 (Alfa Aesar, 99.999%), K_2CO_3 (Alfa Aesar, 99.0%) and $BaCO_3$ (Alfa Aesar, 99.8%). The homogenized stoichiometric mixtures of the reagents for the compositions with x=1, 0.6≤x≤0.9 and 0≤x≤0.4 were calcined at 700–950°C, 700–970°C and 800–1000°C, respectively. The duration of the isothermal annealing during each calcination step was 12 hours, after which the reaction product was ground and milled. The $K_xBa_{1-x}Ga_{2-x}Ge_{2+x}O_8$ (x=1 and 0.9) ceramics, which when synthesized produced the C2/m modification, were transformed to $P2_1/a$ by prolonged annealing (2-100 hours) at 840–860°C. The preparation of C2/m for $K_xBa_{1-x}Ga_{2-x}Ge_{2+x}O_8$ (x=0.6 and 0.67) required annealing at 1020°C. The $P2_1/a$ phase of these two compositions could be obtained again by annealing at 900–940°C. Prior to sintering the $P2_1/a$ and C2/m powders were milled with Y-stabilized ZrO_2 balls in ethanol to yield a median particle size of 0.8 μm. The milled powders were isostatically (at ~560 MPa) pressed into pellets, which were sintered in air at 880–1100°C and then cooled slowly to room temperature with a cooling rate of 1°C/min.

The C2/m powders for the kinetic studies on the C2/m⇒$P2_1/a$ phase transition were uniaxially pressed at ~130 MPa into small pellets of uniform size. The specimens were then heated to the temperature of the fastest C2/m⇒$P2_1/a$ transformation rate. After isothermal firing for a specific period of time, one of these pellets was taken out of the furnace and quenched directly in air to preserve the phase constitution. Meanwhile, the temperature inside the furnace was kept constant. The volume fraction of the formed $P2_1/a$ phase was determined by quantitative powder X-ray diffraction (XRD) analysis, which is described in detail in reference 11.

The powder XRD data were collected with a Bruker AXS D4 Endeavor diffractometer using Cu-Kα radiation. The differential thermal analyses (DTA) of the powders were performed on a simultaneous thermal analysis instrument (Jupiter 449, Netzsch, Selb, Germany).

The aspect densities of the sintered specimens were measured by Archimedes' method using ethanol. The theoretical densities were evaluated from the formula weight and the unit-cell volume, and the relative densities were obtained by comparing the aspect densities with the theoretical values.

The microwave dielectric properties were characterized with a network analyzer (HP 8719C, Agilent, Palo Alto, CA) using the $TE_{01\delta}$-mode dielectric-resonator method, which has been described by Krupka et al.[13] The temperature coefficients of the resonant frequency (τ_f) were measured in the temperature range 20–60°C.

RESULTS AND DISCUSSION

Calcination of the reagents in the stoichiometric ratio of $KGaGe_3O_8$ in the temperature range from 700 to 900°C led to the formation of several potassium-germanium and potassium-gallium oxides with overlapping diffraction lines, which made it difficult to determine the reaction mechanism. $KGaGe_3O_8$ with the C2/m structure started to appear above 900°C and became the dominant phase at 930°C. Single-phase C2/m $KGaGe_3O_8$ was obtained at 950°C. The reaction pathway of the composition with x=0.9 was similar to that of x=1. The difference appeared due to the secondary $BaGe_4O_9$ phase, which required additional calcination and a higher calcination temperature of 970°C for its complete elimination. In the compositions with x=0.6 and 0.67 the $P2_1/a$ modification of the $K_xBa_{1-x}Ga_{2-x}Ge_{2+x}O_8$ solid-solutions started to form at 900–930°C. Additional calcinations with intermediate milling between 950–970°C were required to eliminate the $BaGe_4O_9$ phase and get pure $P2_1/a$ $K_xBa_{1-x}Ga_{2-x}Ge_{2+x}O_8$ solid solutions. The gradual transformation of the $P2_1/a$ phase to the C2/m phase for these two compositions was observed above 980°C, whereas a complete transformation was achieved with a 24-hour isothermal annealing at 1020°C. For compositions with a small K content ($0 \leq x < 0.2$), only the $P2_1/a$ phase could be synthesized, and no phase transition was observed in this case. The powder XRD patterns of the $P2_1/a$ and C2/m phases of $K_{0.67}Ba_{0.33}Ga_{1.33}Ge_{2.67}O_8$ are shown in Fig. 1. An X-ray Rietveld analysis was performed for a structural interpretation of the two phases.[12] Unlike the displacive phase transition in the $CaAl_2Si_2O_8$ anorthite, the frameworks of the $P2_1/a$ and C2/m structures of the $K_xBa_{1-x}Ga_{2-x}Ge_{2+x}O_8$ feldspars are totally different in terms of the linkage between the fundamental tetrahedral sheets. Consequently, the bonds between the atoms in the crystal structure have to be partially disassociated and rearranged to form the new structure. In principle this leads to an intensive thermal effect during the phase transition, which can be detected using thermal analyses.

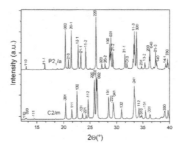

Fig. 1: X-ray powder diffraction patterns of $K_{0.67}Ba_{0.33}Ga_{1.33}Ge_{2.67}O_8$ ceramics with the $P2_1/a$ and C2/m structures

It could already be inferred from the above description of the isothermal annealings that the temperature of the $P2_1/a \Leftrightarrow C2/m$ phase transition increased with a decrease of x. This was additionally confirmed by the DTA, which revealed two endothermic peaks for all compositions with $0.67 \leq x \leq 1$ (Fig. 2). The examinations of the phase compositions of the samples obtained from the interrupted DTA experiments, below and above the on-set temperature and at the peak temperature, revealed that the first peak was associated with the $P2_1/a \Rightarrow C2/m$ phase transition, while the second endothermic peak belonged to melting. A comparison of the first peak on-set temperatures and the temperatures that are required for the isothermal sintering of the $P2_1/a$ $K_xBa_{1-x}Ga_{2-x}Ge_{2+x}O_8$ below the $P2_1/a \Rightarrow C2/m$ phase transition revealed that the first appeared at considerably higher temperatures (Fig. 2, Table I). The different morphology of the powders used for these two types of experiments was one of the reasons for this discrepancy. Namely, the powders examined by the DTA were not milled, while milled powders were used in the sintering process. The other reason lay in the different nature of the experiment. During dynamic DTA measurements the sample was exposed to certain temperatures for shorter times than during the isothermal annealing and therefore detectable heat release due to the phase transition was not observed until high temperatures, when the kinetics became faster. With regards to the sintering of the $P2_1/a$ phase, the maximum temperature at which the $P2_1/a$ modification of a particular composition was still preserved during isothermal annealing was more important information than the on-set temperature on the DTA curve. The critical temperatures summarized in Table I are important data for realizing the synthesis and densification of $K_xBa_{1-x}Ga_{2-x}Ge_{2+x}O_8$ ceramics with the required crystal structure.

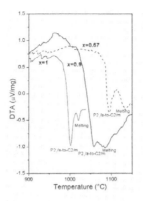

Fig. 2: Differential thermal analysis curves of the non-milled P2₁/a powders of the $K_xBa_{1-x}Ga_{2-x}Ge_{2+x}O_8$ solid solutions, measured at a heating rate of 25°C/min.

Table I: The lowest temperature for the P2₁/a⇒C2/m phase transition, the temperatures of the fastest C2/m ⇒ P2₁/a transformation kinetics and melting temperatures for the $K_xBa_{1-x}Ga_{2-x}Ge_{2+x}O_8$ solid solutions

Composition x	T (°C) (P2₁/a⇒C2/m)	T(°C) (C2/m⇒ P2₁/a)	Melting temperature (°C)
1	900	840	1005
0.9	920	860	1056
0.67	980	900	1158
0.6	990	940	1170

It is clear that the solid-state synthesis of the $K_xBa_{1-x}Ga_{2-x}Ge_{2+x}O_8$ solid solutions with 0.9≤x≤1 resulted in a product with the C2/m structure, while the compositions with 0.4≤x≤0.67 at first formed in the P2₁/a structure and could be transformed to the C2/m symmetry by an increase in the temperature. We followed the idea to sinter $K_xBa_{1-x}Ga_{2-x}Ge_{2+x}O_8$ compositions for 0≤x≤1 in both crystal modifications and characterized them in terms of their dielectric properties. This could be done when certain conditions were fulfilled. In order to sinter the P2₁/a phase the composition had to firstly exist in the P2₁/a modification and secondly to densify below the P2₁/a⇒C2/m phase transition. The preparation of dense $K_xBa_{1-x}Ga_{2-x}Ge_{2+x}O_8$ ceramics with the C2/m structure required the P2₁/a⇔C2/m phase transition and densification below the melting point. We found that all compositions with 0.6≤x≤1 easily transformed to the C2/m modification, while the transformation for x=0.4

occurred above 1130°C, where large losses of potassium occurred. The compositions with $0 \leq x < 0.2$ did not show the $P2_1/a \Leftrightarrow C2/m$ phase transition.

For the compositions ($0.9 \leq x \leq 1$), which formed the $C2/m$ structure directly, the conditions for the transformation from $C2/m$ to $P2_1/a$ were investigated. In analogy with the reversible phase-transition behavior of the $BaAl_2Ge_2O_8$ feldspar, observed by Malcherek et al.,[8,9] where a disordered $C2/m$ modification was transformed into an ordered $I2/c$ modification by annealing at a temperature 400–500°C below the $I2/c \Rightarrow C2/m$ phase-transition temperature, we expected that the $P2_1/a$ phase of the $K_xBa_{1-x}Ga_{2-x}Ge_{2+x}O_8$ ($x=0.9$ and 1) could also be obtained by a post-annealing route from the $C2/m$ phase. Indeed, the $C2/m \Rightarrow P2_1/a$ transformation took place at only 50–80°C below the lowest $P2_1/a \Rightarrow C2/m$ phase-transition temperature (Table I). The temperatures of the fastest $C2/m \Rightarrow P2_1/a$ transition rates were determined by separate experiments and are summarized in Table I.[11] The kinetics of the $C2/m \Rightarrow P2_1/a$ phase transitions were studied at these temperatures. The thermal history was found to have an important effect on the phase-transition kinetics. This was particularly true for the composition with $KGaGe_3O_8$ ($x=1$), for which the transition rate decreased rapidly with aging time. For example, freshly prepared $C2/m$ powder transformed completely to $P2_1/a$ at 840°C in less than 25 hours, while in the kinetics study performed after 1 week the $P2_1/a$ phase started to form after 100 hours. The transition rate increased again when the aged powders were heat treated at 950°C for 12 hours before a repeated kinetic study. However, the transition rate was still considerably slower compared to that of fresh powders. The powders that were additionally annealed before the kinetic study were denoted as "refreshed". We should point out that no obvious structural differences between the fresh, the refreshed and the aged $C2/m$ powders were detected with XRD and Raman spectroscopy. The remarkable slowdown of the $C2/m$-to-$P2_1/a$ kinetics with aging time, which was observed for $KGaGe_3O_8$, could be related to a large increase in the unit-cell volume accompanying this phase transition.[10,11,12] We assume that the strains that were thermally produced in the material could help the lattice to expand and therefore make the phase transition easier. The release of strains over time is believed to cause the observed decrease in the phase-transition rate with aging.[11,12] Compared to $KGaGe_3O_8$, the aging effect on the kinetics of the phase transition was not so pronounced for $K_{0.9}Ba_{0.1}Ga_{1.1}Ge_{2.9}O_8$, $K_{0.67}Ba_{0.33}Ga_{1.33}Ge_{2.67}O_8$ and $K_{0.6}Ba_{0.4}Ga_{1.4}Ge_{2.6}O_8$, which also showed smaller differences between the $C2/m$ and $P2_1/a$ unit-cell volumes. However, the refreshed powders of these compositions still showed faster transition rates compared to the aged ones. The comparison between the $C2/m \Rightarrow P2_1/a$ phase-transition kinetics of the refreshed $KGaGe_3O_8$ and $K_{0.67}Ba_{0.33}Ga_{1.33}Ge_{2.67}O_8$ powders, which were performed at 840 and 900°C, respectively, is represented in Fig. 3.

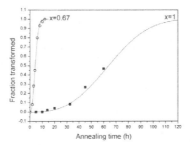

Fig.3: Volume fraction of the $P2_1/a$ phase of the $KGaGe_3O_8$ (x=1) and $K_{0.67}Ba_{0.33}Ga_{1.33}Ge_{2.67}O_8$ (x=0.67), formed from the refreshed powders. The kinetics study was performed at 840°C for the $KGaGe_3O_8$ and at 900°C for the $K_{0.67}Ba_{0.33}Ga_{1.33}Ge_{2.67}O_8$ ceramics.

In order to preserve the $P2_1/a$ phase the ceramics had to be sintered below the $P2_1/a \Rightarrow C2/m$ phase-transition temperature. The $KGaGe_3O_8$ ceramics sintered at 890°C, that is 10°C below the phase-transition temperature, exhibited a relative density of 0.69 and correspondingly poor dielectric properties. The high density of the $KGaGe_3O_8$ was achieved by the addition of 0.3 wt. % H_3BO_3 (Table II). The $K_{0.9}Ba_{0.1}Ga_{1.1}Ge_{2.9}O_8$ ceramics still required a small amount of H_3BO_3 (0.1 wt. %) for sintering at 910°C with a relative density of 0.97, while the compositions with $0 \le x \le 0.67$ sintered in the $P2_1/a$ modification without the sintering aid. The $K_{0.67}Ba_{0.33}Ga_{1.33}Ge_{2.67}O_8$ sintered at 970°C exhibited a Qxf value of 119000 GHz, which was increased to 134000 GHz by an additional 40-hour annealing at 900°C (Table II). It was also noticed for other $P2_1/a$ compositions that prolonged annealing at the temperatures of the fastest $C2/m \Rightarrow P2_1/a$ transition rate led to an improvement in the Qxf-values (Table II). The sintering temperature of the $K_{0.67}Ba_{0.33}Ga_{1.33}Ge_{2.67}O_8$ ceramics could be further lowered to 920°C by the addition of 0.1 wt. % of H_3BO_3, but this caused a decrease in the Qxf-values to 96700 GHz. All the $P2_1/a$ compositions with $0 \le x \le 0.6$ showed high Qxf values around 100000 GHz. The measured permittivity (ε) of the $P2_1/a$ compositions tended to increase with a decrease in the value of x. The same trend, but with lower values, was also observed for the theoretically calculated permittivity ($\varepsilon_{calc.}$) (Table II).[14] In contrast to the $P2_1/a$ phase, the high-temperature-stable $C2/m$ phase of the compositions with $0.67 \le x \le 1$ easily sintered to a high density without any sintering aids. The measured and theoretically determined dielectric constants of the $C2/m$ compositions were higher, while the Qxf values were significantly lower than those of the $P2_1/a$ phase (Table II, III). The crystal-structure refinements of both phases using the X-ray Rietveld method, the results of which are published in reference 12, reveal that the higher permittivity of the $C2/m$ phase is most probably related to a longer M-O distance (M= K, Ba), which means that the M ions have more space in the $C2/m$ structure to vibrate under the periodic force of the electric field. This then results in a higher permittivity. Both modifications, regardless of the composition, exhibited a low temperature coefficient of resonant frequency of around -20 ppm/°C (Table II and III).

Table II: Dielectric properties of the **$P2_1/a$** $K_xBa_{1-x}Ga_{2-x}Ge_{2+x}O_8$ solid solutions determined in the microwave frequency range (at ~13 GHz)

Composition $P2_1/a$	Ts (°C)	Additional annealing	Relative density	ε	$\varepsilon_{calc.}$*	Qxf (GHz)	τ_f (ppm/°C)
$KGaGe_3O_8$	890	-	0.69	4.0	4.49	10210	-
$KGaGe_3O_8$+0.3 wt. % H_3BO_3	880	840°C/40h	0.97	5.5	-	56240	-23
$K_{0.9}Ba_{0.1}Ga_{1.1}Ge_{2.9}O_8$	900	-	0.84	4.7	4.58	10660	-18
$K_{0.9}Ba_{0.1}Ga_{1.1}Ge_{2.9}O_8$+0.1wt.%$H_3BO_3$	910	-	0.97	5.6	-	104500	
$K_{0.67}Ba_{0.33}Ga_{1.33}Ge_{2.67}O_8$	970	-	0.98	6.0	4.76	119000	-25
$K_{0.67}Ba_{0.33}Ga_{1.33}Ge_{2.67}O_8$	970	900°C/40h	0.98	6.0	-	134000	
$K_{0.67}Ba_{0.33}Ga_{1.33}Ge_{2.67}O_8$+0.1wt.% H_3BO_3	920		0.97	6.0	-	96700	
$K_{0.6}Ba_{0.4}Ga_{1.4}Ge_{2.6}O_8$	970	-	0.98	6.1	4.83	109000	
$K_{0.6}Ba_{0.4}Ga_{1.4}Ge_{2.6}O_8$	970	900°C/70h	0.98	6.1	-	120900	-23
$K_{0.4}Ba_{0.6}Ga_{1.6}Ge_{2.4}O_8$	1040	-	0.97	6.4	4.99	94700	-23
$BaGa_2Ge_2O_8$	1100	-	0.98	7.0	5.34	106400	-25

*The theoretical permittivity of $P2_1/a$ $K_xBa_{1-x}Ga_{2-x}Ge_{2+x}O_8$ solid solutions were calculated with the Clausius-Mosotti equation using the ionic polarizability values determined by Shannon.[14]

Table III: Dielectric properties of the **C2/m** $K_xBa_{1-x}Ga_{2-x}Ge_{2+x}O_8$ solid solutions determined in the microwave frequency range (at ~13 GHz)

Composition C2/m	T_s (°C)	ε	$\varepsilon_{calc.}$*	Qxf (GHz)	τ_f (ppm/°C)
$KGaGe_3O_8$	970	6.2	4.53	19800	-21
$K_{0.9}Ba_{0.1}Ga_{1.1}Ge_{2.9}O_8$	990	6.6	4.6	12680	-21
$K_{0.67}Ba_{0.33}Ga_{1.33}Ge_{2.67}O_8$	1020	6.9	4.78	32660	-27

*The theoretical permittivity of C2/m $K_xBa_{1-x}Ga_{2-x}Ge_{2+x}O_8$ solid solutions were calculated with the Clausius-Mosotti equation using the ionic polarizability values determined by Shannon.[14]

CONCLUSIONS

The P2$_1$/a modification of the $K_xBa_{1-x}Ga_{2-x}Ge_{2+x}O_8$ solid solutions was successfully synthesized either by a direct solid-state reaction ($0 \leq x \leq 0.67$) or by post-annealing techniques via the C2/m$\Rightarrow$P2$_1$/a phase transition ($0.6 \leq x \leq 1$). The P2$_1$/a phases of $KGaGe_3O_8$ and $K_{0.9}Ba_{0.1}Ga_{1.1}Ge_{2.9}O_8$ were obtained by isothermal annealing of the C2/m phase at 840 and 860°C, respectively. Annealing above the P2$_1$/a$\Rightarrow$C2/m phase transition led to the C2/m phase for $0.4 \leq x \leq 1$. A dielectric characterization revealed that the P2$_1$/a modification exhibited a slightly lower permittivity, but significantly higher Qxf values compared to the same composition with the C2/m structure. Both phases showed an increase of the permittivity with a decrease of x. The $K_{0.9}Ba_{0.1}Ga_{1.1}Ge_{2.9}O_8$ and $K_{0.67}Ba_{0.33}Ga_{1.33}Ge_{2.67}O_8$ compositions met the requirements of a low permittivity (ε=5.6–6), high Qxf values (96700–134000 GHz), a small temperature coefficient of resonant frequency (~-20 ppm/°C) and a low sintering temperature (970°C (x=0.67), 910°C (x=0.9)), which was for x=0.67 further lowered to 910–920°C by the addition of 0.1 wt. % of H_3BO_3. The latter did not cause a significant deterioration in the dielectric properties.

REFERENCES:

[1] N. Mori, Y. Sugimoto, J. Harada and Y. Higuchi, "Dielectric properties of new glass-ceramics for LTCC applied to microwave or millimeter-wave frequencies", J. Eur. Ceram. Soc., 26 (2006) 1925-1928.

[2] M. Kono, H. Takagi, T. Tatekawa and H.Tamura, "High Q dielectric resonator material with low dielectric constant for millimeter-wave applications", J. Eur. Ceram. Soc., 26 (2006) 1909-1912.

[3] S. X. Dai, R. F. Huang and Sr. D. L Wilcox, "Use of titanates to achieve a temperature-stable low-temperature cofired ceramic dielectric for wireless applications, " J. Am. Ceram. Soc., 85 (2002) 828-832.

THE EFFECT OF A-SITE VACANCIES ON CELL VOLUME AND TOLERANCE FACTOR OF PEROVSKITES

Rick Ubic, Kevin Tolman, Kokfoong Chan, Nicole Lundy
Boise State University
Boise, Idaho, USA

Steven Letourneau and Waltraud Kriven
University of Illinois at Urbana-Champaign
Urbana, IL, USA

ABSTRACT

Point defects like vacancies can have a profound effect on the structure of perovskite ceramics, but the exact mechanisms by which they do this are unclear. A predictive model for the pseudocubic lattice constant of perovskites, based solely on published ionic radii data, has been developed and adapted as a model for tolerance factor. These models more consistently predict both pseudocubic lattice constant, hence cell volume, and the tolerance factor than existing methods, thus also more accurately modeling the temperature coefficient of resonant frequency (TCF). The relationship between tolerance factor and TCF is revisited.

INTRODUCTION

Perovskites abound both in nature and in the laboratory, and their wide compositional range renders a variety of useful properties such that perovskites are encountered in applications as disparate as electroceramics, superconductors, refractories, catalysts, magnetoresistors, and proton conductors. They are also of interest for use as substrates or buffer layers for compound semiconductor heteroepitaxy. The design of such advanced materials requires an understanding of the relationship between chemical composition and crystal structure. Perovskite oxides have the general formula ABO_3, and complex perovskites of the type $A(B'B'')O_3$ are now well-established materials for microwave applications because of their high quality factors (Qf) and low temperature coefficients of resonant frequency (τ_f). These materials are typically engineered with various dopants on both A and B sites and often contain unwanted defects such as vacancies, especially on the anion sublattice.

A recent study[1] established an empirical model relating the ionic radii to the pseudocubic lattice constant with an average absolute error of just 0.60%:

$$a_{pc} = 0.06741 + 0.49052(r_A + r_X) + 1.29212(r_B + r_X) \tag{1}$$

where r_A, r_B, and r_X are the various ionic radii assuming sixfold coordination. A later study[2] used a similar approach in modeling the lattice constants of orthorhombic perovskites in space group *Pbnm*, which accounts for >90% of observed orthorhombic perovskites. The equations so derived yield results with absolute relative errors of 0.616%, 1.089%, and 0.714% for lattice constants a, b, and c, respectively.

The perovskite tolerance factor is generally defined geometrically as:

$$t = \frac{r_A + r_O}{\sqrt{2}(r_B + r_O)} \tag{2}$$

where r_A, r_B, and r_O are the ionic radii of the ionic species on the A, B, and O sites, respectively. By assuming that the cation and anion sublattices are closely packed, the tolerance factor can be viewed as a measure of distortion of oxygen octahedra. The relationship between tolerance factor and the pseudocubic perovskite lattice constant is discussed at some length in reference 1. Reaney *et al.*[3] established a relationship between tolerance factor and the temperature coefficient of permittivity (τ_ε) by which it was determined that changes in τ_ε were closely correlated to octahedral tilt transitions in niobate and tantalate perovskites. They observed that perovskites with $t < {\sim}0.985$ contained axes about which oxygen octahedra were tilted in an anti-phase arrangement, causing cell doubling in the three pseudocubic directions. Similarly, perovskites for which $t < {\sim}0.965$ undergo a further tilt transition whereby octahedra are tilted in-phase about one or more axes as well. Perovskites for which $t > {\sim}0.985$ were not observed to contain a tilt superlattice.

Ubic *et al.*[4] derived an equation for tolerance factor based solely on pseudocubic lattice constant (which can either be experimentally measured or calculated via equation 1) and published values of r_B and r_X:

$$t = \frac{a_{pc} - 0.05444}{0.660460361(r_B + r_X)} \tag{3}$$

This expression, with r_X now in its proper twofold coordination, can be used to derive the correct tilt structure for complex perovskites like $Ba(Sm_{1/2}Sb_{1/2})O_3$, for which equation 2 fails, or simple perovskites like $CaTiO_3$ and $MgSiO_3$, the structures of both of which are incorrectly predicted via equation 2.

It has already been shown[5,6] that charge compensation in rare-earth-doped $SrTiO_3$ occurs via A-site vacancy formation via the reaction:

$$Ln_2O_3 \xrightarrow{\;SrTiO_3\;} 2Ln_{Sr}^{\bullet\bullet} + V_{Sr}'' + 3O_O^{x} \tag{4}$$

Ubic *et al.*[6] have shown by electron and neutron diffraction that the structure of $Sr_{1-3x/2}Ce_xTiO_3$ $(0.1333 \leq x < 0.4)$ is $R\bar{3}c$. Oxygen octahedra are tilted about the pseudocubic [111] by up to 4.7°. Extrapolation of these results suggests that octahedral tilting might start to occur at $0 < x \leq 0.013$. The incorporation of the trivalent species in particular is thought to stabilize the tilted structure despite the fact that the effective tolerance factor $t \geq 0.9895$ for the entire compositional range and only dips to 0.9833 at $x = 0.4$, at which point there is a structural phase transition, possibly to $C2/c$.[7]

In theory, equation 1 is only limited by the accuracy of r_A, r_B, and r_X values used, and equations 2 and 3 should yield equivalent results; however, the question then arises of how to calculate ionic radii when sites are shared by more than one species or are partially vacant. The concept of an "average" cation size seems at first meaningless; however, as the strains caused by each species will be averaged over the whole structure, it is not unreasonable to expect local relaxations to allow for the stabilization of an "average" structure. The assumption that vacancies are zero-dimensional defects inevitably leads to large errors in equations 1 and 2, especially at large vacancy concentrations. In order to understand and correct this problem, compositions have been engineered with exact concentrations of A-site vacancies. When Rietveld refinements of x-ray diffraction data are performed on these compounds, it is possible to back-calculate actual average bond lengths, from which the effective contribution (size) of vacancies to the average A-site ionic radius, $\bar{r}_A$, can be determined.

PROCEDURE

Five compositions in the system $Sr_{1-3x}La_{2x}TiO_3$ ($x = 0.01, 0.05, 0.1667, 0.2222$, and 0.25) were prepared via the mixed-oxide route. Stoichiometric amounts of $SrCO_3$, TiO_2 (99.9%, Aldrich Chemical Co., Milwaukee, WI), and La_2O_3 (99.9%, Alfa-Aesar, Ward Hill, MA) were ball milled in distilled water for four hours, using YSZ media in a high-density nylon pot. Slurries were dried, ground, and calcined at 1300°C for four hours. Approximately 2 wt% of polyethylene glycol (PEG 10,000, Alfa-Aesar, Ward Hill, MA) was added to the dried powders, which were then ground again into fine powder. Cylindrical pellets of about 3-4 mm in height and 10 mm in diameter were made by applying a pressure of 63 MPa. These compacts were then sintered for four hours at temperatures ranging from 1450 to 1650°C.

In order to avoid the appearance of the parasitic $La_2Ti_2O_7$ pyrochlore phase, compositions for which $x > 0.15$ were produced in a three-step process whereby precursor phases La_4SrO_7 and $SrTiO_3$ were produced via the mixed-oxide route:

$$2xLa(OH)_3 + 0.5xSrCO_3 \xrightarrow{1300°C,12hrs} 0.5xLa_4SrO_7 + 0.5xCO_2\uparrow$$

$$(1 - 3.5x)SrCO_3 + (1 - 3.5x)TiO_2 \xrightarrow{1300°C,4hrs} (1-3.5x)SrTiO_3 + (1-3.5x)CO_2\uparrow$$

$$0.5xLa_4SrO_7 + (1-3.5x)SrTiO_3 + 3.5xTiO_2 \xrightarrow{1300°C,4hrs} Sr_{1-3x}La_{2x}TiO_3$$

In order to achieve a phase-pure La_4SrO_7 product, starting powders were first milled for four hours, dried, and calcined at 1300°C for four hours. Then the process was repeated twice with the same parameters. In the case of $SrTiO_3$, starting powders were milled for six hours prior to drying and calcining.

Powder samples were prepared for x-ray diffraction (D8 Discover, Bruker AXS, Madison, Wisconsin, USA) from post-calcined batches. Le Bail fits to x-ray diffraction data were conducted using DiffracPLUS TOPAS 4.2 (Bruker).

RESULTS

All the compositions processed were single-phase perovskites, as illustrated in Fig. 1. Although octahedral tilting was expected in all compositions, Le Bail fitting was conducted assuming an untilted, pseudocubic perovskite structure in $Pm\overline{3}m$. In this setting, it is assumed that r_B ($r_{Ti} = 0.605$)[8] is unaffected by A-site doping and, consequently, both r_O and $\bar{r}_A$ values can be extracted as simple functions of a_{pc}; for example:

$$r_O = 0.5a_{pc} - r_B$$
$$\bar{r}_A = \frac{\sqrt{2}a_{pc} - 2r_O}{2} = \left(\frac{\sqrt{2}-1}{2}\right)a_{pc} + r_B$$

When effective values of r_O are calculated from refined a_{pc} values, only a negligible, nonsystematic change is observed, attributed to measuring error. Oxygen ions are coordinated to two B-site cations and four A-sites; therefore, an increase in the number of A-site vacancies will necessarily lower the coordination of second-nearest neighbors and so might be expected to cause a slight decrease in the average oxygen ionic radius; however, the maximum decrease observed throughout this series is only 1.03% from the ideal value given by Shannon[8] for oxygen in twofold coordination (1.35 Å), so this effect can fairly safely be ignored.

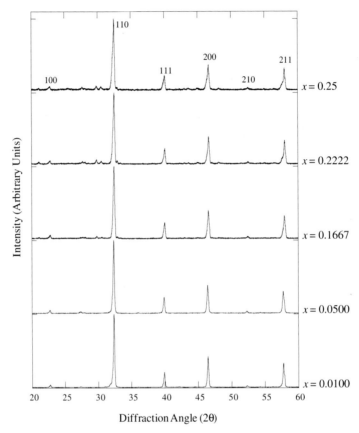

FIG. 1 – X-ray diffraction results of all compositions indexed according to the structure of pure cubic SrTiO$_3$.

Once values of $\bar{r}_A$ are obtained, it is possible to back-calculate the effective size of vacancies, r_V, from the stoichiometry by assuming the ionic sizes of Sr^{2+} (r_{Sr} = 1.44 Å) and La^{3+} (r_{La} = 1.36 Å) in twelve-fold coordination are as published by Shannon.[8] Two additional data points can be included from literature values of lattice constants corresponding $x = 0$ (SrTiO$_3$)[9] and $x = \frac{1}{3}$ (La$_{2/3}$TiO$_3$).[10] Such calculations reveal that the effective vacancy size increases with x and is negative for $x \leq 0.01$ (Table 1).

Table I. Measured pseudocubic latticed constants and calculated ionic radii values

x	a_{pc} (Å)	$\bar{r}_A$ (Å)	r_V (Å)	t (eqn. 3)	a_{pc} calc (Å)
0	3.9050	1.4138	--	1.0011	3.8983
0.0100	3.9010	1.4129	-1.1076	0.9980	3.8961
0.0500	3.9015	1.4130	1.0605	0.9984	3.8954
0.1667	3.8950	1.4117	1.4301	0.9934	3.8937
0.2222	3.8823	1.4091	1.4607	0.9836	3.8906
0.2500	3.8993	1.4126	1.4903	0.9967	3.8879
0.3333	3.9384	1.4207	1.5420	1.0270	3.8733

As Table 1 shows, vacancies have an appreciable effective size in this system. This effect might be explained by the electrostatic repulsion of the oxygen anions surrounding the vacancy. Considering that r_{Sr} = 1.44 Å and r_{La} = 1.36 Å, the values of r_V are generally a considerable fraction of the average size of the A-site cations, climbing even to 113% of the cation size for x = 0.3333 (although the unique layered structure of this compound almost certainly makes this value physically meaningless). The negative value for x = 01 corresponds to the relaxation of oxygen ions towards the vacant site, whereas the positive values correspond to mutual Coulombic repulsion of oxygen ions across the vacant site. These values fit well with the mathematical model published by Ubic et al.[4] in 2009. Without accounting for the finite size of vacancies, the relative errors in a_{pc} values generated via equation 1 range from $0.55\% \leq |\Delta a_{pc}| \leq 3.42\%$ for $0 \leq x \leq \frac{1}{4}$; however, when one accounts for the finite sizes of vacancies, the new predicted values of a_{pc} (Table 1) have relative errors of just $0.03\% \leq |\Delta a_{pc}| \leq 0.29\%$.

Using these refinement results, equation 3 would yield tolerance factors $0.9836 \leq t \leq 1.0270$ throughout the series, with the minimum occurring at x = 0.2222. In all but the case of x = 0.2222, the calculated tolerance factors would predict an untilted structure; however, it is known[11] that the rhombohedral $a^-a^-a^-$ tilt system is stabilized by highly charged A-site cations (e.g., La^{3+}) and small tilt angles.

CONCLUSIONS
Five compositions in the $Sr_{2-3x/2}La_xTiO_3$ homologous series corresponding to x = 0.01, 0.05, 0.1667, 0.2222, and 0.25 have been produced and characterized via x-ray diffraction. LeBail refinements show that A-site vacancies in this system have an effective size due to both bond relaxation and mutual repulsion of coordinating oxygen ions. Such vacancies also cause a very slight reduction in the radius of oxygen anions as a result of the lowering of their secondary coordination.

REFERENCES

[1]R. Ubic, Revised Method for the Prediction of Lattice Constants in Cubic and Pseudocubic Perovskites, *J. Am. Ceram. Soc.*, **90**, 3326-30 (2007).

[2]R. Ubic and G. Subodh, The Prediction of Lattice Constants in Orthorhombic Perovskites, *J. Alloys Compd.*, **488**, 374-379 (2010).

[3]I.M. Reaney, E.L. Colla, and N. Setter, Dielectric and Structural Characteristics of Ba- and Sr-Based Complex Perovskites as a Function of Tolerance Factor, *Jpn. J. Appl. Phys., Part 1*, **33**, 3984-90 (1994).

[4]R. Ubic, G. Subodh, M.T. Sebastian, D. Gout, and T. Proffen, Effective Size of Vacancies in the $Sr_{1-3x/2}Ce_xTiO_3$ Superstructure, *Ceram. Trans.*, **204**, 177-185 (2009).

[5]G. Subodh, J. James, M.T. Sebastian, R. Paniago, A. Dias, and R.L. Moreira, Structure and Microwave Dielectric Properties of $Sr_{2+n}Ce_2Ti_{5+n}O_{15+3n}$ ($n \leq 10$) Homologous Series, *Chem. Mater.*, **19**, 4077-82 (2007).

[6]R. Ubic, G. Subodh, M.T. Sebastian, D. Gout, and T. Proffen, Structure of Compounds in the $Sr_{1-3x/2}Ce_xTiO_3$ Homologous Series, *Chem.Mater.*, **20**, 3127-33 (2008).

[7]R. Ubic, G. Subodh, M.T. Sebastian, D. Gout, and T. Proffen, Structure of $Sr_{0.4}Ce_{0.4}TiO_3$, *Chem. Mater.*, **21**, 4706-4710 (2009).

[8]R.D. Shannon, Revised Effective Ionic Radii and Systematic Studies of Interatomic Distances in Halides and Chalcogenides, *Acta Cryst.*, **A32**, 751-767 (1976).

[9]H.E. Swanson and R.K. Fuyat, Natl. Bur. Stand. (U.S.), Circ. **539**, 3 44 (1954).

[10] Z.S. Gönen, D. Paluchowski, P.Yu Zavalii, B.W. Eichhorn, and J. Gopalakrishnan, Reversible Cation/Anion Extraction from $K_2La_2Ti_3O_{10}$: Formation of New Layered Titanates, $KLa_2Ti_3O_{9.5}$ and $La_2Ti_3O_9$, *Inorg. Chem.*, **45**, 8736-8742 (2006).

[11]P.M. Woodward, Octahedral Tilting in Perovskites. II. Structure Stabilizing Forces, *Acta Cryst.*, **B53**, 44-66 (1997).

ACKNOWLEDGEMENTS

This work has been supported by the National Science Foundation through the Major Research Instrumentation Program, Award Number 0619795, and DMR 1052788.

SINTERING EFFECTS ON MICROSTRUCTURE AND ELECTRICAL PROPERTIES OF CaCu$_3$Ti$_4$O$_{12}$ CERAMICS

S. Marković[1], M. Lukić[1], Č. Jovalekić[2], S.D. Škapin[3], D. Suvorov[3] and D. Uskoković[1]

[1]Institute of Technical Sciences of the Serbian Academy of Sciences and Arts, Belgrade, Serbia
[2]Institute for Multidisciplinary Research, Belgrade, Serbia
[3]Jožef Stefan Institute, Ljubljana, Slovenia

ABSTRACT

CCTO powders were prepared by solid state reaction and mechanochemically, respectively. Synthesized powders were characterized by XRD, FE–SEM and PSA techniques. The sinterability of CCTO powders was investigated by heating microscopy. Powders were uniaxially pressed into pellets and sintered up to 1100 °C, with heating rates of 2, 5, 10 and 20 °/min. The recorded shrinkage curves were used for choosing conventional and two step sintering (TSS) conditions. By TSS the samples were heated up to 1070 °C and after retention for 10 min cooled down to 1020 °C and kept for 20 h. The microstructure of CCTO ceramics sintered by conventional and TSS techniques was examined by FE–SEM method; the electrical properties were investigated in medium frequency (MF) range (42 Hz–5 MHz) and in the microwave (MW) range of frequencies. Electrical properties of the sintered CCTO ceramics were correlated to the samples microstructure. Finally, we have shown that appropriate choice of sintering conditions is important for preparation of high-quality CCTO ceramics with high dielectric permittivity in the kilohertz range as well as at the resonant frequency.

INTRODUCTION

Calcium copper titanate (CaCu$_3$Ti$_4$O$_{12}$, CCTO) belongs to a group of ACu$_3B_4$O$_{12}$ perovskite-type compounds,[1] which attracted ever-increasing attention for its practical applications in microelectronics, especially for preparation of capacitors and memory devices. CCTO ceramics are very attractive because of their giant dielectric constant ($\sim10^4$–10^5) in the kilohertz region at room temperature, and their good stability over a wide temperature range from 100 to 600 K. Reasons for interesting electrical properties of CCTO-type materials are not fully understood, and during the years different theoretical models and suggestions have been proposed to explain this behavior. However, since the giant dielectric properties primarily depend on the grain boundary resistivity while an intrinsic mechanism is excluded, ceramic microstructure (i.e. the average grain size and pellet density) and processing conditions (such as the oxygen partial pressure, sintering temperature, and cooling rate) are very important.[2,3] It is shown that varying sintering parameters desired electrical properties could be tailored.[4] Solid state reaction and mechanochemical treatment have been widely used as simple and fast synthesis methods. Having this in mind, designing appropriate processing route for such CCTO powders to achieve dense electroceramics with high dielectric permittivity and its stability in wide temperature range is desired. Until now, TSS method has been used for processing of sol-gel derived CCTO ceramics and method applicability for grain growth suppression at level of 6 μm is shown.[5]

In this study, CCTO powders were prepared by solid-state reaction and mechanochemically, respectively. Synthesized powder was characterized by XRD, FE–SEM and PSA techniques. The sinterability of CCTO powders was investigated by heating microscopy. Powder was uni-axially pressed into pellets ($\varnothing$ 6.7 mm) and sintered up to 1100 °C, with heating rate of 2, 5, 10 and 20

°/min. The recorded shrinkage curves were used for choosing of further sintering conditions, conventional as well as two-step sintering. The microstructure and electrical properties of sintered ceramics were examined and obtained results were correlated.

EXPERIMENTAL PROCEDURE

The polycrystalline calcium copper titanate ($CaCu_3Ti_4O_{12}$, CCTO) powders were prepared by two different techniques: (1) solid-state reaction and (2) mechanochemically by high-energy ball milling. In both cases the starting materials were commercially available powders of high purity: calcium carbonate ($CaCO_3$) (>99 %, Centrohem, Serbia), copper oxide (CuO) (>99%, Kemika, Zagreb) and titanium oxide (TiO_2) (>99.8%, Ventron GmbH, Germany) in a stoichiometric ratio $CaCO_3$–$3CuO$–$4TiO_2$. For solid-state reaction synthesis, mixture of the starting powders was homogenized during 24 hours in isopropanol under constant stirring at 1000 rpm. After that powder slurry was filtered, dried and calcined at 1000 °C for 12 hours. Calcined powder was grinded in agate mortar with addition of isopropanol and XRD pattern was recorded. The process of calcination and grinding was repeated several times until the formation of CCTO compound was confirmed by XRD studies. For the mechanochemical synthesis the powders mixture was grinded in a Fritsch Pulverisette 5 planetary ball mill. Milling was performed with tungsten carbide balls (Ø 10 mm) and sealed vials (250 ml). The mass of the powder was 6.5 g and the balls-to-powder mass ratio was 40:1. The milling was done in air atmosphere without any additives for 5 h. The angular velocities of the supporting disk and vials were 317 and 396 rpm, respectively. The milling vessels were opened every hour for removing the carbon dioxide overpressure, since the overpressure of gases in the vessel reduced the efficiency of milling.

The reaction occurring at high temperature or during milling can be summarized as:

$$CaCO_3 + 3\, CuO + 4\, TiO_2 \rightarrow CaCu_3Ti_4O_{12} + CO_2$$

CCTO powders synthesized by solid-state reaction and mechanochemically were denoted as CCTO_ss and CCTO_mc, respectively.

The phase composition of the synthesized CCTO powders and sintered ceramics was determined according to X-ray diffraction (XRD) data. XRD patterns were obtained at room temperature using a Philips PW-1050 automated diffractometer with Cu tube ($\lambda_{CuK\alpha}$ = 1.54178 Å); the X-ray generator operated at 40 kV and 20 mA. The diffraction measurements were done over scattering angle 2θ from 10 to 70 ° with a scanning step size of 0.05 °. JCPDS database[6] was used for phase identification. The particle size distribution was determined by the particle size analyzer Mastersizer 2000 (Malvern Instruments Ltd., UK).

The synthesized CCTO powders were uniaxially pressed in die (Ø 6.7 mm) under a pressure of 400 MPa; each of compacts has the thickness *circa* 2 mm. Green compacts with 60±2 % TD were prepared. The sintering of the green bodies was carried out in a Protherm tube furnace in air atmosphere, by both, conventional (CS) and two-step sintering (TSS) methods, respectively. At first, in the aim of finding the appropriate conditions for sintering, the non-isothermal sintering of the CCTO compacts was done in a heating microscope with automatic image analysis (New Heating Microscope EM201, Hesse Instruments, Germany). The non-isothermal experiments were performed in air up to 1100 °C, using a heating rate of 2, 5, 10 and

20 °/min. The recorded shrinkage curves were used for choosing sintering conditions for further conventional (CS) and two-step sintering (TSS) experiments in the aim to prepare dense ceramics with good dielectric properties. The CS was done by the heating rate of 5 °/min up to 1050 °C; the dwell time was 12 hours. Furthermore, for TSS, the samples were heated up to T_1 (1070 °C) and after retention for 10 min at T_1, the samples were cooled down to T_2 (1020 °C) and, subsequently, kept in the second-step temperature for 20 h. The heating rate of TSS was 5 °/min, while the cooling rate, between T_1 and T_2, was 50 °C min^{-1} and after T_2, samples were naturally cooled down with the furnace to room temperature. The density of the sintered samples was estimated according to Archimedes' principle, and listed in Table I.

The morphology of the used CCTO powders and microstructure of the sintered ceramics were analyzed by field emission scanning electron microscopy (FE–SEM, Supra 35 VP, Carl Zeiss). Before the analysis, the powders were dispersed in ethanol, filtered, and carbon coated, while the sintered samples were polished and thermally etched at 1000 °C for 10 min, and afterwards carbon coated. The obtained micrographs were used for the estimation of the average grain size with a SemAfore digital slow scan image recording system (JEOL, version 4.01 demo).

The electrical characterization of CCTO ceramics was performed in medium frequency (MF) and microwave (MW) regions. For impedance measurements in MF region HIOKI 3532-50 LCR HiTESTER was used. The measurements were done in frequency interval 42 Hz–5 MHz, in air atmosphere, during cooling from 500 to 25 °C; an applied ac voltage was 1 V. As electrodes, high conductivity silver paste was applied onto both sides of the samples. The resonance measurements in the 10 MHz–67 GHz range were done in a conventional set up using a HP E8361C PNA Network Analyzer; measurements were done on samples without electrodes.

RESULTS AND DISCUSSION
XRD patterns of CCTO powders synthesized by solid-state reaction and mechanochemically are presented in Figure 1. The XRD measurements confirmed a perovskite phase which has the centrosymmetric structure with cubic space group *Im-3* (according to JCPDS 75-2188);[6,7] besides perovskite phase, sample prepared by solid state reaction (CCTO_ss) possesses small amount of CuO and TiO_2 while CCTO_mc sample is single-phased. The broad reflections in the XRD pattern of CCTO_mc powder indicate low crystallinity and small crystallite size, what is typical for powders prepared mechanochemically i.e. by high energy ball-milling;[8] quite contrary, high temperature of the solid-state reaction, 1000 °C, yields XRD pattern with narrower diffraction maximums which indicates better crystallinity and increase in crystallite size of CCTO_ss powder.

Furthermore, CCTO powders morphology was examined by FE–SEM analysis, Figure 2. CCTO_ss powder contained nonuniform particles with sizes from 1 to 10 μm, consolidated in large condensed agglomerates with sizes rising up to 50 μm. Large, strong agglomerates with partially sintered particles are consequence of high temperature of solid-state reaction. FE–SEM micrograph of CCTO_mc powder shows nonuniform soft agglomerates with sizes from 200 to 800 nm. Evidently, the agglomerates are consisted of particles smaller of 100 nm in average size.

While FE–SEM was used for the observation of powders morphology and estimation of particles and agglomerates sizes, the average particle size and particle size distribution were studied in details by a laser particle size analyzer (PSA). However, it is known that the success of PSA technique depends on the dispersion of the powder. Since the synthesized powders are agglomerates of primary nanoparticles (found according to the FE–SEM micrographs), it was difficult to disperse them as individual particles; and hence, in spite of real particle size, the results presented in Figure 3 indicate size (and distribution) of agglomerates.

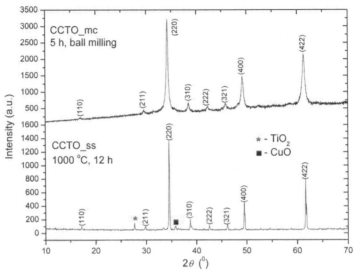

Figure 1. XRD patterns of CCTO powders synthesized by solid state reaction (CCTO_ss) and mechanochemically (CCTO_mc).

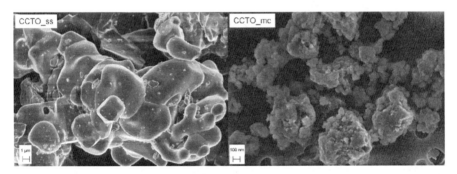

Figure 2. FE–SEM micrographs of CCTO powders synthesized by: solid state reaction (CCTO_ss) and mechanochemically (CCTO_mc).

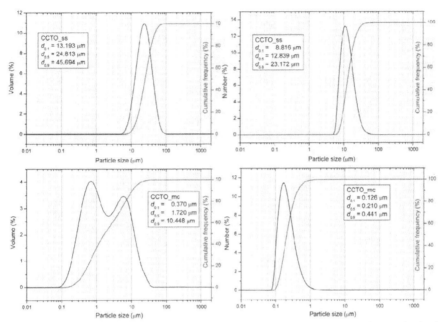

Figure 3. Particle size distribution (based on volume and number) of CCTO powders synthesized by solid state reaction (CCTO_ss) and mechanochemically (CCTO_mc).

From Figure 3, for CCTO_ss, it can be seen that particle size distributions, according to volume and number, are very narrow (*span* = 1.30 and 1.17, respectively), indicating uniform powder. From volume distribution can be seen that CCTO_ss powder possesses particles of about 25 μm in average ($d_{0.5}$), 10 % of particles ($d_{0.1}$) is smaller than 13 μm, while 90 % of particles ($d_{0.9}$) is smaller than 46 μm. Furthermore, particle size distribution according to the number shows that 10 % of particles is smaller than 8.8 μm, average particle size is 12.8 μm, while 90 % of particles is smaller than 23 μm. Thus, CCTO powder synthesized by solid-state reaction has large particles; furthermore, it can be supposed that, in average, agglomerates are consisted of two interconnected particles (grains). CCTO_mc powder shows wide (*span* = 5.86) bimodal particle size distribution based on volume. Average agglomerate size of one fraction is about 600 nm while second agglomerate fraction is about 7 μm in average. Generally, 10 % of particle agglomerates is smaller than 0.370 μm, 90 % is smaller than 10.5 μm while average size is about 1.7 μm. Quite contrary, particle size distribution according to the number is narrow (*span* = 1.50); 10 % of particles is smaller than 0.120 μm, average particle size is 0.210 μm, while 90 % of particles is smaller than 0.440 μm. Such difference between distribution according to the volume and number can be explained by a large number of small particles in CCTO_mc powder. It can be supposed that small particles (below 100 nm) are arranged in agglomerates that are basically soft in nature; besides, attractive forces between particles are weak van der Waals forces that can be easily broken down by low-intensity ultrasound.

In the second step of our investigation, the synthesized CCTO powders were pressed and sintered to prepare dense ceramics with good dielectric properties. At first, in the aim of finding the appropriate conditions for sintering, the non-isothermal sintering of the CCTO compacts was done in a heating microscope. A heating microscope was used for detailed quantitative studies of sintering as well for *in situ* monitoring of the shrinkage process; the sintering shrinkage of cylindrical compact area (A) was recorded. From the experimental data for the area recorded at 2-s time intervals during non-isothermal sintering and using equation (1), the percentage of shrinkage was calculated:

$$shrinkage \ (\%) = \frac{\Delta A}{A_o} \times 100 \qquad (1)$$

where ΔA ($=A_o-A_i$) denotes the difference between the initial value of area A_o at time t_o and the values A_i at time t_i. The calculated values of shrinkage were used for the determination of the samples' sintering behavior.

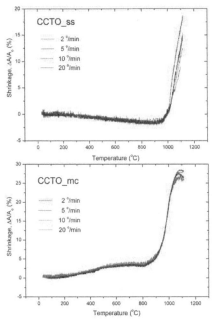

The densification of CCTO cylindrical compacts during non-isothermal sintering up to 1100 °C, using a heating rate of 2, 5, 10 and 20 °/min is represented by shrinkage curves of the samples area *versus* temperature, Figure 4.

It can be noticed that, each of the powders shows relatively similar shrinkage curves, independent on heating rate. For CCTO_ss is obvious that sintering starts at different temperatures depending on heating regime; actually, densification starts at 950 °C for slower heating and at 1050 °C for faster heating regimes i.e. delayed onset on densification with increased heating rate exists. Furthermore, CCTO_ss powder does not reach final densification stage until 1100 °C. Quite contrary, sintering curves of CCTO_mc powder are practically overlapped yielding that densification of the powder is independent on heating rate. This fact implicate good powder' sinterability. Besides, shrinkage curves show that the main densification starts at the 800 °C and reaches final stage around 1070 °C when starts to expand, probably due to forming of some eutectic melting with alumina substrate.

Figure 4. Shrinkage curves for CCTO_ss and CCTO_mc.

Thus, it can be concluded that CCTO_mc has better sinterability comparing to CCTO_ss powder, which is a consequence of powders characteristics, actually, crystallinity, average particle size and nature of agglomerates.

In the following part, according to the results of non-isothermal sintering, the conventional and two-step sintering experiments were done. Conditions for CS are chosen in the manner to obtain dense ceramics avoiding as much as possible final stage grain growth; that temperature was 1050 °C with dwell time of 12 h. Additionally, TSS sintering conditions were chosen based on sintering curves. CCTO_ss powder has very narrow temperature range of densification from 1000 to 1100 °C, while CCTO_mc powder reaches final sintering stage around 1070 °C. Since TSS method can be used for microstructural refinement and grain growth suppression, after first heating step samples should not reach final sintering stage where accelerated and uncontrolled grain growth may occur. The first sintering step should provide critical density at which interconnected pores start to collapse, while second step enables conditions for slow diffusion kinetics.[9] So, we chose 1070 °C with dwell time of 10 min for the first and 1020 °C for 20 h for the second sintering step. Heating rate of 5 °/min is chosen considering sintering curves of both CCTO_ss and CCTO_mc powders. The list of all experiments as well as characteristics of final ceramics are given in Table I.

Table I. Performed heating cycles and characteristics of final ceramics.

Powder	Heating cycle	T_1 (°C)	t_1 (min)	T_2 (°C)	t_2 (min)	Density (g/cm^3)	Average grain size (μm)
CCTO_ss	CS	1050	720	-	-	4.26	4.73
	TSS	1070	10	1020	1200	4.26	2.93
CCTO_mc	CS	1050	720	-	-	4.46	3.32
	TSS	1070	10	1020	1200	4.56	1.80

Furthermore, we used complex impedance spectroscopy (IS) to determine the electrical characteristics of CCTO ceramics tailored by different heating cycles, as well as to distinguish the grain-interior and grain boundary resistivity of the ceramics. It is known that *ac* impedance spectroscopy is a powerful tool in separating the grain-interior, grain boundary and electrode process of the ceramics.[10] In particular, a typical complex impedance diagram, in the so-called Nyquist presentation (the plot of imaginary, Z' *versus* real impedance, Z'', with the frequency f as an independent parameter) of a sintered, low-conducting material between blocking electrodes, consists of three parts (a bulk semicircle, a grain boundary semicircle, and an electrode arc), ending at the origin point of coordinate system at infinite frequency. For sintered perovskite materials, the high-frequency semicircle of the impedance spectra is attributed to the grain-interior impedance, the middle one is attributed to the grain boundary response, whereby the low-frequency arc (<0.1 Hz) corresponds to the electrode response.[11,12] The grain-interior and grain boundary resistance may be read as the diameter of appropriate arcs.

The room temperature complex impedance plots (Z^*, Z' *versus* Z'') for CCTO ceramic sintered in different heating cycles are shown in Figure 5. All Z^* plots exhibit only one arc with a large slope and the zero high frequency intercepts (see inset). Only one arc in the impedance plot is the consequence of the predominant grain boundary effect. It is known that the room temperature resistivity of perovskite materials depends mainly on the microstructural development associated with grain growth; precisely, resistivity increases with an increase of density and decrease of average grain size.[13]

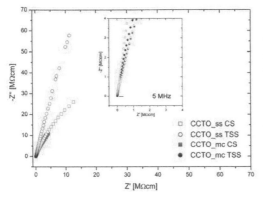

Here, it can be emphasized that IS spectra without point dissipation, indicating high-quality ceramics. This is especially important for electronic ceramics meaning no insulator interfaces (cracks and/or delamination) were produced during powders processing and high-temperature sintering.

Figure 5. Complex impedance spectra for CCTO sintered ceramics at room temperature.

From the complex impedance spectroscopy experimental data and using equation (2), the complex dielectric permittivity (ε^*) was calculated:

$$\varepsilon^* = \varepsilon' - i\varepsilon'' = \frac{1}{i\omega C_0 Z^*} \qquad (2)$$

where ω is the angular frequency, and C_0 is the empty cell capacitance.

Figure 6 shows behavior of complex dielectric permittivity at 1 kHz in temperature range from room temperature to 400 °C. For CCTO_mc ceramics, we found high dielectric permittivity, above 10000, at 1 kHz that is nearly constant from room temperature to 225 °C. It can be emphasized that CCTO_mc ceramics, especially TSS sample, could be very promising for capacitor applications. CCTO_ss ceramics show smaller dielectric permittivity but also stable in a wide temperature region. As it is previously stressed, a difference in dielectric permittivity between the CCTO ceramics is influenced by different microstructure.

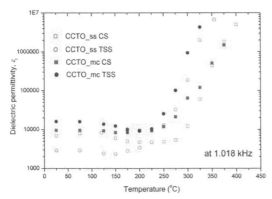

Figure 6. Dependence of dielectric permittivity on temperature, at frequency of 1.018 kHz.

Figure 7 shows the dielectric permittivity of the CCTO samples in the range from 42 Hz to 5 MHz measured at room temperature. It can be observed that dielectric constant decrease with increase of frequency to 1 MHz, while in the range of MHz is quite stable.

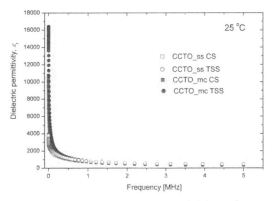

Figure 7. Dependence of dielectric permittivity on frequency measured at room temperature.

The main electrical characteristics of sintered ceramics at room temperature, in MF as well at resonant frequencies are listed in Table II.

Table II. Electrical characteristics of CCTO ceramics.

Sample	Dielectric permittivity at room temperature		Resonant frequency [GHz]	Quality factor [GHz]
	At 1 kHz	At resonant frequency		
CCTO_ss CS	6870	62.2	7.778	4256
CCTO_ss TSS	2917	61.4	6.872	1885
CCTO_mc CS	9566	62.5	7.968	2405
CCTO_mc TSS	16273	64.2	6.773	1310

Thus, according to the IS results, high specific resistivity of the CCTO at room temperature (without leakage currents) in addition to relative permittivity above 16000 at 1 kHz in a wide temperatures interval, make this sintered CCTO ceramics suitable for practical application as capacitors. Besides, the combination of capacitance behavior and high values of dielectric permittivity (above 60) in the MW range confirm the potential use of such materials for preparation of high dielectric planar antennas, with applicability in microelectronics or for microwave devices (for example cell mobile phones), where the miniaturization of the devices is crucial.[14]

In the final stage of this study, we studied the influence of heating regime on microstructure which is correlated with electrical properties. Figure 8 shows microstructure of CCTO ceramics prepared by different heating regimes. It is obvious that heating regimes affected microstructure; it impacts final density as well as average grain size. Estimated values of average grain sizes and final densities are shown in Table I. Here, it can be emphasized that CCTO_mc exhibits bimodal grain size distribution: for CS regime smaller fraction is around 2 μm and larger is around 5 μm; for the TSS regime one fraction is below 1 μm while there are larger grains of several micrometers. In both systems, TSS achieved microstructural refinement, decreasing average grains size for almost 50 %, while density improvement was obtained in the case of CCTO_mc powder. Better response of this powder to TSS method is correlated to powder morphology, much smaller average particle size as well as softer agglomerates, compared to CCTO_ss powder obtained by more severe synthesis conditions. Smaller average grain size of the CCTO_mc compared to CCTO_ss ceramics obtained by both CS and TSS methods influenced the value of dielectric permittivity at 1.018 kHz, which is several times higher. Furthermore, considering CCTO_mc system, average grain size reduction and density improvement obtained in TSS regime yielded to significant increase of dielectric permittivity at 1 kHz value.

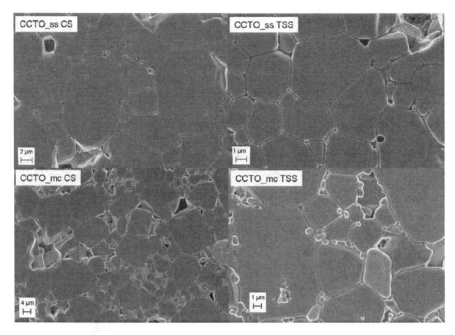

Figure 8. Microstructure of CCTO ceramics prepared from CCTO_ss and CCTO_mc powders by CS and TSS sintering regimes.

In order to confirm that differences in electrical properties of CCTO ceramics are determined by created microstructure rather than by different phase composition, XRD patterns of sintered ceramics are recorded, Figure 9. Obviously, all of the examined electronic ceramics possess the same phase composition corresponding to pure CCTO crystal phase.

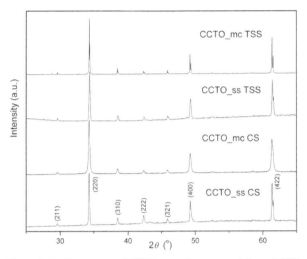

Figure 9. XRD patterns of CCTO ceramics prepared from CCTO_ss and CCTO_mc powders by CS and TSS sintering regimes.

CONCLUSION

CCTO powders with different characteristics were prepared by solid-state (CCTO_ss) and mechanochemical (CCTO_mc) methods. Their sintering behavior was investigated by heating microscopy with different heating rates and it is found that CCTO_mc powder possesses better sinterability which is connected with its more uniform powder morphology and existence of smaller, softly agglomerated particles in spite of CCTO_ss powder, consisting of hard aggregation of a large micrometer sized particles. Based on sintering behavior, experiments of conventional (CS) and two-step sintering (TSS) were carried out. It is shown that in the case of both powders, TSS yielded to microstructural refinement; in addition, for CCTO_mc density increase is found. Furthermore, electrical properties of sintered ceramics were investigated in MF as well in MW region. According to the results of MF measurements, the sintered CCTO ceramics has both, the high specific resistivity at room temperature and high relative permittivity at 1 kHz in a wide temperatures interval, which makes these materials suitable for practical application as capacitors. Besides, high values of dielectric permittivity (above 60) at resonant frequency are measured. The combination of capacitance behavior and high dielectric permittivity in the MW range promote the potential use of such materials for preparation of high dielectric planar antennas, with applicability in microelectronics

XRD measurements of investigated samples confirmed that difference in electrical characteristics is a consequence of microstructural changes rather than that of phase composition since only present phase in all samples corresponds to pure CCTO crystal phase. This fact emphasized the responsibility of sintering strategy development to prepare appropriate microstructure with desirable electrical properties.

ACKNOWLEDGEMENT

This study was supported by the Ministry of Education and Science of the Republic of Serbia under Grant No. III45004, and the bilateral cooperation program between the Republic of Serbia and the Republic of Slovenia under Grant No. 651-03-1251/2012-09/06.

REFERENCES:
[1] M.A. Subramanian, D. Li, N. Duan, B.A. Reisner, A.W. Sleight, High dielectric constant in $ACu_3Ti_4O_{12}$ and $ACu_3Ti_3FeO_{12}$ phases, Journal of Solid State Chemistry 151 (2000) 323–325.

[2] J. Liu, R.W. Smith, W.-N. Mei, Synthesis of the giant dielectric constant material $CaCu_3Ti_4O_{12}$ by wet-chemistry methods, Chemistry of Materials 19 (2007) 6020–6024.

[3] M.A. De la Rubia, P. Leret, J.bDe Frutos, J.F. Fernández, Effect of the synthesis route on the microstructure and the dielectric behavior of $CaCu_3Ti_4O_{12}$ ceramics, Journal of the American Ceramic Society 95 (2012) 1866–1870

[4] D.-L. Sun, A.-Y. Wu, S.-T. Yin, Structure, properties, and impedance spectroscopy of $CaCu_3Ti_4O_{12}$ ceramics prepared by sol–gel process, Journal of the American Ceramic Society 91 (2008) 169–173.

[5] S. K. Jo, Y. H. Han, Sintering behavior and dielectric properties of polycrystalline $CaCu_3Ti_4O_{12}$. Journal of Materials Science: Materials in Electronics 20 (2009) 680–684.

[6] JCPDS Database on CD-ROM, International Centre for Diffraction Data, Newton Square, PA, 1999.

[7] B. Bochu, M.N. Deschizeaux, J.C. Joubert, Synthèse et caractérisation d'une série de titanates pérowskites isotypes de $[CaCu_3](Mn_4)O_{12}$, Journal of Solid State Chemistry 29 (2) (1979) 291–298.

[8] S.K. Manik, S.K. Pradhan, Microstructure characterization of ball-mill-prepared nanocrystalline $CaCu_3Ti_4O_{12}$ by Rietveld method, Physica E 33 (2006) 160–168.

[9] I.-W. Chen, X.-H. Wang, Sintering dense nanocrystalline ceramics without final-stage grain growth. Nature 404 (2000) 168–171.

[10] J.E. Bauerle, Study of solid electrolyte polarization by a complex admittance method, J Phys Chem Solids 30 (1969) 2657–70.

[11] N. Hirose, A.R.West, Impedance spectroscopy of undoped BaTiO3 ceramics. J Am Ceram Soc 79 (1996) 1633–41.

[12] J.R. Macdonald, editor. Impedance Spectroscopy. New York/Chichester/ Brisbane/Toronto/Singapore: John Wiley & Sons; 1987.

[13] S. Marković, Č. Jovalekić, Lj. Veselinović, S. Mentus and D. Uskoković, Electrical properties of barium titanate stannate functionally graded materials, Journal of the European Ceramic Society 30 (2010) 1427–1435.

[14] A.F.L. Almeida, P.B.A. Fechine, L.C. Kretly, A.S.B. Sombra, $BaTiO_3$ (BTO)–$CaCu_3Ti_4O_{12}$ (CCTO) substrates for microwave devices and antennas, Journal of Materials Science 41 (2006) 4623–4631.

CRYSTAL CHEMISTRY AND PHASE DIAGRAMS OF THREE THERMOELECTRIC
ALKALINE-EARTH COBALTATE (Ca-M-Co-O, M=Sr, Zn and La) SYSTEMS

W. Wong-Ng
Materials Measurement Science Division, Material Measurement Laboratory, National Institute
of Standards and Technology, Gaithersburg, MD 20899.

ABSTRACT
    Research interest in waste heat conversion using thermoelectric materials has increased
substantially in recent years partly because of the increasing price of fuel and also due to the
effect of greenhouse gas emissions. For waste heat energy conversion applications, oxide
materials which have high temperature stability are potential candidates. Three phase diagrams
of selected systems that contain low dimensional alkaline cobaltates (layers or chains), as well as
the crystal chemistry and thermoelectric properties of selected thermoelectric oxides in these
systems are summarized. Examples of compounds in these systems that are of interest as
potential thermoelectric materials include $(Ca,M)_3Co_4O_9$ (M=Sr, La), $Ca_3(Co, M)_4O_9$ (M=Zn),
$A_{n+2}Co_nCo'O_{3n+3}$ (A= Sr, Ca), and $Ca_{n+2}(Co,Zn)_n(Co,Zn)'O_{3n+3}$.

INTRODUCTION
    Over the last decade, the increasing global interest in research and development of
thermoelectric materials was partly due to the soaring demand for energy and partly due to the
need to create a sustainable energy future. The efficiency and performance of thermoelectric
energy conversion or cooling is related to the dimensionless figure of merit (ZT) of
thermoelectric (TE) materials, given by $ZT=S^2\sigma T/\kappa$, where T is the mean absolute temperature, S
is the Seebeck coefficient or thermoelectric power, σ is the electrical conductivity, and k is the
thermal conductivity [1]. ZT is directly related to the coefficient of performance of a
thermoelectric material. Optimization of ZT values is not a straight-forward process because S,
σ, and κ are interrelated. For almost half a decade, only a small number of materials have been
found to have practical industrial applications and they all have ZT values around or below 1.0.
Recent reports that relatively high ZT values are possible in both thin film and bulk forms [2-7]
have revitalized interest in thermoelectric materials development.
    The stability of thermoelectric oxides at high temperature has rendered these materials
relevant to waste heat conversion applications. For example, the low dimensional cobaltates that
include $NaCoO_x$ [8], $Ca_2Co_3O_6$ [9, 10], and $Ca_3Co_4O_9$ [11-14] exhibit the coexistence of a large
Seebeck coefficient and relatively low thermal conductivity. Consequently in recent years,
considerable research has been conducted on thermoelectric oxides.
    Phase equilibrium diagrams provide road maps for processing and facilitate an
understanding of materials properties. Because of the promising properties of calcium-containing
cobaltate, Sr-, Zn- and La-doped calcium cobaltates may also offer desirable properties. This
paper summarizes the phase compatibility relationships, crystal chemistry, and crystallography
of three Ca-Co-O containing systems, namely, Ca-Sr-Co-O, Ca-Co-Zn-O, and Ca-La-Co-O
systems [15-18] that have been prepared at NIST. Thermoelectric property measurements have
also been conducted on selected phases.

CRYSTAL CHEMISTRY, CRYSTALLOGRAPHY, AND PHASE DIAGRAMS OF THE Ca-M-Co-O SYSTEMS

The Ca-Sr-Co-O system [15]

The Sr-Ca-Co-O system is an important system for thermoelectric research because it consists of two low dimension phases, namely, $(Ca,Sr)_3Co_4O_9$ with misfit layered structure, and $(Sr, Ca)_{n+2}Co_nCo'O_{3n+3}$ with one dimensional cobalt oxide chains that offer interesting thermoelectric properties. Figure 1 gives the phase diagram of the Sr-Ca-Co-O system that was determined at 850 °C in air [15]. The phase relationships between solid solutions and other phases are expressed as tie-line bundles. Because of the presence of various solid solutions in the Sr-Ca-Co-O system, series of tie-line bundles are constructed. The $(Ca,Sr)_3Co_2O_6$ phase is compatible with $(Ca, Sr)O$, $Ca_3Co_4O_9$ and with $(Sr_{0.7}Ca_{0.3})_4Co_3O_9$. The $(Ca, Sr)_3Co_4O_9$ phase is in equilibrium with $(Sr_{0.7}Ca_{0.3})_4Co_3O_9$, $(Sr_{2/3}Ca_{1/3})_5Co_4O_{12}$, $(Sr_{0.725}Ca_{0.275})_6Co_5O_{15}$, and CoO_x. The $Sr_{n+2}Co_nCo'O_{3n+3}$ series is in equilibrium with $(Sr, Ca)O$ and CoO_x.

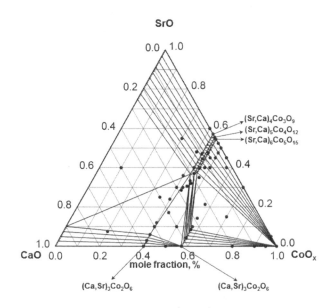

Fig. 1. Phase diagram of the Ca-Sr-Co-O system prepared at 850 °C [15].

The $Ca_3Co_4O_9$ phase has been extensively studied [11-14]. It is a misfit layered oxide that has two monoclinic subsystems with identical a, c, β, but different b [12]. The 1st subsystem consists of triple rock-salt layers of Ca_2CoO_3 in the ab plane while the second subsystem consists of a single CoO_2 layer, which has the CdI_2-type structure. This phase exhibits strong anisotropic thermoelectric properties in the ab-plane. The chemical formula can be written as $[Ca_2CoO_3]_{RS}[CoO_2]_{1.61}$, where RS represents the rock salt structure and 1.61 expresses the incommensurate character for the b parameter of the rock salt and the CdI_2-type structure.

$Co_3Co_2O_6$ (R-$3c$, $a = 9.0793(7)$ Å, and $c = 10.381(1)$ Å [20]; Fig. 2a) is the n = 1 member of the homologous series with the general formula of $A_{n+2}B_nB'O_{3n+3}$, where A is an alkali-earth element such as Ca, Sr, and Ba. B describes the cobalt ion inside the octahedral cage, and B' is the cobalt ion inside a trigonal prism. $Ca_{n+2}Co_nCoO_{3n+3}$ consists of 1-dimensional linear parallel $Co_2O_6^{6-}$ chains, built by successive alternating face-sharing CoO_6 trigonal prisms and CoO_6 octahedra along the hexagonal c-axis [21]. This face-sharing feature is in contrast with $Ca_3Co_4O_9$ [12] and $NaCo_2O_4$ [8] which consist of edge-sharing CoO_6 octahedra. The linear $Co_2O_6^{6-}$ chains of $Ca_3Co_2O_6$ consist of one CoO_6 octahedron unit alternating with one CoO_6 trigonal prism. Each $Co_2O_6^{6-}$ chain is surrounded by six other chains which form a hexagonal arrangement. These $Co_2O_6^{6-}$ chains are separated by octahedral-coordinated Ca^{2+} ions (Fig. 2a). The compounds $A_{n+2}Co_nCo'O_{3n+3}$ can also be considered as ordered intergrowths between the n = ∞ ($ACoO_3$) and n = 1 ($A_3Co_2O_6$) end members [22, 23]. We found that when A = Ca, only the n = 1 member, namely, $Ca_3Co_2O_6$, can be made.

The only phase found in the SrO-CoO$_x$ system is the homologous series $Sr_{n+2}Co_nCo'O_{3n+3}$. The stable $Sr_{n+2}Co_nCo'O_{3n+3}$ compounds for the relatively larger Sr (as compared to Ca in $Ca_{n+2}Co_nCoO_{3n+3}$) are those with $2 \leq n < 5$. The n = 1 member ("$Sr_3Co_2O_6$") cannot be prepared. The n = 2 member, $Sr_4Co_3O_{10}$, was reported to be isostructural with $Sr_4Ni_3O_{10}$ [24] (Fig. 2b).

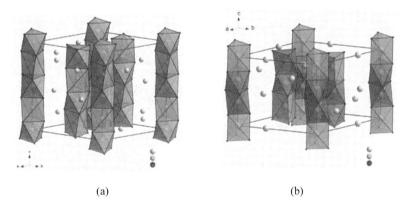

(a)   (b)

Fig. 2. General structure of the (a) n=1, and (b) n=2 members of the homologous series $(A,A')_{n+2}(Co,M)_n(Co',M')O_{3n+3}$ (A, A'= alkaline earth or rare-earth elements; M and M'=transition elements).

From X-ray diffraction, the space group and unit cell parameters of $Sr_4Co_3O_{10}$ are determined to be $P321$, $a = 9.5074$ Å, and $c = 7.9175$ Å. In the $Co_3O_{10}^{8-}$ chains, there are two units of CoO_6 octahedra alternating with a CoO_6 prism [23]. The structures for the n = 3 and n = 4 members, namely, $Sr_5Co_4O_{11}$ ($P3c1$, $a = 9.4$ Å, $c = 20.2$ Å [23], Fig. 3a) and $Sr_6Co_5O_{15}$ ($R32$, $a = 9.5035(2)$ Å, $c = 12.3966(4)$ Å [25], Fig. 3b) feature 3 and 4 octahedra interleaving with one trigonal prism along the c-axis, respectively. We were not able to make the n≥5 phases; however, the $(Ba_{0.5}Sr_{0.5})_7Co_6O_{18}$, $Ba_8Co_7O_{21}$ and $Ba_9Co_8O_{24}$ phases were reported by Boulahya et al. [22, 23]. Although the $BaCoO_3$ (n = ∞) structure with linear chains exists (Fig. 4) [22], the n = ∞ hexagonal Sr-analogue does not form under the conditions tested.

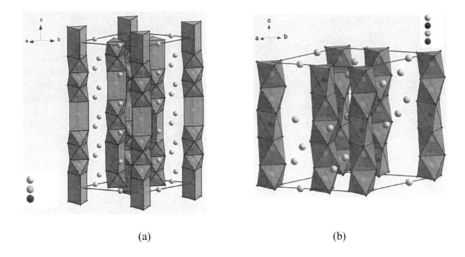

(a) (b)

Fig. 3. General structure of the (a) n=3, and (b) n=4 members of the homologous series $(A,A')_{n+2}(Co,M)_n(Co',M')O_{3n+3}$ (A, A'= alkaline earth or rare-earth elements; M and M'=transition elements).

There appears to be a general trend that larger alkaline-earth metals (A) are required in order to from stable $A_{n+2}Co_nCo'O_{3n+3}$ phases as n increases. According to Boulahya et al. [22, 23], these compounds can be considered as making up of A_3O_9 layers as well as linear parallel chains of cobalt oxides. In order to stabilize the n ≥ 5 members of the $A_{n+2}Co_nCo'O_{3n+3}$ series, it is necessary to increase the distance between the AO_3 layers by increasing the size of the cations in A positions.

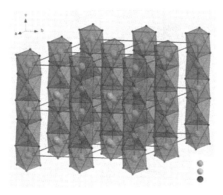

Fig. 4. General structure of the n=∞ member of the homologous series
$(A,A')_{n+2}(Co,M)_n(Co',M')O_{3n+3}$ (A, A'= alkaline earth or rare-earth elements; M and M'=transition elements).

No ternary oxide phase was found in the SrO-CaO-CoO_x system other than the three solid solution series as a result of the substitution of Ca into the Sr-Co-O compounds or Sr into the Ca sites of the Ca-Co-O compounds. The solid solution, $(Ca, Sr)_3Co_4O_9$, also features a misfit layered oxide that has an incommensurate structure [12]. Sr substitutes in the Ca site of $Ca_3Co_4O_9$ to the limit of $(Ca_{0.8}Sr_{0.2})_3Co_4O_9$. The second phase, $(Ca,Sr)_3Co_2O_6$, is the n = 1 member of the homologous series $(Ca,Sr)_{n+2}Co_nCo'O_{3n+3}$, where Sr substitutes in the Ca site to the limit of $(Ca_{0.9}Sr_{0.1})_3Co_2O_6$. In the third series, $(Sr, Ca)_{n+2}Co_nCo'O_{3n+3}$, Ca substitutes in the Sr site of the n = 2, 3 and 4 members to the limit of $(Sr_{0.7}Ca_{0.3})_4Co_3O_9$, $(Sr_{0.67}Ca_{0.33})_5Co_4O_{11}$ and $(Sr_{0.725}Ca_{0.275})_6Co_5O_{13.5}$, respectively. The detailed structure of the phases $(Sr_{0.8}Ca_{0.2})_5Co_4O_{12}$ and $Sr_6Co_5O_{15}$ and their reference x-ray powder diffraction patterns was reported [16].

The Ca-Co–Zn-O system [17]
Figure 5 gives the phase diagram of the Ca-Co-Zn-O system that was determined at 885 °C in air [17]. This system consists of $Ca_3(Co,Zn)_4O_9$ with misfit layered structure, and the n=1 phase of the $Ca_{n+2}(Co,Zn)_n(Co,Zn)'O_{3n+3}$ homologous series (Fig. 2a). These phases are isostructural to the $Ca_3Co_4O_9$ and $Ca_3Co_2O_6$ phases, respectively. The solid solution limits for these series were found to be extremely small, with end point at $Ca_3(Co_{1.95}Zn_{0.05})O_6$. Extended X-ray Absorption Fine Structure (EXAFS) experiments (provide local bonding environments), found that the main concentration of Zn is in the 6-coordination site with 2.11(2) Å (which is longer than the average octahedral Zn-O distance of 1.9111(5) Å). The 2.11(2) Å of Zn-O would suggest a bond length of a trigonal prism environment [17].

No binary compound was found in the CaO-ZnO system. The only phases found in the ZnO-CoO_z system are the two solid solutions formed in the vicinity of ZnO and CoO_z. Co substitutes

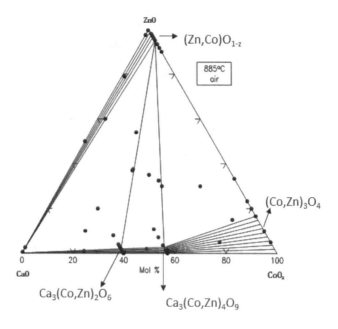

Fig. 5. Phase diagram of the Ca-Co-Zn-O system prepared at 885 °C.

into the Zn site of ZnO ($P6_3mc$, a=3.250 Å, c=5.207 Å [26]) to form $(Zn_{1-x}Co_x)O_{1-z}$ (0 <x≤ 0.06)), and Zn substitutes into the tetrahedral Co site of Co_3O_4 ($Fd\overline{3}m$, a=8.066 (3) Å [27]) to form $(Co_{1-x}Zn_x)_3O_{4-z}$ (0<x≤0.17, with the maximum substitution composition of $(Co_{2.49}Zn_{0.51})O_4$). The green-colored $(Zn_{1-x}Co_x)O$ (x=0.02) phase crystallizes in the hexagonal wurtzite type structure in which an extensive network of corner-shared ZnO_4 tetrahedra was found. Essentially Co was found to dope in the Zn site. The black-colored $(Co_{1-x}Zn_x)_3O_4$ (x=0.3) spinel phase has a cubic closed-packed spinel-structure. $(Co_{1-x}Zn_x)_3O_4$ is isostructural with Co_3O_4 which crystallizes in the space group $Fd\overline{3}m$ (a=8.087558(8) Å). In $(Co,Zn)_3O_4$, there are eight filled tetrahedral sites and 16 octahedral sites (8a and 16d) per unit cell. In one unit cell, there are 4 layers of CoO_6 octahedral chains along the c-axis, with neighboring layers perpendicular to each other. The CoO_4 tetrahedra are found between the CoO_6 octahedral chains (Fig. 6). As the common coordination of Zn is tetrahedral, Zn was indeed found to substitute solely in the tetrahedral 8a site in the structure of $(Co_{2.7}Zn_{0.3})O_4$.

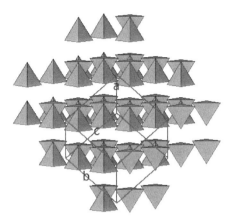

Fig. 6. Crystal structure of $(Co_{1-x}Zn_x)_3O_4$ showing the arrangement of the CoO_4 tetrahedra.

The phase relationships between various solid solutions and other phases in the Ca-Co-Zn-O system are expressed as tie-lines and tie-line bundles. The limit of the two solid solution series, $Ca_3(Co,Zn)_2O_6$ and $Ca_3(Co,Zn)_4O_9$, are $Ca_3Co_{1.9}Zn_{0.1}O_6$ and $Ca_3(Co_{3.85}Zn_{0.15})O_9$, respectively. Tie-line bundles are found to be between $Ca_3(Co,Zn)_4O_9$ and $(Co,Zn)_3O_4$, between $Ca_3(Co,Zn)_2O_6$ and $Ca_3(Co,Zn)_4O_9$, and between $Ca_3(Co,Zn)_2O_6$ and CaO. The end members of the $Ca_3(Co,Zn)_2O_6$ and $Ca_3(Co,Zn)_4O_9$ series were found to be compatible with the $(Zn_{1-x}Co_x)O_{1-z}$ solid solution series.

The Ca-La-Co-O system [18]

Figure 7 gives the phase diagram of the Ca-La-Co-O system that was determined at 885 °C in air. Our phase diagram is substantially different from that reported by Cherepanov et al. [28] which was determined at 1100 °C. At 885 °C, no intermediate phases form in the La_2O_3-CaO system other than the $(La,Ca)_2O_{3-z}$ solid solution. At 1100 °C [28] the solubility of CaO in La_2O_3 does not exceed 0.1 % mole fraction, and at 885 °C, the solubility limit was found to be $(La_{0.92}Ca_{0.08})_2O_3$. The only phase found in the La_2O_3-CoO_z system is $LaCoO_3$. There is no solid solution of the end members of this binary system, namely, $(La,Co)_2O_{3-z}$ or $(Co, La)_3O_4$.

The only two ternary phases in the Ca-La-Co-O system are $(Ca_{1-x}La_x)_3Co_4O_9$ (a small homogeneity range of $0 \leq x \leq 0.07$) and the pervoskite-type solid solution, $La_{1-x}Ca_xCoO_{3-z}$; however, attempts to substitute La ions for Ca in $Ca_3Co_2O_6$, failed in air at 885 °C. As the extent of solid solution depends on temperature, at 885 °C, the value of x in $La_{1-x}Ca_xCoO_{3-z}$ was determined to be about 0.2, whereas at 1100 °C, it was reported to be between 0.3 and 0.4 [28].

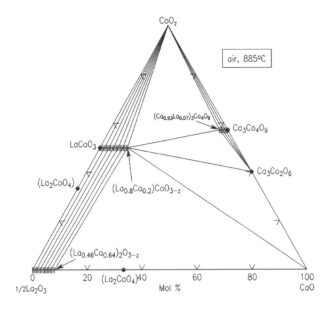

Fig. 7. Phase diagram of the Ca-La-Co-O system prepared at 885 °C [18].

There has been controversy about the structure of $La_{1-x}Ca_xCoO_{3-z}$. It has been reported to be rhombohedral $R-3c$ and the rhombohedral distortion decreases while x increases [29]. Recently, Wang et al. [30], using high resolution X-ray diffraction (XRD), reported that when the monoclinic cell $I2/a$ was used for the refinement of the structure, not only the refinement result was improved, but the space group was compatible with the Jahn-Teller distortion of Co^{3+}. Figure 7 gives the tie-line relationships of the Ca-La-Co-O system. There are a total of three solid solution series, two in the Ca-La-Co-O system, namely, $(Ca,La)_3Co_4O_9$ and $(La,Ca)CoO_{3-z}$, and one in the La_2O_3-CaO system, namely, $(La, Ca)_2O_{3-z}$. There are a total of four 3-phase regions, namely, CoO_z-$(La,Ca)CoO_{3-z}$-$(Ca,La)_3Co_4O_9$, $(La,Ca)CoO_{3-z}$-$(Ca,La)_3Co_4O_9$-$Ca_3Co_2O_6$, $(La,Ca)CoO_{3-z}$-$Ca_3Co_2O_6$-CaO, and $(La,Ca)CoO_{3-z}$-$(La,Ca)_2O_{3-z}$-CaO. The tie-line bundles extended from CoO_z to $(La,Ca)CoO_{3-z}$, from $(La,Ca)CoO_{3-z}$ to $(La, Ca)_2O_{3-z}$, from CoO_z to $(Ca,La)_3Co_4O_9$, and from $(Ca,La)_3Co_4O_9$ to $Ca_3Co_2O_6$.

THERMOELECTRIC PROPERTIES

Thermoelectric properties of a number of compounds in the Sr-Co-O and Ca-Co-O systems have been reported in literature [31-38]. In the Ca-Co-O system, the $Ca_3Co_4O_9$ phase was reported to exhibit strong anisotropic properties and has good thermoelectric property on the ab-plane. It has very low resistivity. ZT was reported to be ≈ 1 at 1000 K. It is thought that the increased scattering of phonons at the interface of misfit layers leads to the lowering of the lattice thermal conductivity. The Seebeck coefficient for single crystal $Ca_3Co_2O_6$ has been reported by Mikami and Funahashi to be relatively high and positive, and the thermal conductivity is relatively low at high temperature [10]. The transport properties are dominated mainly by p-type carriers. ZT was determined to be about 0.15 at 1000 K for a single crystal of $Ca_3Co_2O_6$ [10].

The Seebeck coefficient, thermal conductivity, and resistivity data of selected n = 2, 3 and n = 4 members of the $(Sr, Ca)_{n+2}Co_nCo'O_{3n+3}$ series, namely, $((Sr_{0.7}Ca_{0.3})_4Co_3O_9$, $(Sr_{0.8}Ca_{0.2})_5Co_4O_{12}$, and $(Sr_{0.87}Ca_{0.13})_6Co_5O_{15})$ as a function of temperature are given in Fig. 8a, 8b, and 8c. In general, the thermal conductivity of all three compounds appears to be low. The resistivity values indicate an activated dependence with temperature, similar to out-of-plane carrier transport observed in $Ca_3Co_4O_9$ [12] but with larger magnitudes in part due to the high porosity. Below 200 K the magnitudes are too large for the apparatus to acquire a measurement. The positive Seebeck coefficients suggest hole dominated conduction, with room temperature magnitudes between 110 μV/K and 140 μV/K (Fig. 8a). The magnitudes and temperature dependencies of the transport properties are similar in all 3 compounds. Although the $(Ca,Sr)_{n+2}Co_nCo'O_{3n+3}$ oxides show relatively high Seebeck coefficients and low thermal conductivities in general, their electrical resistivity values are relatively high.

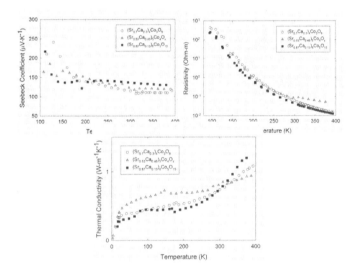

Fig. 8. (a) Seebeck coefficient, (b) resistivity, and (c) thermal conductivity for the n = 2 $((Sr_{0.7}Ca_{0.3})_4Co_3O_9)$, n = 3 $(Sr,Ca)_5Co_4O_{13})$, and n = 4 $((Sr_{0.87}Ca_{0.13})_6Co_5O_{15})$ members of $(Sr,Ca)_{n+2}Co_nCoO_{3n+3}$ [15].

The Seebeck coefficient, thermal conductivity, resistivity, and ZT values of $Ca_3Co_2O_6$, $Ca_3(Co, Zn)_2O_{6-z}$, $Ca_3(Co, Zn)_4O_{9-z}$ and $Ca_3(Co, Zn)_4O_{9-z}$ as a function of temperature are given in Fig. 9a to Fig. 9d for comparison. The transport properties are dominated mainly by p-type carriers, with high temperature magnitudes between 200 µV/K at 900 K and (400 µV/K to 700 µV/K) at 300 K for the $Ca_3(Co, Zn)_2O_{6-z}$ phase, and about 190 µV/K at between 300 K and 900 K for the $Ca_3(Co, Zn)_4O_{9-z}$ phase. The thermal conductivity data of the two solid solution series are similar and are reasonably low in the temperature range of 300 K to 900 K. Zn substitution does not seem to have much effect on the thermoelectric properties of the $Ca_3(Co, Zn)_4O_{9-z}$ series. Although Zn-doping lowers the resistivity of the $Ca_3(Co, Zn)_2O_{6-z}$ series between 300 K to 600 K, presumably due to the changes of carrier concentration, at 900 K the values converge. The $Ca_3(Co, Zn)_4O_{9-z}$ series has a higher ZT value than that of the $Ca_3(Co, Zn)_2O_{6-z}$ series. At present, compounds with the $Ca_3Co_4O_9$ structure type still give the highest ZT values among the oxides studied.

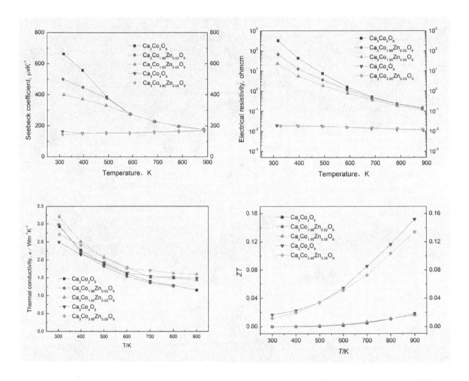

Fig. 9. (a) Seebeck coefficient,(b)resistivity, (c)thermal conductivity, and (d) figure of merit, ZT, of $Ca_3Co_2O_6$, $Ca_3(Co, Zn)_2O_6$, $Ca_3Co_4O_9$ and $Ca_3(Co, Zn)_4O_9$ as a function of temperature [17].

The Ca-La-Co-O system contains a remarkable lanthanum cobaltate phase, $La_{1-x}Ca_xCoO_{3-z}$. Wang et al. [30] found that there is a correlation between the perovskite structural distortion and the thermoelectric response in the solid solution, $La_{1-x}Ca_xCoO_{3-z}$. The ZT value exhibits a maximum at x=0.06 at room temperature (Fig. 10), which is a relatively high value.

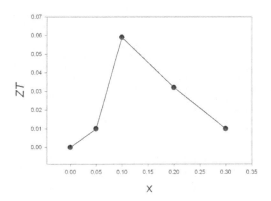

Fig. 10. ZT value of $(La_{1-x}Ca_x)CoO_{3-z}$ as a function of x [30].

SUMMARY AND FUTURE WORK

The three Ca-M-Co-O (M=Sr, Zn, and La) systems provide an important basis for thermoelectric research because they include two low dimensional phases, namely, $(Ca,M)_3Co_4O_9$ (M=Sr, La), $Ca_3(Co, M)_4O_9$ (M'=Zn) with misfit layered structure, and $(Ca, Sr)_{n+2}Co_nCo'O_{3n+3}$ and $Ca_{n+2}(Co,Zn)_n(Co,Zn)'O_{3n+3}$ phases with one dimensional cobalt oxide chains that offer promising thermoelectric properties. These three phase diagrams were determined at < 900 °C in air (below the melting temperatures of the two low dimensional phases) and they offer detailed compatibility relationships in the binary as well as the ternary oxide systems. The homogeneity range of various solid solutions have also been determined.

Although the homologous series, $(Ca, Sr)_{n+2}(Co,Zn)_n(Co,Zn)'O_{3n+3}$, which consists of linear chain structures shows relatively high Seebeck coefficient and low thermal conductivity in general, the resistivity values are relatively high; therefore unless one can decrease the resistivity of these linear chain type phases via either substitution or improved processing, the best cobaltate material for thermoelectric applications, at present, still has the $Ca_3Co_4O_9$ type of misfit layered structure. Continuation to determine the phase diagrams for additional R-Ca-Co-O systems, with R=Nd, Sm, Eu and Gd, in the near future has been planned.

REFERENCES

[1] G.S. Nolas, J. sharp, and H.J. Goldsmid, *Thermoelectric: Basic Principles and New Materials Developments* (Springer, New York, 2001).

[2] T. M. Tritt and M.A. Subramanian, guest editors, *Harvesting Energy through thermoelectrics: Power Generation and Cooling*, pp. 188-195, MRS Bulletin, Materials Research Soc. (2006).

[3] T. M. Tritt, *Science* **272**, 1276 (1996).

[4] Kuei Fang Hsu, Sim Loo, Fu Guo, Wei Chen, Jeffrey S. Dyck, Ctirad Uher, Tim Hogan, E. K. Polychroniadis, and Mercouri G. Kanatzidis, *Science* **303**, 818 (2004).

[5] R. Venkatasubramanian, E. Siivola, T. Colpitts, and B. O'Quinn, *Nature* **413**, 597(2001).

[6] S. Ghamaty and N.B. Eisner, Proceeding of Interpack: ASME Technical Conference on Packaging of MEMS, NEWS and Electric Systems, July 17-22, San Francisco, CA (2005).

[7] M.S. Dresselhaus, G. Chen, M.Y. Tang, R.G. Yang, H. Lee, D.Z. Wang, Z.F. Ren, J.P. Fleurial, and P. Gogna, Mater. Res. Soc. Symp. Proc. **886**, *Materials and Technologies for Direct Thermal-to-Electric Energy Conversion*, pp. 3-12 (2006).

[8] I. Terasaki, Y. Sasago, K. Uchinokura, *Phys. Rev.* B **56** 12685-12687 (1997).

[9] M. Mikami, R. Funashashi, M. Yoshimura, Y Mori, and T. Sasaki, *J. Appl. Phys.* **94** (10) 6579 - 6582 (2003).

[10] M. Mikami, R. Funahashi, *J. Solid State Chem.* **178** 1670-1674 (2005)

[11] D. Grebille, S. Lambert, F. Bouree, and V. Petricek, *J. Appl. Crystallogr.* **37** 823-831 (2004).

[12] A.C. Masset, C. Michel, A. Maignan, M. Hervieu, O. Toulemonde, F. Studer, and B. Raveau, *Phys. Rev* B **62** 166-175 (2000).

[13] H. Minami, K. Itaka, H. Kawaji, Q.J. Wang, H. Koinuma, and M. Lippmaa, *Appl. Surface Sci.* **197** 442-447 (2002).

[14] Y.F. Hu, W.D. Si, E. Sutter, and Q. Li, *Appl Phys. Lett.* **86** 082103 (2005).

[15] W. Wong-Ng, G. Liu, J. Martin, E. Thomas, N. Lowhorn, and M. Otani, J. Appl. Phys., **107** 033508 (2010).

[16] W. Wong-Ng, J. A. Kaduk, and G. Liu, Powder Diff. **26**, 22 (2011).

[17] W. Wong-Ng, T. Luo, M. Tang, M. Xie, J.A. Kaduk, Q. Huang, Y. Yang, M. Tang, and T. Tritt, J. Solid State Chem., **184**(8), 2159 (2011).

[18] W. Wong-Ng, W. Laws, Y.G. Yan, Solid State Sciences, in press (2012).

[19] W. Wong-Ng, Y.F. Hu, M.D. Vaudin, B. He, M. Otani, N.D. Lowhorn, and Q. Li, J. Appl. Phys, **102**(3) 33520 (2007).

[20] T. Takami, H. Ikuta, and U. Mizutani, Jap. J. Appl. Phys. **43** (22) 8208-8212 (2004).

[21] H. Fjellvag, E. Gulbrandsen, S. Aasland, A. Olsem, B. C. Hauback, J. Solid State Chem. **124** 190 (1996).

[22] K. Boulahya, M. Parras, and J.M. González-Calbet, J. Solid State Chem. **142** 419-427 (1999)

[23] K. Boulahya, M. Parras, and J.M. González-Calbet, J. Solid State Chem. **145** 116-127 (1999).

[24] M. Huvê, C. Renard, F. Abraham, G. van Tendeloo, and S. Amelinckx, J. Solid State Chem. **135**, 1 (1998).

[25] W.T.A. Harrison, S.I. Hegwood, and A.J. Jacobson, Chem. Commun. 1953 (1995).

[26] H. McMurdie, M. Morris, E. Evans, B. Paretzkin, W. Wong-Ng, L. Ettinger, C. R. Hubard, *Powd. Diff.* **1** (1988) 76.

[27] W.L. Roth, *J. Phys. Chem. Solids* **25**, 1 (1964).

[28] V.A. Cherepanov, L. Ya. Gavrilova, L. Yu. Barkhatova, V.I. Voronon, M.V. Trifonova, and O.A. Bukhner, Ionics **4** 309 (1998).

[29] P.G. Radaclli, and S.W. Cheong, Phys. Rev. B **66** 094408 (2002).

[30] Y. Wang, Y. Sui, P. Ren, L. Wang, X. Wang, W. Su, and H.J. Fan, Inorg. Chem. **49** (2010) 3216.

[31] T. Takami, H. Ikuta, and U. Mizutani, Jap. J. Appl. Phys. **43** (22) 8208-8212 (2004).

[32] M. Mikami and R. Funahashi, IEEE Proceedings on 22nd International Conference on Thermoelectrics, pp. 200-202, (2003).

[33] S. Hébert, D. Flahaut, C. Martin, S. Lemonnier, J. Noudem, C. Goupil, A. Maifnan, and J. Hejtmanek, Prog. in Solid State Chem. **35** 457-467 (2007).

[34] K. Iwasaki, H. Yamane, S. Kubota, J. Takahashi, and M. Shimada, J. Alloys and Compounds **358** 210-215 (2003).

[35] Y. Miyazaki, Solid State Ionics **172** 463-467 (2004).

[36] S. Li, R. Funahashi, I. Matsubara, H. Yamada, K. Ueno, and S. Sodeoka, Ceram. International **27** 321-324 (2001).

[37] S. Li, R. Funahashi, I. Matsubara, H. Yamada, K. Ueno, S. Sodeoka and H. Yamada, Chem . Mater. **12** 2424-2427 (2000).

[38] M. Prevel, O. Perez, and J.G. Noudem, Solid State Sciences **9** 231-235 (2007).

Author Index